SOLUTIONS TO RED EXERCISES

ROXY WILSON

THE UNIVERSITY OF ILLINOIS, URBANA-CHAMPAIGN

TENTH EDITION

CHEMISTRY

THE CENTRAL SCIENCE

BROWN | LeMAY | BURSTEN

PEARSON

Prentice
Hall

Upper Saddle River, New Jersey 07458

Project Manager: Kristen Kaiser
Executive Editor: Nicole Folchetti
Executive Managing Editor: Kathleen Schiaparelli
Assistant Managing Editor: Becca Richter
Production Editor: Diane Hernandez
Supplement Cover Designer: Elizabeth Wright
Manufacturing Buyer: Ilene Kahn

© 2006 Pearson Education, Inc.
Pearson Prentice Hall
Pearson Education, Inc.
Upper Saddle River, NJ 07458

Pearson Prentice Hall™ is a trademark of Pearson Education, Inc.

The author and publisher of this book have used their best efforts in preparing this book. These efforts include the development, research, and testing of the theories and programs to determine their effectiveness. The author and publisher make no warranty of any kind, expressed or implied, with regard to these programs or the documentation contained in this book. The author and publisher shall not be liable in any event for incidental or consequential damages in connection with, or arising out of, the furnishing, performance, or use of these programs.

Printed in the United States of America

10 9 8 7 6 5 4 3

ISBN 0-13-146486-8

Pearson Education Ltd., *London*
Pearson Education Australia Pty. Ltd., *Sydney*
Pearson Education Singapore, Pte. Ltd.
Pearson Education North Asia Ltd., *Hong Kong*
Pearson Education Canada, Inc., *Toronto*
Pearson Educación de Mexico, S.A. de C.V.
Pearson Education—Japan, *Tokyo*
Pearson Education Malaysia, Pte. Ltd.

Contents

Introduction

Chemistry: The Central Science, 10th edition, contains nearly 2600 end-of-chapter exercises. Considerable attention has been given to these exercises because one of the best ways for students to master chemistry is by solving problems. Grouping the exercises according to subject matter is intended to aid the student in selecting and recognizing particular types of problems. Within each subject matter group, similar problems are arranged in pairs. This provides the student with an opportunity to reinforce a particular kind of problem. There are also a substantial number of general exercises in each chapter to supplement those grouped by topic. Visualizing Concepts, a new feature of the 10th edition, is a set of general exercises that requires students to analyze visual data in order to formulate conclusions about chemical concepts. Integrative exercises, which require students to integrate concepts from several chapters, are a continuing feature of the 10th edition. Answers to the odd numbered topical exercises plus selected general and integrative exercises, about 1200 in all, are provided in the text. These appendix answers help to make the text a useful self-contained vehicle for learning.

This manual, **Solutions to Red Exercises** in **Chemistry: The Central Science, 10th edition**, was written to enhance the end-of-chapter exercises by providing documented solutions for those problems answered in the appendix of the text. The manual assists the instructor by saving time spent generating solutions for assigned problem sets and aids the student by offering a convenient independent source to check their understanding of the material. Most solutions have been worked in the same detail as the in-chapter sample exercises to help guide students in their studies.

To reinforce the '*Analyze, Plan, Solve, Check*' problem-solving method used extensively in the text, this strategy has also been incorporated into the Solution Manual. Solutions to most red topical exercises and selected Additional and Integrative exercises feature this four-step approach. We strongly encourage students to master this powerful and totally general method.

When using this manual, keep in mind that the numerical result of any calculation is influenced by the precision of the numbers used in the calculation. In this manual, for example, atomic masses and physical constants are typically expressed to four significant figures, or at least as precisely as the data given in the problem. If students use slightly different values to solve problems, their answers will differ slightly from those listed in the appendix of the text or this manual. This is a normal and a common occurrence when comparing results from different calculations or experiments.

Rounding methods are another source of differences between calculated values. In this manual, when a solution is given in steps, intermediate results will be rounded to the correct number of significant figures; however, unrounded numbers will be used in subsequent calculations. By following this scheme, calculators need not be cleared to re-enter rounded intermediate results in the middle of a calculation sequence. The final answer will appear with the correct number of significant figures. This may result in a small discrepancy in the last significant digit between student-calculated answers and those given in this manual. Variations due to rounding can occur in any analysis of numerical data.

The first step in checking your solution and resolving differences between your answer and the listed value is to look for similarities and differences in problem-solving methods. Ultimately, resolving the small numerical differences described above is less important than understanding the general method for solving a problem. The goal of this manual is to provide a reference for sound and consistent problem-solving methods in addition to accurate answers to text exercises.

Extraordinary efforts have been made to keep this manual as error-free as possible. All exercises were worked and proof-read by at least three chemists to ensure clarity in methods and accuracy in mathematics. The work and advice of Dr. Angela Manders Cannon and Mr. David Shinn have been invaluable to this project. However, in a written work as technically challenging as this manual, typos and errors inevitably creep in. Please help us find and eliminate them. We hope that both instructors and students will find this manual accurate, helpful and instructive.

Roxy B. Wilson
University of Illinois
School of Chemical Sciences
505 S. Mathews Ave., Box 49-1
Urbana, IL 61801
rbwilson@uiuc.edu

1 Introduction: Matter and Measurement

Visualizing Concepts

1.1   *Pure elements* contain only one kind of atom. Atoms can be present singly or as tightly bound groups called molecules. *Compounds* contain two or more kinds of atoms bound tightly into molecules. *Mixtures* contain more than one kind of atom and/or molecule, not bound into discrete particles.

   (a)   pure element: i, v

   (b)   mixture of elements: vi

   (c)   pure compound: iv

   (d)   mixture of an element and a compound: ii, iii

1.5   (a)   7.5 cm. There are two significant figures in this measurement; the number of cm can be read precisely, but there is some estimating (uncertainty) required to read tenths of a centimeter. Listing two significant figures is consistent with the convention that measured quantities are reported so that there is uncertainty in only the last digit.

   (b)   140°C. The temperature can be read to the nearest 50°C and estimated to the nearest 5–10°C. Since there is uncertainty in the tens digit, the measurement has two significant figures.

1.8   Given: mi/hr   Find: km/s

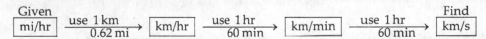

Classification and Properties of Matter

1.9   (a)   heterogeneous mixture

   (b)   homogeneous mixture (If there are undissolved particles, such as sand or decaying plants, the mixture is heterogeneous.)

   (c)   pure substance

   (d)   homogeneous mixture

1.11   (a) S   (b) K   (c)   Cl   (d) Cu   (e) Si   (f) N   (g) Ca   (h)   He

1.13   (a)   lithium   (b) aluminum   (c) lead   (d) sulfur   (e) bromine

   (f)   tin   (g)   chromium   (h) zinc

1

1.15 $A(s) \xrightarrow{\text{heat}} B(s) + C(g)$

When solid carbon is burned in excess oxygen gas, the two elements combine to form a gaseous compound, carbon dioxide. Clearly substance C is this compound. Since C is produced when A is heated in the absence of oxygen (from air), both the carbon and oxygen in C must have been present in A originally. A is, therefore, a compound composed of two or more elements chemically combined. Without more information on the chemical or physical properties of B, we cannot determine absolutely whether it is an element or a compound. However, few if any elements exist as white solids, so B is probably also a compound.

1.17   *Physical properties*: silvery white (color); lustrous; melting point = 649°C; boiling point = 1105°C; density at 20°C = 1.738 g/cm^3; pounded into sheets (malleable); drawn into wires (ductile); good conductor. *Chemical properties*: burns in air to give intense white light; reacts with Cl_2 to produce brittle white solid.

1.19   (a)   chemical   (b)   physical   (c)   physical   (d)   chemical   (e)   chemical

1.21   (a)   Take advantage of the different water solubilities of the two solids. Add water to dissolve the sugar; filter this mixture, collecting the sand on the filter paper and the sugar water in the flask. Evaporate the water from the flask to reproduce solid sugar.

           (b)   Either the melting-point difference or magnetism difference between iron and sulfur can be used to separate these two elements. Heat the mixture until the sulfur melts, then decant (pour off) the liquid sulfur. Or use a magnet to attract the iron particles, leaving the solid sulfur behind.

Units and Measurement

1.23   (a) 1×10^{-1} (b) 1×10^{-2} (c) 1×10^{-15} (d) 1×10^{-6} (e) 1×10^6

           (f) 1×10^3 (g) 1×10^{-9} (h) 1×10^{-3} (i) 1×10^{-12}

1.25   (a) $25.5 \, \text{mg} \times \dfrac{1 \times 10^{-3} \, \text{g}}{1 \, \text{mg}} = 0.0255 \, \text{g} \; (2.55 \times 10^{-2} \, \text{g})$

           (b) $4.0 \times 10^{-10} \, \text{m} \times \dfrac{1 \, \text{nm}}{1 \times 10^{-9} \, \text{m}} = 0.40 \, \text{nm}$

           (c) $0.575 \, \text{mm} \times \dfrac{1 \times 10^{-3} \, \text{m}}{1 \, \text{mm}} \times \dfrac{1 \, \mu\text{m}}{1 \times 10^{-6} \, \text{m}} = 575 \, \mu\text{m}$

1.27   (a) $\text{density} = \dfrac{\text{mass}}{\text{volume}} = \dfrac{39.73 \, \text{g}}{25.0 \, \text{mL}} = 1.59 \, \text{g/mL or } 1.59 \, \text{g/cm}^3$

           (The units cm^3 and mL will be used interchangeably in this manual.)

           Carbon tetrachloride, 1.59 g/mL, is more dense than water, 1.00 g/mL; carbon tetrachloride will sink rather than float on water.

(b) $75.00 \, \text{cm}^3 \times 21.45 \dfrac{\text{g}}{\text{cm}^3} = 1.609 \times 10^3 \, \text{g} \, (1.609 \, \text{kg})$

(c) $87.50 \, \text{g} \times \dfrac{1 \, \text{cm}^3}{1.738 \, \text{g}} = 50.3452 = 50.35 \, \text{cm}^3 = 50.35 \, \text{mL}$

1.29   (a) $\text{density} = \dfrac{38.5 \, \text{g}}{45 \, \text{mL}} = 0.86 \, \text{g/mL}$

        The substance is probably toluene, density = 0.866 g/mL.

    (b) $45.0 \, \text{g} \times \dfrac{1 \, \text{mL}}{1.114 \, \text{g}} = 40.4 \, \text{mL}$ ethylene glycol

    (c) $(5.00)^3 \, \text{cm}^3 \times \dfrac{8.90 \, \text{g}}{1 \, \text{cm}^3} = 1.11 \times 10^3 \, \text{g} \, (1.11 \, \text{kg})$ nickel

1.31   thickness = volume/area

    $\text{volume} = 200 \, \text{mg} \times \dfrac{1 \times 10^{-3} \, \text{g}}{1 \, \text{mg}} \times \dfrac{1 \, \text{cm}^3}{19.32 \, \text{g}} = 0.01035 = 0.0104 \, \text{cm}^3$

    $\text{area} = 2.4 \, \text{ft} \times 1.0 \, \text{ft} \times \dfrac{12^2 \, \text{in}^2}{1 \, \text{ft}^2} \times \dfrac{2.54^2 \, \text{cm}^2}{\text{in}^2} = 2.23 \times 10^3 = 2.2 \times 10^3 \, \text{cm}^2$

    $\text{thickness} = \dfrac{0.01035 \, \text{cm}^3}{2{,}230 \, \text{cm}^2} \times \dfrac{1 \times 10^{-2} \, \text{m}}{1 \, \text{cm}} = 4.6 \times 10^{-8} \, \text{m}$

    $4.6 \times 10^{-8} \, \text{m} \times \dfrac{1 \, \text{nm}}{1 \times 10^{-9} \, \text{m}} = 46 \, \text{nm thick}$

1.33   (a) $^{\circ}\text{C} = 5/9 \, (^{\circ}\text{F} - 32^{\circ}); \, 5/9 \, (62 - 32) = 17^{\circ}\text{C}$

    (b) $^{\circ}\text{F} = 9/5 \, (^{\circ}\text{C}) + 32^{\circ}; \, 9/5 \, (216.7) + 32 = 422.1^{\circ}\text{F}$

    (c) $\text{K} = {}^{\circ}\text{C} + 273.15; \, 233^{\circ}\text{C} + 273.15 = 506 \, \text{K}$

    (d) $315 \, \text{K} - 273 = 42^{\circ}\text{C}; \, 9/5 \, (42^{\circ}\text{C}) + 32 = 108^{\circ}\text{F}$

    (e) $^{\circ}\text{C} = 5/9 \, (^{\circ}\text{F} - 32^{\circ}); \, 5/9 \, (2500 - 32) = 1371^{\circ}\text{C}; \, 1371^{\circ}\text{C} + 273 = 1644 \, \text{K}$
        (assuming 2500 C has 4 sig figs)

Uncertainty In Measurement

1.35   Exact: (c), (d), and (f) (All others depend on measurements and standards that have margins of error, e.g., the length of a week as defined by the earth's rotation.)

1.37   (a)   3   (b) 2   (c) 5   (d) 3   (e) 5

1.39   (a) 1.025×10^2 (b) 6.570×10^5 (c) 8.543×10^{-3}

    (d) 2.579×10^{-4} (e) -3.572×10^{-2}

1.41   (a)   12.0550 + 9.05 = 21.105 = 21.11 (For addition and subtraction, the minimum number of decimal places, here two, determines decimal places in the result.)

    (b)   257.2 – 19.789 = 237.4

(c) $(6.21 \times 10^3)(0.1050) = 652$ (For multiplication and division, the minimum number of significant figures, here three, determines sig figs in the result.)

(d) $0.0577/0.753 = 7.66 \times 10^{-2}$

Dimensional Analysis

1.43   (a) $cm \rightarrow ft : \dfrac{1\,in}{2.54\,cm} \times \dfrac{1\,ft}{12\,in}$ (b) $in^3 \rightarrow cm^3 : \dfrac{(2.54)^3\,cm^3}{1^3\,in^3}$

1.45   (a) $0.076\,L \times \dfrac{1000\,mL}{1\,L} = 76\,mL$

(b) $5.0 \times 10^{-8}\,m \times \dfrac{1\,nm}{1 \times 10^{-9}\,m} = 50.\,nm$

(c) $6.88 \times 10^5\,ns \times \dfrac{1 \times 10^{-9}\,s}{1\,ns} = 6.88 \times 10^{-4}\,s$

(d) $0.50\,lb \times \dfrac{453.6\,g}{1\,lb} = 226.8 = 2.3 \times 10^2\,g$ The data and the result have 2 sig figs.

(e) $\dfrac{1.55\,kg}{m^3} \times \dfrac{1000\,g}{1\,kg} \times \dfrac{1\,m^3}{(10)^3\,dm^3} \times \dfrac{1\,dm^3}{1\,L} = 1.55\,g/L$

(f) $\dfrac{5.850\,gal}{hr} \times \dfrac{3.7854\,L}{1\,gal} \times \dfrac{1\,hr}{60\,min} \times \dfrac{1\,min}{60\,s} = 6.151 \times 10^{-3}\,L/s$

Estimated answer: $6 \times 4 = 24$; $24/60 = 0.4$; $0.4/60 = 0.0066 = 7 \times 10^{-3}$. This agrees with the calculated answer of $6.151 \times 10^{-3}\,L/s$.

1.47   (a) $5.00\,days \times \dfrac{24\,hr}{1\,day} \times \dfrac{60\,min}{1\,hr} \times \dfrac{60\,s}{1\,min} = 4.32 \times 10^5\,s$

(b) $0.0550\,mi \times \dfrac{1.6093\,km}{mi} \times \dfrac{1000\,m}{1\,km} = 88.5\,m$

(c) $\dfrac{\$1.89}{gal} \times \dfrac{1\,gal}{3.7854\,L} = \dfrac{\$0.499}{L}$

(d) $\dfrac{0.510\,in}{ms} \times \dfrac{2.54\,cm}{1\,in} \times \dfrac{1 \times 10^{-2}\,m}{1\,cm} \times \dfrac{1\,km}{1000\,m} \times \dfrac{1\,ms}{1 \times 10^{-3}\,s} \times \dfrac{60\,s}{1\,min} \times \dfrac{60\,min}{1\,hr} = 46.6\,\dfrac{km}{hr}$

Estimate: $0.5 \times 2.5 = 1.25$; $1.25 \times 0.01 \approx 0.01$; $0.01 \times 60 \times 60 \approx 36\,km/hr$

(e) $\dfrac{22.50\,gal}{min} \times \dfrac{3.7854\,L}{gal} \times \dfrac{1\,in}{60\,s} = 1.41953 = 1.420\,L/s$

Estimate: $20 \times 4 = 80$; $80/60 \approx 1.3\,L/s$

(f) $0.02500\,ft^3 \times \dfrac{12^3\,in^3}{1\,ft^3} \times \dfrac{2.54^3\,cm^3}{1\,in^3} = 707.9\,cm^3$

Estimate: $10^3 = 1000$; $3^3 = 27$; $1000 \times 27 = 27{,}000$; $27{,}000/0.04 \approx 700\,cm^3$

1.49   (a) $31\,\text{gal} \times \dfrac{4\,\text{qt}}{1\,\text{gal}} \times \dfrac{1\,\text{L}}{1.057\,\text{qt}} = 1.2 \times 10^2\,\text{L}$

          Estimate: $(30 \times 4)/1 \approx 120\,\text{L}$

       (b) $\dfrac{6\,\text{mg}}{\text{kg (body)}} \times \dfrac{1\,\text{kg}}{2.205\,\text{lb}} \times 150\,\text{lb} = 4 \times 10^2\,\text{mg}$

          Estimate: $6/2 = 3; 3 \times 150 = 450\,\text{mg}$

       (c) $\dfrac{254\,\text{mi}}{11.2\,\text{gal}} \times \dfrac{1.609\,\text{km}}{1\,\text{mi}} \times \dfrac{1\,\text{gal}}{4\,\text{qt}} \times \dfrac{1.057\,\text{qt}}{1\,\text{L}} = \dfrac{9.64\,\text{km}}{\text{L}}$

          Estimate: $250/10 = 25; 1.6/4 = 0.4; 25 \times 0.4 \times 1 \approx 10\,\text{km/L}$

       (d) $\dfrac{50\,\text{cups}}{1\,\text{lb}} \times \dfrac{1\,\text{qt}}{4\,\text{cups}} \times \dfrac{1\,\text{L}}{1.057\,\text{qt}} \times \dfrac{1000\,\text{mL}}{1\,\text{L}} \times \dfrac{1\,\text{lb}}{453.6\,\text{g}} = \dfrac{26\,\text{mL}}{\text{g}}$

          Estimate: $50/4 = 12; 1000/500 = 2; (12 \times 2)/1 \approx 24\,\text{mL/g}$

1.51 $12.5\,\text{ft} \times 15.5\,\text{ft} \times 8.0\,\text{ft} = 1580 = 1.6 \times 10^3\,\text{ft}^3$ (2 sig figs)

      $1550\,\text{ft}^3 \times \dfrac{(1\,\text{yd})^3}{(3\,\text{ft})^3} \times \dfrac{(1\,\text{m})^3}{(1.0936)^3\,\text{yd}^3} \times \dfrac{10^3\,\text{dm}^3}{1\,\text{m}^3} \times \dfrac{1\,\text{L}}{1\,\text{dm}^3} \times \dfrac{1.19\,\text{g}}{\text{L}} \times \dfrac{1\,\text{kg}}{1000\,\text{g}} = 52\,\text{kg air}$

      Estimate: $1550/30 = 50; (50 \times 1)/1 \approx 50\,\text{kg}$

1.53   Select a common unit for comparison, in this case the cm.

      $1\,\text{in} \approx 2.5\,\text{cm}, 1\,\text{m} = 100\,\text{cm}$

      $57\,\text{cm} = 57\,\text{cm}$

      $14\,\text{in} \approx 35\,\text{cm}$

      $1.1\,\text{m} = 110\,\text{cm}$

      The order of length from shortest to longest is 14-in shoe < 57-cm string < 1.1-m pipe.

1.55   (a) $26.73\,\text{g total} \times \dfrac{0.90\,\text{g Ag}}{1\,\text{g total}} \times \dfrac{1\,\text{tr oz}}{31.1\,\text{g Ag}} \times \dfrac{\$1.18}{1\,\text{tr oz}} = \$0.91$

       (b) $\$25.00 \times \dfrac{1\,\text{tr oz}}{\$5.30} \times \dfrac{31.1\,\text{g}}{1\,\text{tr oz}} \times \dfrac{1\,\text{g total}}{0.90\,\text{g Ag}} \times \dfrac{1\,\text{coin}}{26.73\,\text{g}} = 6.1\,\text{coins}$

      Since coins come in integer numbers, 7 coins are required.

Additional Exercises

1.57   *Composition* is the contents of a substance, the kinds of elements that are present and their relative amounts. *Structure* is the arrangement of these contents.

1.60   Any sample of vitamin C has the same relative amount of carbon and oxygen; the ratio of oxygen to carbon in the isolated sample is the same as the ratio in synthesized vitamin C.

$$\frac{2.00\,\text{g O}}{1.50\,\text{g C}} = \frac{x\,\text{g O}}{6.35\,\text{g C}}; \quad x = \frac{(2.00\,\text{g O})(6.35\,\text{g C})}{1.50\,\text{g C}} = 8.47\,\text{g O}$$

This calculation assumes the law of constant composition.

1.63   (a)   volume   (b) area   (c) volume   (d) density

(e)   time   (f) length   (g) temperature

1.66   (a) $\dfrac{40\,\text{lb peat}}{14 \times 20 \times 30\,\text{in}^3} \times \dfrac{1\,\text{in}^3}{(2.54)^3\,\text{cm}^3} \times \dfrac{453.6\,\text{g}}{1\,\text{lb}} = 0.13\,\text{g/cm}^3\ \text{peat}$

$\dfrac{40\,\text{lb soil}}{1.9\,\text{gal}} \times \dfrac{1\,\text{gal}}{4\,\text{qt}} \times \dfrac{1.057\,\text{qt}}{1\,\text{L}} \times \dfrac{1 \times 10^{-3}\,\text{L}}{1\,\text{mL}} \times \dfrac{1\,\text{mL}}{1\,\text{cm}^3} \times \dfrac{453.6\,\text{g}}{1\,\text{lb}} = 2.5\,\text{g/cm}^3\ \text{soil}$

No. Volume must be specified in order to compare mass. The densities tell us that a certain volume of peat moss is "lighter" (weighs less) than the same volume of top soil.

(b)   1 bag peat = $14 \times 20 \times 30 = 8.4 \times 10^3\,\text{in}^3$

$10.\,\text{ft} \times 20.\,\text{ft} \times 2.0\,\text{in} \times \dfrac{12^2\,\text{in}^2}{\text{ft}^2} = 57{,}600 = 5.8 \times 10^4\,\text{in}^3\ \text{peat needed}$

$57{,}600\,\text{in}^3 \times \dfrac{1\,\text{bag}}{8.4 \times 10^3\,\text{in}^3} = 6.9\ \text{bags needed}\quad\text{(Buy 7 bags of peat.)}$

1.69   (a)   density = (35.66 g – 14.23 g)/4.59 cm^3 = 4.67 g/cm^3

(b) $34.5\,\text{kg} \times \dfrac{1000\,\text{g}}{1\,\text{kg}} \times \dfrac{1\,\text{mL}}{13.6\,\text{g}} \times \dfrac{1\,\text{L}}{1000\,\text{mL}} = 2.54\,\text{L}$

(c)   V = $4/3\,\pi\,r^3$ = $4/3\,\pi\,(28.9\,\text{cm})^3$ = 1.0111×10^5 = 1.01×10^5 = $1.01 \times 10^5\,\text{cm}^3$

$1.011 \times 10^5\,\text{cm}^3 \times \dfrac{19.3\,\text{g}}{\text{cm}^3} = 1.95 \times 10^6\,\text{g}$

The sphere weighs 1950 kg or 4300 pounds. The student is unlikely to be able to carry the sphere.

1.73   (a)   The distance is exactly 10 mi (infinite sig figs).

$25\,\text{min} \times \dfrac{1\,\text{hr}}{60\,\text{min}} = 0.41667 = 0.42\,\text{hr};$

$14\,\text{s} \times \dfrac{1\,\text{min}}{60\,\text{s}} \times \dfrac{1\,\text{hr}}{60\,\text{min}} = 0.00389 = 0.0039\,\text{hr}$

Time can be measured to the nearest second, or 0.0003 hr, so the total time is reported with 4 decimal places and 5 sig figs.

total time = 1 hr + 0.4167 hr + 0.0039 hr = 1.4206 hr

avg. speed = 10 mi/1.4206 hr = 7.0393 mi/hr

(b) $\dfrac{1.4206\,\text{hr}}{10\,\text{mi}} \times \dfrac{60\,\text{min}}{1\,\text{hr}} = \dfrac{8.5236\,\text{min}}{\text{mi}}$

$\dfrac{1.4206\,\text{hr}}{10\,\text{mi}} \times \dfrac{60\,\text{min}}{1\,\text{hr}} \times \dfrac{60\,\text{s}}{1\,\text{min}} = \dfrac{511.42\,\text{s}}{\text{mi}}$

1.76 (a) $\dfrac{\$2480}{\text{acre} \cdot \text{ft}} \times \dfrac{1\,\text{acre}}{4840\,\text{yd}^2} \times \dfrac{3\,\text{ft}}{1\,\text{yd}} \times \dfrac{(1.094\,\text{yd})^3}{(1\,\text{m})^3} \times \dfrac{(1\,\text{m})^3}{(10\,\text{dm})^3} \times \dfrac{(1\,\text{dm})^3}{1\,\text{L}} =$

$\$2.013 \times 10^{-3}/\text{L}$ or $0.2013\ \text{¢}/\text{L}$ ($0.201\ \text{¢}/\text{L}$ to 3 sig figs)

(b) $\dfrac{\$2480}{\text{acre} \cdot \text{ft}} \times \dfrac{1\,\text{acre} \cdot \text{ft}}{2\,\text{households} \cdot \text{year}} \times \dfrac{1\,\text{year}}{365\,\text{days}} \times 1\,\text{household} = \dfrac{\$3.397}{\text{day}} = \dfrac{\$3.40}{\text{day}}$

1.79 (a) Let x = mass of Au in jewelry

9.85 - x = mass of Ag in jewelry

The total volume of jewelry = volume of Au + volume of Ag

$0.675\,\text{cm}^3 = x\,\text{g} \times \dfrac{1\,\text{cm}^3}{19.3\,\text{g}} + (9.85 - x)\text{g} \times \dfrac{1\,\text{cm}^3}{10.5\,\text{g}}$

$0.675 = \dfrac{x}{19.3} + \dfrac{9.85 - x}{10.5}$ (To solve, multiply both sides by (19.3)(10.5))

$0.675\,(19.3)(10.5) = 10.5\,x + (9.85 - x)(19.3)$

$136.79 = 10.5\,x + 190.105 - 19.3\,x$

$-53.315 = -8.8\,x$

x = 6.06 g Au; 9.85 g total - 6.06 g Au = 3.79 g Ag

$\text{mass \% Au} = \dfrac{6.06\,\text{g Au}}{9.85\,\text{g jewelry}} \times 100 = 61.5\%\ \text{Au}$

(b) 24 carats × 0.615 = 15 carat gold

1.81 The separation is successful if two distinct spots are seen on the paper. To quantify the characteristics of the separation, calculate a reference value for each spot that is

$$\dfrac{\text{distance travelled by spot}}{\text{distance travelled by solvent}}$$

If the values for the two spots are fairly different, the separation is successful. (One could measure the distance between the spots, but this would depend on the length of paper used and be different for each experiment. The values suggested above are independent of the length of paper.)

2 Atoms, Molecules, and Ions

Visualizing Concepts

2.1   (a)   The path of the charged particle bends because it is repelled by the negatively charged plate and attracted to the positively charged plate.

(b)   Like charges repel and opposite charges attract, so the sign of the electrical charge on the particle is negative.

(c)   The greater the magnitude of the charges, the greater the electrostatic repulsion or attraction. As the charge on the plates is increased, the bending will increase.

(d)   As the mass of the particle increases and speed stays the same, linear momentum (mv) of the particle increases and bending decreases. (See **A Closer Look**: The Mass Spectrometer.)

2.3   Since the number of electrons (negatively charged particles) does not equal the number of protons (positively charged particles), the particle is an ion. The charge on the ion is 2–.

Atomic number = number of protons = 16. The element is S, sulfur.

Mass number = protons + neutrons = 32

$$^{32}_{16}\text{S}^{2-}$$

2.5   Formula: IF_5 Name: iodine pentafluoride
Since the compound is composed of elements that are all nonmetals, it is molecular.

Atomic Theory and the Discovery of Atomic Structure

2.7   Postulate 4 of the atomic theory is the *law of constant composition*. It states that the relative number and kinds of atoms in a compound are constant, regardless of the source. Therefore, 1.0 g of pure water should always contain the same relative amounts of hydrogen and oxygen, no matter where or how the sample is obtained.

2.9   (a) $\dfrac{17.60\,\text{g oxygen}}{30.82\,\text{g nitrogen}} = \dfrac{0.5711\,\text{g O}}{1\,\text{g N}}$; $0.5711/0.5711 = 1.0$

$\dfrac{35.20\,\text{g oxygen}}{30.82\,\text{g nitrogen}} = \dfrac{1.142\,\text{g O}}{1\,\text{g N}}$; $1.142/0.5711 = 2.0$

8

$$\frac{70.40\,\text{g oxygen}}{30.82\,\text{g nitrogen}} = \frac{2.284\,\text{g O}}{1\,\text{g N}}; \quad 2.284/0.5711 = 4.0$$

$$\frac{88.00\,\text{g oxygen}}{30.82\,\text{g nitrogen}} = \frac{2.855\,\text{g O}}{1\,\text{g N}}; \quad 2.855/0.5711 = 5.0$$

2.11   Evidence that cathode rays were negatively charged particles was (1) that electric and magnetic fields deflected the rays in the same way they would deflect negatively charged particles and (2) that a metal plate exposed to cathode rays acquired a negative charge.

2.13   (a)   If the positive plate were lower than the negative plate, the oil drops "coated" with negatively charged electrons would be attracted to the positively charged plate and would descend much more quickly.

         (b)   The more times a measurement is repeated, the better the chance of detecting and compensating for experimental errors. That is, if a quantity is measured five times and four measurements agree but one does not, the measurement that disagrees is probably the result of an error. Also, the four measurements that agree can be averaged to compensate for small random fluctuations. Millikan wanted to demonstrate the validity of his result via its reproducibility.

Modern View of Atomic Structure; Atomic Weights

2.15   (a) $1.9\,\text{Å} \times \dfrac{1 \times 10^{-10}\,\text{m}}{1\,\text{Å}} \times \dfrac{1\,\text{nm}}{1 \times 10^{-9}\,\text{m}} = 0.19\,\text{nm}$

                 $1.9\,\text{Å} \times \dfrac{1 \times 10^{-10}\,\text{m}}{1\,\text{Å}} \times \dfrac{1\,\text{pm}}{1 \times 10^{-12}\,\text{m}} = 1.9 \times 10^{2} \text{ or } 190\,\text{pm}\ (1\,\text{Å} = 100\,\text{pm})$

         (b)   Aligned Kr atoms have **diameters** touching. $d = 2r = 2(1.9\,\text{Å}) = 3.8\,\text{Å}$

                 $1.0\,\text{mm} \times \dfrac{1\,\text{m}}{1000\,\text{mm}} \times \dfrac{1\,\text{Å}}{1 \times 10^{-10}\,\text{m}} \times \dfrac{1\,\text{Kr atom}}{3.8\,\text{Å}} = 2.6 \times 10^{6}\,\text{Kr atoms}$

         (c) $V = 4/3\,\pi r^{3}. \quad r = 1.9\,\text{Å} \times \dfrac{1 \times 10^{-10}\,\text{m}}{1\,\text{Å}} \times \dfrac{100\,\text{cm}}{\text{m}} = 1.9 \times 10^{-8}\,\text{cm}$

                 $V = (4/3)(\pi)(1.9 \times 10^{-8})^{3}\,\text{cm}^{3} = 2.9 \times 10^{-23}\,\text{cm}^{3}$

2.17   (a)   proton, neutron, electron

         (b)   proton = +1, neutron = 0, electron = –1

         (c)   The neutron is most massive, the electron least massive. (The neutron and proton have very similar masses).

2.19   (a)   *Atomic number* is the number of protons in the nucleus of an atom. *Mass number* is the total number of nuclear particles, protons plus neutrons, in an atom.

         (b)   The mass number can vary without changing the identity of the atom, but the atomic number of every atom of a given element is the same.

2.21   p = protons, n = neutrons, e = electrons

   (a) ^{40}Ar has 18 p, 22 n, 18 e   (b) ^{65}Zn has 30 p, 35 n, 30 e

   (c) ^{70}Ga has 31 p, 39 n, 31 e   (d) ^{80}Br has 35 p, 45 n, 35 e

   (e) ^{184}W has 74 p, 110 n, 74 e   (f) ^{243}Am has 95 p, 148 n, 95 e

2.23

Symbol	^{52}Cr	^{55}Mn	^{112}Cd	^{222}Rn	^{207}Pb
Protons	24	25	48	86	82
Neutrons	28	30	64	136	125
Electrons	24	25	48	86	82
Mass no.	52	55	112	222	207

2.25   (a) $^{196}_{78}$Pt   (b) $^{84}_{36}$Kr   (c) $^{75}_{33}$As   (c) $^{24}_{12}$Mg

2.27   (a) $^{12}_{6}$C

   (b)   Atomic weights are really average atomic masses, the sum of the mass of each naturally occurring isotope of an element times its fractional abundance. Each B atom will have the mass of one of the naturally occurring isotopes, while the "atomic weight" is an average value. The naturally occurring isotopes of B, their atomic masses, and relative abundances are:

   ^{10}B, 10.012937, 19.9%; ^{11}B, 11.009305, 80.1%.

2.29   Atomic weight (average atomic mass) = Σ fractional abundance × mass of isotope

   Atomic weight = 0.6917(62.9296) + 0.3083(64.9278) = 63.5456 = 63.55 amu

2.31   (a)   Compare Figures 2.4 and 2.13, referring to Solution 2.12. In Thomson's cathode ray experiments and in mass spectrometry a stream of charged particles is passed through a magnetic field. The charged particles are deflected by the magnetic field according to their mass and charge. For a constant magnetic field strength and speed of the particles, the lighter particles experience a greater deflection.

   (b)   The x-axis label (independent variable) is atomic weight and the y-axis label (dependent variable) is signal intensity.

   (c)   Uncharged particles are not deflected in a magnetic field. The effect of the magnetic field on moving, *charged* particles is the basis of their separation by mass.

2.33   (a)   Average atomic mass = 0.7899(23.98504) + 0.1000(24.98584) + 0.1101(25.98259)

   = 24.31 amu

(b)

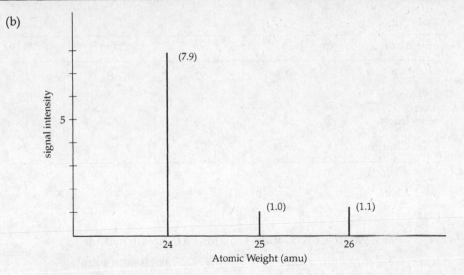

The relative intensities of the peaks in the mass spectrum are the same as the relative abundances of the isotopes. The abundances and peak heights are in the ratio ^{24}Mg: ^{25}Mg: ^{26}Mg as 7.8 : 1.0 : 1.1.

The Periodic Table; Molecules and Ions

2.35   (a)   Cr (metal)   (b) He (nonmetal)   (c) P (nonmetal)   (d) Zn (metal)

(e)   Mg (metal)   (f) Br (nonmetal)   (g) As (metalloid)

2.37   (a)   K, alkali metals (metal)   (b) I, halogens (nonmetal)

(c)   Mg, alkaline earth metals (metal)   (d) Ar, noble gases (nonmetal)

(e)   S, chalcogens (nonmetal)

2.39   An *empirical formula* shows the simplest ratio of the different atoms in a molecule. A *molecular formula* shows the exact number and kinds of atoms in a molecule. A *structural formula* shows how these atoms are arranged.

2.41   (a) $AlBr_3$ (b) C_4H_5 (c) C_2H_4O (d) P_2O_5

(e) C_3H_2Cl (f) BNH_2

2.43   (a)   6   (b) 6   (c) 12

2.45   (a) C_2H_6O
$$H-\underset{\underset{H}{|}}{\overset{\overset{H}{|}}{C}}-O-\underset{\underset{H}{|}}{\overset{\overset{H}{|}}{C}}-H$$
(b) C_2H_6O
$$H-\underset{\underset{H}{|}}{\overset{\overset{H}{|}}{C}}-\underset{\underset{H}{|}}{\overset{\overset{H}{|}}{C}}-O-H$$

(c) CH_4O
$$H-\underset{\underset{H}{|}}{\overset{\overset{H}{|}}{C}}-O-H$$
(d) PF_3
$$F-\underset{\underset{F}{|}}{P}-F$$

2.47

Symbol	$^{59}Co^{3+}$	$^{80}Se^{2-}$	$^{192}Os^{2+}$	$^{200}Hg^{2+}$
Protons	27	34	76	80
Neutrons	32	46	116	120
Electrons	24	36	74	78
Net Charge	3+	2-	2+	2+

2.49   (a) Mg^{2+} (b) Al^{3+} (c) K^+

        (d) S^{2-} (e) F^-

2.51   (a) GaF_3, gallium(III) fluoride   (b) LiH, lithium hydride

        (c) AlI_3, aluminum iodide   (d) K_2S, potassium sulfide

2.53   (a) $CaBr_2$ (b) K_2CO_3 (c) $Al(C_2H_3O_2)_3$

        (d) $(NH_4)_2SO_4$ (e) $Mg_3(PO_4)_2$

2.55   Molecular (all elements are nonmetals):

        (a) B_2H_6 (b) CH_3OH (f) $NOCl$ (g) NF_3

        Ionic (formed by a cation and an anion, usually contains a metal cation):

        (c) $LiNO_3$ (d) Sc_2O_3 (e) $CsBr$ (h) Ag_2SO_4

Naming Inorganic Compounds; Organic Molecules

2.57   (a) ClO_2^- (b) Cl (c) ClO_3^-

        (d) ClO_4^- (e) ClO^-

2.59   (a)   magnesium oxide   (b) aluminum chloride

        (c)   lithium phosphate   (d) barium perchlorate

        (e)   copper(II) nitrate (cupric nitrate)   (f) iron(II) hydroxide (ferrous hydroxide)

        (g)   calcium acetate   (h) chromium(III) carbonate (chromic carbonate)

        (i) potassium chromate   (j) ammonium sulfate

2.61   (a) $Al(OH)_3$ (b) K_2SO_4 (c) Cu_2O (d) $Zn(NO_3)_2$

        (e) $HgBr_2$ (f) $Fe_2(CO_3)_3$ (g) $NaBrO$

2.63   (a)   bromic acid   (b) hydrobromic acid   (c) phosphoric acid

        (d) $HClO$ (e) HIO_3 (f) H_2SO_3

2.65   (a)   sulfur hexafluoride   (b) iodine pentafluoride   (c) xenon trioxide

        (d) N_2O_4 (e) HCN (f) P_4S_6

2.67   (a) $ZnCO_3$, ZnO, CO_2 (b) HF, SiO_2, SiF_4, H_2O (c) SO_2, H_2O, H_2SO_3

        (d) PH_3 (e) $HClO_4$, Cd, $Cd(ClO_4)_2$ (f) VBr_3

2.69 (a) A hydrocarbon is a compound composed of the elements hydrogen and carbon only.

(b)

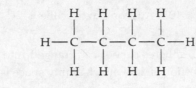

molecular: C_4H_{10}

empirical: C_2H_5

2.71 (a) *Functional groups* are groups of specific atoms that are constant from one molecule to the next. For example, the alcohol functional group is an –OH. Whenever a molecule is called an alcohol, it contains the –OH group.

(b) ——OH   (c)

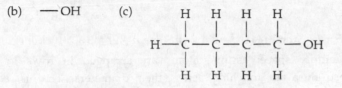

Additional Exercises

2.74 *Radioactivity* is the spontaneous emission of radiation from a substance. Becquerel's discovery showed that atoms could decay, or degrade, *implying* that they are not indivisible. However, it wasn't until Rutherford and others characterized the nature of radioactive emissions, especially the particle nature of α and β rays, that the full significance of the discovery was apparent.

2.77 (a) ^{3}He has 2 protons, 1 neutron, and 2 electrons.

(b) ^{3}H has 1 proton, 2 neutrons, and 1 electron.

^{3}He: $2(1.6726231 \times 10^{-24} \text{ g}) + 1.6749286 \times 10^{-24} \text{ g} + 2(9.1093897 \times 10^{-28} \text{ g})$

$$= 5.021996 \times 10^{-24} \text{ g}$$

^{3}H: $1.6726231 \times 10^{-24} \text{ g} + 2(1.6749286 \times 10^{-24} \text{ g}) + 9.1093897 \times 10^{-28} \text{ g}$

$$= 5.023391 \times 10^{-24} \text{ g}$$

Tritium, ^{3}H, is more massive.

(c) The masses of the two particles differ by 0.0014×10^{-24} g. Each particle loses 1 electron to form the +1 ion, so the difference in the masses of the ions is still 1.4×10^{-27}. A mass spectrometer would need precision to 1×10^{-27} g to differentiate ^{3}He$^+$ and ^{3}H.

2.79 (a) Calculate the mass of a single gold atom, then divide the mass of the cube by the mass of the gold atom.

$$\frac{197.0 \text{ amu}}{\text{gold atom}} \times \frac{1 \text{ g}}{6.022 \times 10^{23} \text{ amu}} = 3.2713 \times 10^{-22} = 3.271 \times 10^{-22} \text{ g/gold atom}$$

$$\frac{19.3 \text{ g}}{\text{cube}} \times \frac{1 \text{ gold atom}}{3.271 \times 10^{-22} \text{ g}} = 5.90 \times 10^{22} \text{ Au atoms in the cube}$$

(b) The shape of atoms is spherical; spheres cannot be arranged into a cube so that there is no empty space. The question is, how much empty space is there? We can calculate the two limiting cases, no empty space and maximum empty space. The true diameter will be somewhere in this range.

No empty space: volume cube/number of atoms = volume of one atom

$V = 4/3\pi r^3$; $r = (3\pi V/4)^{1/3}$; $d = 2r$

$$\text{vol. of cube} = (1.0 \times 1.0 \times 1.0) = \frac{1.0 \text{ cm}^3}{5.90 \times 10^{22} \text{ Au atoms}} = 1.695 \times 10^{-23}$$

$$= 1.7 \times 10^{-23} \text{ cm}^3$$

$r = [\pi (1.695 \times 10^{-23} \text{ cm}^3)/4]^{1/3} = 3.4 \times 10^{-8} \text{ cm}; d = 2r = 6.8 \times 10^{-8} \text{ cm}$

Maximum empty space: assume atoms are arranged in rows in all three directions so they are touching across their diameters. That is, each atom occupies the volume of a cube, with the atomic diameter as the length of the side of the cube. The number of atoms along one edge of the gold cube is then

$(5.90 \times 10^{22})^{1/3} = 3.893 \times 10^7 = 3.89 \times 10^7$ atoms/1.0 cm.

The diameter of a single atom is $1.0 \text{ cm}/3.89 \times 10^7$ atoms $= 2.569 \times 10^{-8}$

$$= 2.6 \times 10^{-8} \text{ cm.}$$

The diameter of a gold atom is between 2.6×10^{-8} cm and 6.8×10^{-8} cm (2.6 – 6.8 Å).

(c) Some atomic arrangement must be assumed, since none is specified. The solid state is characterized by an orderly arrangement of particles, so it isn't surprising that atomic arrangement is required to calculate the density of a solid. A more detailed discussion of solid-state structure and density appears in Chapter 11.

2.82 (a) $^{16}_{8}\text{O}$, $^{17}_{8}\text{O}$, $^{18}_{8}\text{O}$

(b) All isotopes are atoms of the same element, oxygen, with the same atomic number (Z = 8), 8 protons in the nucleus and 8 electrons. Elements with similar electron arrangements have similar chemical properties (Section 2.5). Since the 3 isotopes all have 8 electrons, we expect their electron arrangements to be the same and their chemical properties to be very similar, perhaps identical. Each has a different number of neutrons (8, 9, or 10), a different mass number (A = 16, 17, or 18) and thus a different atomic mass.

2.86 (a)   A Br_2 molecule could consist of two atoms of the same isotope or one atom of each of the two different isotopes. This second possibility is twice as likely as the first. Therefore, the second peak (twice as large as peaks 1 and 3) represents a Br_2 molecule containing different isotopes. The mass numbers of the two isotopes are determined from the masses of the two smaller peaks. Since $157.836 \approx 158$, the first peak represents a $^{79}Br - ^{79}Br$ molecule. Peak 3, $161.832 \approx 162$, represents a $^{81}Br - ^{81}Br$ molecule. Peak 2 then contains one atom of each isotope, $^{79}Br - ^{81}Br$, with an approximate mass of 160 amu.

(b)   The mass of the lighter isotope is 157.836 amu/2 atoms, or 78.918 amu/atom. For the heavier one, 161.832 amu/2 atoms = 80.916 amu/atom.

(c)   The relative size of the three peaks in the mass spectrum of Br_2 indicates their relative abundance. The average mass of a Br_2 molecule is

$0.2569(157.836) + 0.4999(159.834) + 0.2431(161.832) = 159.79$ amu.

(Each product has four significant figures and two decimal places, so the answer has two decimal places.)

(d) $\dfrac{159.79 \text{ amu}}{\text{avg. } Br_2 \text{ molecule}} \times \dfrac{1 \, Br_2 \text{ molecule}}{2 \, Br \text{ atoms}} = 79.895 \text{ amu}$

(e)   Let x = the abundance of ^{79}Br, 1 - x = abundance of ^{81}Br. From (b), the masses of the two isotopes are 78.918 amu and 80.916 amu, respectively. From (d), the mass of an average Br atom is 79.895 amu.

$x(78.918) + (1 - x)(80.916) = 79.895, \; x = 0.5110$

$^{79}Br = 51.10\%, \; ^{81}Br = 48.90\%$

2.89   (a)   an alkali metal: K   (b) an alkaline earth metal: Ca   (c) a noble gas: Ar

(d)   a halogen: Br   (e)   a metalloid: Ge   (f) a nonmetal in 1A: H

(g)   a metal that forms a 3+ ion: Al   (h) a nonmetal that forms a 2- ion: O

(i)   an element that resembles Al: Ga

2.92   (a)   nickel(II) oxide, 2+   (b) manganese(IV) oxide, 4+

(c)   chromium(III) oxide, 3+   (d) molybdenium(VI) oxide, 6+

2.95   (a)   sodium chloride   (b) sodium bicarbonate (or sodium hydrogen carbonate)

(c)   sodium hypochlorite   (d) sodium hydroxide

(e)   ammonium carbonate   (f) calcium sulfate

2.97   (a) $CaS, Ca(HS)_2$ (b) $HBr, HBrO$ (c) $AlN, Al(NO_2)_3$ (d) FeO, Fe_2O_3

(e) NH_3, NH_4^+ (f) $K_2SO_3, KHSO_3$ (g) $Hg_2Cl_2, HgCl_2$ (h) $HClO_3, HClO_4$

3 Stoichiometry: Calculations with Chemical Formulas and Equations

Visualizing Concepts

3.1 Reactant A = red, reactant B = blue

Overall, 4 red A_2 molecules + 4 blue B atoms $\rightarrow$ 4 A_2B molecules

Since 4 is a common factor, this equation reduces to equation (a).

3.4 The box contains 4 C atoms and 16 H atoms, so the empirical formula of the hydrocarbon is CH_4.

3.7 *Analyze.* Given a box diagram and formulas of reactants, draw a box diagram of products.

Plan. Write and balance the chemical equation. Determine combining ratios of elements and decide on limiting reactant. Draw a box diagram of products, containing the correct number of product molecules and only excess reactant.

Solve. $N_2 + 3H_2 \longrightarrow 3NH_3$. $N_2 = $ ●● , $NH_3 = $ ○●○

Each N atom (1/2 of an N_2 molecule) reacts with 3 H atoms (1.5 H_2 molecules) to form an NH_3 molecule. Eight N atoms (4 N_2 molecules) require 24 H atoms (12 H_2 molecules) for complete reaction. Only 9 H_2 molecules are available, so H_2 is the limiting reactant. Nine H_2 molecules (18 H atoms) determine that 6 NH_3 molecules are produced. One N_2 molecule is in excess.

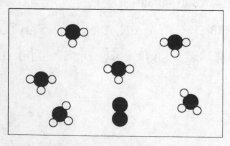

Check. Verify that mass is conserved in your solution, that the number and kinds of atoms are the same in reactant and product diagrams. In this example, there are 8 N atoms and 18 H atoms in both diagrams, so mass is conserved.

Balancing Chemical Equations

3.9   (a)   In balancing chemical equations, the law of conservation of mass, that atoms are neither created nor destroyed during the course of a reaction, is observed. This means that the **number** and **kinds** of atoms on both sides of the chemical equation must be the same.

        (b)   Subscripts in chemical formulas should not be changed when balancing equations, because changing the subscript changes the identity of the compound (law of constant composition).

        (c)   gases, (g); liquids, (l); solids, (s); aqueous solutions, (aq)

3.11   (a) $2CO(g) + O_2(g) \rightarrow 2CO_2(g)$

        (b) $N_2O_5(g) + H_2O(l) \rightarrow 2HNO_3(aq)$

        (c) $CH_4(g) + 4Cl_2(g) \rightarrow CCl_4(l) + 4HCl(g)$

        (d) $Al_4C_3(s) + 12H_2O(l) \rightarrow 4Al(OH)_3(s) + 3CH_4(g)$

        (e) $2C_5H_{10}O_2(l) + 13O_2(g) \rightarrow 10CO_2(g) + 10H_2O(l)$

        (f) $2Fe(OH)_3(s) + 3H_2SO_4(aq) \rightarrow Fe_2(SO_4)_3(aq) + 6H_2O(l)$

        (g) $Mg_3N_2(s) + 4H_2SO_4(aq) \rightarrow 3MgSO_4(aq) + (NH_4)_2SO_4(aq)$

3.13   (a) $CaC_2(s) + 2H_2O(l) \rightarrow Ca(OH)_2(aq) + C_2H_2(g)$

        (b) $2KClO_3(s) \xrightarrow{\Delta} 2KCl(s) + 3O_2(g)$

        (c) $Zn(s) + H_2SO_4(aq) \rightarrow H_2(g) + ZnSO_4(aq)$

        (d) $PCl_3(l) + 3H_2O(l) \rightarrow H_3PO_3(aq) + 3HCl(aq)$

        (e) $3H_2S(g) + 2Fe(OH)_3(s) \rightarrow Fe_2S_3(s) + 6H_2O(g)$

Patterns of Chemical Reactivity

3.15   (a)   When a metal reacts with a nonmetal, an ionic compound forms. The combining ratio of the atoms is such that the total positive charge on the metal cation(s) is equal to the total negative charge on the nonmetal anion(s). All ionic compounds are solids. $2\,Na(s) + Br_2(l) \rightarrow 2NaBr(s)$

        (b)   The second reactant is oxygen gas from the air, $O_2(g)$. The products are $CO_2(g)$ and $H_2O(l)$. $2C_6H_6(l) + 15O_2(g) \rightarrow 12CO_2(g) + 6H_2O(l)$.

3.17   (a) $Mg(s) + Cl_2(g) \rightarrow MgCl_2(s)$

        (b) $BaCO_3(s) \xrightarrow{\Delta} BaO(s) + CO_2(g)$

        (c) $C_8H_8(l) + 10O_2(g) \rightarrow 8CO_2(g) + 4H_2O(l)$

        (d) CH_3OCH_3 is C_2H_6O. $C_2H_6O(l) + 3O_2(g) \rightarrow 2CO_2(g) + 3H_2O(l)$

3.19   (a) $2Al(s) + 3Cl_2(g) \rightarrow 2AlCl_3(s)$ combination

         (b) $C_2H_4(g) + 3O_2(g) \rightarrow 2CO_2(g) + 2H_2O(l)$ combustion

         (c) $6Li(s) + N_2(g) \rightarrow 2Li_3N(s)$ combination

         (d) $PbCO_3(s) \rightarrow PbO(s) + CO_2(g)$ decomposition

         (e) $C_7H_8O_2(l) + 8O_2(g) \rightarrow 7CO_2(g) + 4H_2O(l)$ combustion

Formula Weights

3.21   *Analyze.* Given molecular formula or name, calculate formula weight.

        Plan. If a name is given, write the correct molecular formula. Then, follow the method in Sample Exercise 3.5.   *Solve.*

        (a) N_2O_5: $2(14.0) + 5(16.0) = 108.0$ amu

        (b) $CuSO_4$: $1(63.6) + 1(32.1) + 4(16.0) = 159.6$ amu

        (c) $(NH_4)_3PO_4$: $3(14.0) + 12(1.0) + 1(31.0) + 4(16.0) = 149.0$ amu

        (d) $Ca(HCO_3)_2$: $1(40.1) + 2(1.0) + 2(12.0) + 6(16.0) = 162.1$ amu

        (e) Al_2S_3: $2(27.0) + 3(32.1) = 150.3$ amu

        (f) $Fe_2(SO_4)_3$: $2(55.8) + 3(32.1) + 12(16.0) = 399.9$ amu

        (g) Si_2Br_6: $2(28.1) + 6(79.9) = 535.6$ amu

3.23   *Plan.* Calculate the formula weight (FW), then the mass % oxygen in the compound. *Solve.*

        (a) SO_3: FW = $1(32.1) + 3(16.0) = 80.1$ amu

$$\% \, O = \frac{3(16.0)\,\text{amu}}{80.1\,\text{amu}} \times 100 = 59.9\%$$

        (b) $CH_3COOCH_3 = C_3H_6O_2$: FW = $3(12.0) + 6(1.0) + 2(16.0) = 74.0$ amu

$$\% \, O = \frac{2(16.0)\,\text{amu}}{74.0\,\text{amu}} \times 100 = 43.2\%$$

        (c) $Cr(NO_3)_3$: FW = $1(52.0) + 3(14.0) + 9(16.0) = 238.0$ amu

$$\% \, O = \frac{9(16.0)\,\text{amu}}{238.0\,\text{amu}} \times 100 = 60.5\%$$

        (d) Na_2SO_4: FW = $2(23.0) + 1(32.1) + 4(16.0) = 142.1$ amu

$$\% \, O = \frac{4(16.0)\,\text{amu}}{142.1\,\text{amu}} \times 100 = 45.0\%$$

        (e) NH_4NO_3: FW = $2(14.0) + 4(1.0) + 3(16.0) = 80.0$ amu

$$\% \, O = \frac{3(16.0)\,\text{amu}}{80.0\,\text{amu}} \times 100 = 60.0\%$$

3.25   *Plan.* Follow the logic for calculating mass % C given in Sample Exercise 3.6.   *Solve.*

(a) C_7H_6O: FW = 7(12.0) + 6(1.0) + 1(16.0) = 106.0 amu

$$\% \, C = \frac{7(12.0)\,amu}{106.0\,amu} \times 100 = 79.2\%$$

(b) $C_8H_8O_3$: FW = 8(12.0) + 8(1.0) + 3(16.0) = 152.0 amu

$$\% \, C = \frac{8(12.0)\,amu}{152.0\,amu} \times 100 = 63.2\%$$

(c) $C_7H_{14}O_2$: FW = 7(12.0) + 14(1.0) + 2(16.0) = 130.0 amu

$$\% \, C = \frac{7(12.0)\,amu}{130.0\,amu} \times 100 = 64.6\%$$

Avogadro's Number and the Mole

3.27   (a) 6.022×10^{23}. This is the number of objects in a mole of anything.

(b)   The formula weight of a substance in amu has the same numerical value as the molar mass expressed in grams.

3.29   *Plan.* Since the mole is a counting unit, use it as a basis of comparison; determine the total moles of atoms in each given quantity.   *Solve.*

23 g Na contains 1 mol of atoms

0.5 mol H_2O contains (3 atoms × 0.5 mol) = 1.5 mol atoms

6.0×10^{23} N_2 molecules contains (2 atoms × 1 mol) = 2 mol atoms

3.31   *Analyze.* Given: 16 lb/ball, Avogadro's number of balls, 6.022×10^{23} balls. Find: mass in kg of Avogadro's number of balls; compare with mass of Earth.

Plan. balls → mass in lb → mass in kg; mass of balls/mass of Earth

Solve. 6.022×10^{23} balls $\times \dfrac{16\,lb}{ball} \times \dfrac{1\,kg}{2.2046\,lb} = 4.370 \times 10^{24} = 4.4 \times 10^{24}$ kg

$\dfrac{4.370 \times 10^{24}\,kg\,of\,balls}{5.98 \times 10^{24}\,kg\,Earth} = 0.73;$ One mole of shot-put balls weighs 0.73 times as much as Earth.

Check. This mass of balls is reasonable since Avogadro's number is large.

Estimate: 16 lb ≈ 7 kg; $6 \times 10^{23} \times 7 = 4.2 \times 10^{24}$ kg

3.33   (a)   *Analyze.* Given: 0.773 mol CaH_2. Find: mass in g.

Plan. Use molar mass (g/mol) of CaH_2 to find g CaH_2

Solve. molar mass = 1(40.08) + 2(1.008) = 42.096 = 42.10 g/mol CaH_2

0.773 mol CaH_2 $\times \dfrac{42.096\,g}{1\,mol} = 32.5$ g CaH_2

Check. 3/4(4) = 30 g. The calculated result is reasonable.

(b)   *Analyze.* Given: mass. Find: moles. *Plan.* Use molar mass of $Mg(NO_3)_2$.

Solve. molar mass = 1(24.31) + 2(14.01) + 6(16.00) = 148.33 = 148.3

$$5.35 \text{ g Mg(NO}_3)_2 \times \frac{1 \text{ mol}}{148.33 \text{ g}} = 0.0361 \text{ mol Mg(NO}_3)_2$$

Check. $5/150 \approx 1/30 = 0.033$ mol

(c)   *Analyze.* Given: moles. Find: molecules. *Plan.* Use Avogadro's number.

Solve. $0.0305 \text{ mol CH}_3\text{OH} \times \dfrac{6.022 \times 10^{23} \text{ molecules}}{1 \text{ mol}} = 1.8367 \times 10^{22}$

$$= 1.84 \times 10^{22} \text{ CH}_3\text{OH molecules}$$

Check. $(0.03 \times 6 \times 10^{23}) = 0.18 \times 10^{23} = 1.8 \times 10^{22}$

(d)   *Analyze.* Given: mol C_4H_{10}. Find: C atoms.

Plan. mol $C_4H_{10} \rightarrow$ mol H atoms $\rightarrow$ H atoms

Solve. $0.585 \text{ mol C}_4\text{H}_{10} \times \dfrac{4 \text{ mol C atoms}}{1 \text{ mol C}_4\text{H}_{10}} \times \dfrac{6.022 \times 10^{23} \text{ atoms}}{1 \text{ mol}}$

$$= 1.41 \times 10^{24} \text{ C atoms}$$

Check. $(0.6 \times 4 \times 6 \times 10^{23}) = 14 \times 10^{23} = 1.4 \times 10^{24}.$

3.35   *Analyze/Plan.* See Solution 3.33 for stepwise problem-solving approach.   *Solve.*

(a) $(NH_4)_3PO_4$ molar mass = 3(14.007) + 12(1.008) + 1(30.974) + 4(16.00) = 149.091

$$= 149.1 \text{ g/mol}$$

$$2.50 \times 10^{-3} \text{ mol (NH}_4)_3\text{PO}_4 \times \frac{149.1 \text{ g (NH}_4)_3\text{PO}_4}{1 \text{ mol}} = 0.373 \text{ g (NH}_4)_3\text{PO}_4$$

(b) $AlCl_3$ molar mass = 26.982 + 3(35.453) = 133.341 = 133.34 g/mol

$$0.2550 \text{ g AlCl}_3 \times \frac{1 \text{ mol}}{133.34 \text{ g AlCl}_3} \times \frac{3 \text{ mol Cl}^-}{1 \text{ mol AlCl}_3} = 5.737 \times 10^{-3} \text{ mol Cl}^-$$

(c) $C_8H_{10}N_4O_2$ molar mass = 8(12.01) + 10(1.008) + 4(14.01) + 2(16.00) = 194.20

$$= 194.2 \text{ g/mol}$$

$$7.70 \times 10^{20} \text{ molecules} \times \frac{1 \text{ mol}}{6.022 \times 10^{23} \text{ molecules}} \times \frac{194.2 \text{ g C}_8\text{H}_{10}\text{N}_4\text{O}_2}{1 \text{ mol caffeine}}$$

$$= 0.248 \text{ g C}_8\text{H}_{10}\text{N}_4\text{O}_2$$

(d) $\dfrac{0.406 \text{ g cholesterol}}{0.00105 \text{ mol}} = 387 \text{ g cholesterol/mol}$

(d) $\dfrac{15.86 \text{ g Valium}}{0.05570 \text{ mol}} = 284.7 \text{ g Valium/mol}$

3.37   (a) $C_6H_{10}OS_2$ molar mass = 6(12.01) + 10(1.008) + 1(16.00) + 2(32.07) = 162.28

$$= 162.3 \text{ g/mol}$$

(b) *Plan.* mg → g → mol   *Solve.*

$$5.00 \text{ mg allicin} \times \frac{1 \times 10^{-3} \text{ g}}{1 \text{ mg}} \times \frac{1 \text{ mol}}{162.3 \text{ g}} = 3.081 \times 10^{-5} = 3.08 \times 10^{-5} \text{ mol allicin}$$

Check. 5.00 mg is a small mass, so the small answer is reasonable.

$(5 \times 10^{-3})/200 = 2.5 \times 10^{-5}$

(c) *Plan.* Use mol from part (b) and Avogadro's number to calculate molecules.

$$\text{Solve. } 3.081 \times 10^{-5} \text{ mol allicin} \times \frac{6.022 \times 10^{23} \text{ molecules}}{\text{mol}} = 1.855 \times 10^{19}$$
$$= 1.86 \times 10^{19} \text{ allicin molecules}$$

Check. $(3 \times 10^{-5})(6 \times 10^{23}) = 18 \times 10^{18} = 1.8 \times 10^{19}$

(d) *Plan.* Use molecules from part (c) and molecular formula to calculate S atoms.

$$\text{Solve. } 1.855 \times 10^{19} \text{ allicin molecules} \times \frac{2 \text{ S atoms}}{1 \text{ allicin molecule}} = 3.71 \times 10^{19} \text{ S atoms}$$

Check. Obvious.

3.39 (a) *Analyze.* Given: $C_6H_{12}O_6$, 1.250×10^{21} C atoms. Find: H atoms.

Plan. Use molecular formula to determine number of H atoms that are present with 1.250×10^{21} C atoms.   *Solve.*

$$\frac{12 \text{ H atoms}}{6 \text{ C atoms}} = \frac{2 \text{ H}}{1 \text{ C}} \times 1.250 \times 10^{21} \text{ C atoms} = 2.500 \times 10^{21} \text{ H atoms}$$

Check. $(2 \times 1 \times 10^{21}) = 2 \times 10^{21}$

(b) *Plan.* Use molecular formula to find the number of glucose molecules that contain 1.250×10^{21} C atoms. *Solve.*

$$\frac{1 \text{ C}_6\text{H}_{12}\text{O}_6 \text{ molecule}}{6 \text{ C atoms}} \times 1.250 \times 10^{21} \text{ C atoms} = 2.0833 \times 10^{20}$$
$$= 2.083 \times 10^{20} \text{ C}_6\text{H}_{12}\text{O}_6 \text{ molecules}$$

Check. $(12 \times 10^{20}/6) = 2 \times 10^{20}$

(c) *Plan.* Use Avogadro's number to change molecules → mol.   *Solve.*

$$2.0833 \times 10^{20} \text{ C}_6\text{H}_{12}\text{O}_6 \text{ molecules} \times \frac{1 \text{ mol}}{6.022 \times 10^{23} \text{ molecules}}$$
$$= 3.4595 \times 10^{-4} = 3.460 \times 10^{-4} \text{ mol C}_6\text{H}_{12}\text{O}_6$$

Check. $(2 \times 10^{20})/(6 \times 10^{23}) = 0.33 \times 10^{-3} = 3.3 \times 10^{-4}$

(d) *Plan.* Use molar mass to change mol → g.   *Solve.*

1 mole of $C_6H_{12}O_6$ weighs 180.0 g (Sample Exercise 3.9)

$$3.4595 \times 10^{-4} \text{ mol C}_6\text{H}_{12}\text{O}_6 \times \frac{180.0 \text{ g C}_6\text{H}_{12}\text{O}_6}{1 \text{ mol}} = 0.06227 \text{ g C}_6\text{H}_{12}\text{O}_6$$

Check. $3.5 \times 180 = 630$; $630 \times 10^{-4} = 0.063$

3.41   *Analyze.* Given: g C_2H_3Cl/L.   Find: mol/L, molecules/L.

Plan. The /L is constant throughout the problem, so we can ignore it. Use molar mass for g → mol, Avogadro's number for mol → molecules.   *Solve.*

$$\frac{2.0\times10^{-6}\ \text{g}\ C_2H_3Cl}{1\ \text{L}}\times\frac{1\ \text{mol}\ C_2H_3Cl}{62.50\ \text{g}\ C_2H_3Cl}=3.20\times10^{-8}=3.2\times10^{-8}\ \text{mol}\ C_2H_3Cl/\text{L}$$

$$\frac{3.20\times10^{-8}\ \text{mol}\ C_2H_3Cl}{1\ \text{L}}\times\frac{6.022\times10^{23}\ \text{molecules}}{1\ \text{mol}}=1.9\times10^{16}\ \text{molecules}/\text{L}$$

Check. $(200\times10^{-8})/60=2.5\times10^{-8}$ mol

$$(2.5\times10^{-8})\times(6\times10^{23})=15\times10^{15}=1.5\times10^{16}$$

Empirical Formulas

3.43   (a)   *Analyze.* Given: moles. Find: empirical formula.

              Plan. Find the **simplest ratio of moles** by dividing by the smallest number of moles present.

              Solve. 0.0130 mol C / 0.0065 = 2

                      0.039 mol H / 0.0065 = 6

                      0.0065 mol O / 0.0065 = 1

              The empirical formula is C_2H_6O.

              Check. The subscripts are simple integers.

     (b)   *Analyze.* Given: grams. Find: empirical formula.

              Plan. Calculate the moles of each element present, then the simplest ratio of moles.

              Solve. $11.66\ \text{g Fe}\times\dfrac{1\ \text{mol Fe}}{55.85\ \text{g Fe}}=0.2088\ \text{mol Fe};\ 0.2088/0.2088=1$

              $5.01\ \text{g O}\times\dfrac{1\ \text{mol O}}{16.00\ \text{g O}}=0.3131\ \text{mol O};\ 0.3131/0.2088\approx1.5$

              Multiplying by two, the integer ratio is 2 Fe : 3 O; the empirical formula is Fe_2O_3.

              Check. The subscripts are simple integers.

     (c)   *Analyze.* Given: mass %. Find: empirical formulas.

              Plan. Assume 100 g sample, calculate moles of each element, find the simplest ratio of moles.

              Solve. $40.0\ \text{g C}\times\dfrac{1\ \text{mol C}}{12.01\ \text{g C}}=3.33\ \text{mol C};\ 3.33/3.33=1$

              $6.7\ \text{g H}\times\dfrac{1\ \text{mol H}}{1.008\ \text{mol H}}=6.65\ \text{mol H};\ 6.65/3.33\approx2$

              $53.3\ \text{g O}\times\dfrac{1\ \text{mol O}}{16.00\ \text{mol O}}=3.33\ \text{mol O};\ 3.33/3.33=1$

The empirical formula is CH_2O.

Check. The subscripts are simple integers.

3.45   *Analyze/Plan.* The procedure in all these cases is to assume 100 g of sample, calculate the number of moles of each element present in that 100 g, then obtain the ratio of moles as smallest whole numbers.   *Solve.*

(a) $10.4 \text{ g C} \times \dfrac{1 \text{ mol C}}{12.01 \text{ g C}} = 0.866 \text{ mol C}; \quad 0.866 / 0.866 = 1$

$27.8 \text{ g S} \times \dfrac{1 \text{ mol S}}{32.07 \text{ g S}} = 0.867 \text{ mol S}; \quad 0.867 / 0.866 \approx 1$

$61.7 \text{ g Cl} \times \dfrac{1 \text{ mol Cl}}{35.45 \text{ g Cl}} = 1.74 \text{ mol Cl}; \quad 1.74 / 0.866 \approx 2$

The empirical formula is $CSCl_2$.

(b) $21.7 \text{ g C} \times \dfrac{1 \text{ mol C}}{12.01 \text{ g C}} = 1.81 \text{ mol C}; \quad 1.81 / 0.600 \approx 3$

$9.6 \text{ g O} \times \dfrac{1 \text{ mol O}}{16.00 \text{ g O}} = 0.600 \text{ mol O}; \quad 0.600 / 0.600 = 1$

$68.7 \text{ g F} \times \dfrac{1 \text{ mol F}}{19.00 \text{ g F}} = 3.62 \text{ mol F}; \quad 3.62 / 0.600 \approx 6$

The empirical formula is C_3OF_6.

(c) $32.79 \text{ g Na} \times \dfrac{1 \text{ mol Na}}{22.99 \text{ g Na}} = 1.426 \text{ mol Na}; \quad 1.426 / 0.4826 \approx 3$

$13.02 \text{ g Al} \times \dfrac{1 \text{ mol Al}}{26.98 \text{ g Al}} = 0.4826 \text{ mol Al}; \quad 0.4826 / 0.4826 = 1$

$54.19 \text{ g F} \times \dfrac{1 \text{ mol F}}{19.00 \text{ g F}} = 2.852 \text{ mol F}; \quad 2.852 / 0.4826 \approx 6$

The empirical formula is Na_3AlF_6.

3.47   *Analyze.* Given: empirical formula, molar mass. Find: molecular formula.

Plan. Calculate the empirical formula weight (FW); divide FW by molar mass (MM) to calculate the integer that relates the empirical and molecular formulas. Check. If FW/MM is an integer, the result is reasonable. *Solve.*

(a) $FW \ CH_2 = 12 + 2(1) = 14. \quad \dfrac{MM}{FW} = \dfrac{84}{14} = 6$

The subscripts in the empirical formula are multiplied by 6. The molecular formula is C_6H_{12}.

(b) $FW \ NH_2Cl = 14.01 + 2(1.008) + 35.45 = 51.48. \quad \dfrac{MM}{FW} = \dfrac{51.5}{51.5} = 1$

The empirical and molecular formulas are NH_2Cl.

3.49   *Analyze.* Given: mass %, molar mass. Find: molecular formula.

           Plan. Use the plan detailed in Solution 3.45 to find an empirical formula from mass % data. Then use the plan detailed in 3.47 to find the molecular formula. Note that some indication of molar mass must be given, or the molecular formula cannot be determined. *Check.* If there is an integer ratio of moles and MM/ FW is an integer, the result is reasonable.   *Solve.*

     (a) $92.3\,g\,C \times \dfrac{1\,mol\,C}{12.01\,g\,C} = 7.685\,mol\,C;\ 7.685/7.639 = 1.006 \approx 1$

            $7.7\,g\,H \times \dfrac{1\,mol\,H}{1.008\,g\,H} = 7.639\,mol\,H;\ 7.639/7.639 = 1$

            The empirical formula is CH, FW = 13.

            $\dfrac{MM}{FW} = \dfrac{104}{13} = 8;$ the molecular formula is C_8H_8.

     (b) $49.5\,g\,C \times \dfrac{1\,mol\,C}{12.01\,g\,C} = 4.12\,mol\,C;\ 4.12/1.03 \approx 4$

            $5.15\,g\,H \times \dfrac{1\,mol\,H}{1.008\,g\,H} = 5.11\,mol\,H;\ 5.11/1.03 \approx 5$

            $28.9\,g\,N \times \dfrac{1\,mol\,N}{14.01\,g\,N} = 2.06\,mol\,N;\ 2.06/1.03 \approx 2$

            $16.5\,g\,O \times \dfrac{1\,mol\,O}{16.00\,g\,O} = 1.03\,mol\,O;\ 1.03/1.03 = 1$

            Thus, $C_4H_5N_2O$, FW = 97. If the molar mass is about 195, a factor of 2 gives the molecular formula $C_8H_{10}N_4O_2$.

     (c) $35.51\,g\,C \times \dfrac{1\,mol\,C}{12.01\,g\,C} = 2.96\,mol\,C;\ 2.96/0.592 = 5$

            $4.77\,g\,H \times \dfrac{1\,mol\,H}{1.008\,g\,H} = 4.73\,mol\,H;\ 4.73/0.592 = 7.99 \approx 8$

            $37.85\,g\,O \times \dfrac{1\,mol\,O}{16.00\,g\,O} = 2.37\,mol\,O;\ 2.37/0.592 = 4$

            $8.29\,g\,N \times \dfrac{1\,mol\,N}{14.01\,g\,N} = 0.592\,mol\,N;\ 0.592/0.592 = 1$

            $13.60\,g\,Na \times \dfrac{1\,mol\,Na}{22.99\,g\,Na} = 0.592\,mol\,Na;\ 0.592/0.592 = 1$

            The empirical formula is $C_5H_8O_4NNa$, FW = 169 g. Since the empirical formula weight and molar mass are approximately equal, the empirical and molecular formulas are both $NaC_5H_8O_4N$.

3.51   (a)   *Analyze.* Given: mg CO_2, mg H_2O Find: empirical formula of hydrocarbon, C_xH_y

Plan. Upon combustion, all C → CO$_2$, all H → H$_2$O.

mg CO$_2$ → g CO$_2$ → mol C; mg H$_2$O → g H$_2$O, mol H

Find simplest ratio of moles and empirical formula.   *Solve.*

$$5.86 \times 10^{-3} \text{ g CO}_2 \times \frac{1 \text{ mol CO}_2}{44.01 \text{ g CO}_2} \times \frac{1 \text{ mol C}}{1 \text{ mol CO}_2} = 1.33 \times 10^{-4} \text{ mol C}$$

$$1.37 \times 10^{-3} \text{ g H}_2\text{O} \times \frac{1 \text{ mol H}_2\text{O}}{18.02 \text{ g H}_2\text{O}} \times \frac{2 \text{ mol H}}{1 \text{ mol H}_2\text{O}} = 1.52 \times 10^{-4} \text{ mol H}$$

Dividing both values by 1.33 × 10^{-4} gives C:H of 1:1.14. This is not "close enough" to be considered 1:1. No obvious multipliers (2, 3, 4) produce an integer ratio. Testing other multipliers (trial and error!), the correct factor seems to be 7. The empirical formula is C$_7$H$_8$.

Check. See discussion of C:H ratio above.

(b)   *Analyze.* Given: g of menthol, g CO$_2$, g H$_2$O, molar mass. Find: molecular formula.

Plan/Solve. Calculate mol C and mol H in the sample.

$$0.2829 \text{ g CO}_2 \times \frac{1 \text{ mol CO}_2}{44.01 \text{ g CO}_2} \times \frac{1 \text{ mol C}}{1 \text{ mol CO}_2} = 0.0064281 = 0.006428 \text{ mol C}$$

$$0.1159 \text{ g H}_2\text{O} \times \frac{1 \text{ mol H}_2\text{O}}{18.02 \text{ g H}_2\text{O}} \times \frac{2 \text{ mol H}}{1 \text{ mol H}_2\text{O}} = 0.012863 = 0.01286 \text{ mol H}$$

Calculate g C, g H and get g O by subtraction.

$$0.0064281 \text{ mol C} \times \frac{12.01 \text{ g C}}{1 \text{ mol C}} = 0.07720 \text{ g C}$$

$$0.012863 \text{ mol H} \times \frac{1.008 \text{ g H}}{1 \text{ mol H}} = 0.01297 \text{ g H}$$

mass O = 0.1005 g sample – (0.07720 g C + 0.01297 g H) = 0.01033 g O

Calculate mol O and find integer ratio of mol C: mol H: mol O.

$$0.01033 \text{ g O} \times \frac{1 \text{ mol O}}{16.00 \text{ g O}} = 6.456 \times 10^{-4} \text{ mol O}$$

Divide moles by 6.456 × 10^{-4}.

$$C : \frac{0.006428}{6.456 \times 10^{-4}} \approx 10; \quad H : \frac{0.01286}{6.456 \times 10^{-4}} \approx 20; \quad O : \frac{6.456 \times 10^{-4}}{6.456 \times 10^{-4}} = 1$$

The empirical formula is C$_{10}$H$_{20}$O.

$$FW = 10(12) + 20(1) + 16 = 156; \quad \frac{M}{FW} = \frac{156}{156} = 1$$

The molecular formula is the same as the empirical formula, C$_{10}$H$_{20}$O.

Check. The mass of O wasn't negative or greater than the sample mass; empirical and molecular formulas are reasonable.

3.53   *Analyze.* Given 2.558 g $Na_2CO_3 \cdot xH_2O$, 0.948 g Na_2CO_3. Find: x.

Plan. The reaction involved is $Na_2CO_3 \cdot xH_2O(s) \rightarrow Na_2CO_3(s) + xH_2O(g)$.

Calculate the mass of H_2O lost and then the mole ratio of Na_2CO_3 and H_2O.

Solve. g H_2O lost = 2.558 g sample – 0.948 g Na_2CO_3 = 1.610 g H_2O

$$0.948 \text{ g } Na_2CO_3 \times \frac{1 \text{ mol } Na_2CO_3}{106.0 \text{ g } Na_2CO_3} = 0.00894 \text{ mol } Na_2CO_3$$

$$1.610 \text{ g } H_2O \times \frac{1 \text{ mol } H_2O}{18.02 \text{ g } H_2O} = 0.08935 \text{ mol } H_2O$$

The formula is $Na_2CO_3 \cdot \underline{\mathbf{10}} \, H_2O$.

Check. x is an integer.

Calculations Based on Chemical Equations

3.55   The mole ratios implicit in the coefficients of a balanced chemical equation express the fundamental relationship between amounts of reactants and products. If the equation is not balanced, the mole ratios will be incorrect and lead to erroneous calculated amounts of products.

3.57 $Na_2SiO_3(s) + 8HF(aq) \rightarrow H_2SiF_6(aq) + 2NaF(aq) + 3H_2O(l)$

   (a)   *Analyze.* Given: mol Na_2SiO_3. Find: mol HF. *Plan.* Use the mole ratio $8HF:1Na_2SiO_3$ from the balanced equation to relate moles of the two reactants.

   Solve.

   $$0.300 \text{ mol } Na_2SiO_3 \times \frac{8 \text{ mol HF}}{1 \text{ mol } Na_2SiO_3} = 2.40 \text{ mol HF}$$

   Check. Mol HF should be greater than mol Na_2SiO_3.

   (b)   *Analyze.* Given: mol HF. Find: g NaF. *Plan.* Use the mole ratio $2NaF:8HF$ to change mol HF to mol NaF, then molar mass to get NaF.   *Solve.*

   $$0.500 \text{ mol HF} \times \frac{2 \text{ mol NaF}}{8 \text{ mol HF}} \times \frac{41.99 \text{ g NaF}}{1 \text{ mol NaF}} = 5.25 \text{ g NaF}$$

   Check. (0.5/4) = 0.125; 0.13 × 42 > 4 g NaF

   (c)   *Analyze.* Given: g HF Find: g Na_2SiO_3.

   Plan. g HF $\rightarrow$ mol HF $\left(\dfrac{\text{mol}}{\text{ratio}}\right) \rightarrow$ mol $Na_2SiO_3 \rightarrow$ g Na_2SiO_3

   The mole ratio is at the heart of every stoichiometry problem. Molar mass is used to change to and from grams.   *Solve.*

   $$0.800 \text{ g HF} \times \frac{1 \text{ mol HF}}{20.01 \text{g HF}} \times \frac{1 \text{ mol } Na_2SiO_3}{8 \text{ mol HF}} \times \frac{122.1 \text{ g } Na_2SiO_3}{1 \text{ mol } Na_2SiO_3} = 0.610 \text{ g } Na_2SiO_3$$

   Check. 0.8 (120/160) < 0.75 mol

3.59   (a) $Al(OH)_3(s) + 3HCl(aq) \rightarrow AlCl_3(aq) + 3H_2O(l)$

(b)   *Analyze.* Given mass of one reactant, find stoichiometric mass of other reactant and products.

Plan. Follow the logic in Sample Exercise 3.16. Calculate mol $Al(OH)_3$ in 0.500 g $Al(OH_3)_3$ separately, since it will be used several times.

Solve. $0.500 \text{ g Al(OH)}_3 \times \dfrac{1 \text{ mol Al(OH)}_3}{78.00 \text{ g Al(OH)}_3} = 6.410 \times 10^{-3} = 6.41 \times 10^{-3} \text{ mol Al(OH)}_3$

$6.410 \times 10^{-3} \text{ mol Al(OH)}_3 \times \dfrac{3 \text{ mol HCl}}{1 \text{ mol Al(OH)}_3} \times \dfrac{36.46 \text{ g HCl}}{1 \text{ mol HCl}} = 0.7012 = 0.701 \text{ g HCl}$

(c) $6.410 \times 10^{-3} \text{ molAl(OH)}^3 \times \dfrac{1 \text{ mol HCl}}{1 \text{ mol Al(OH)}_3} \times \dfrac{133.34 \text{ g AlCl}_3}{1 \text{ mol AlCl}_3} = 0.8547$

$= 0.855 \text{ g AlCl}_3$

$6.410 \times 10^{-3} \text{ mol Al(OH)}_3 \times \dfrac{3 \text{ mol H}_2\text{O}}{1 \text{ mol Al(OH)}_3} \times \dfrac{18.02 \text{ g H}_2\text{O}}{1 \text{ mol H}_2\text{O}} = 0.3465 = 0.347 \text{ g H}_2\text{O}$

(d)   Conservation of mass: mass of products = mass of reactants

reactants: $Al(OH)_3$ + HCl, 0.500 g + 0.701 g = 1.201 g

products: $AlCl_3$ + H_2O, 0.855 g + 0.347 g = 1.202 g

The 0.001 g difference is due to rounding (0.8547 + 0.3465 = 1.2012). This is an excellent *check* of results.

3.61   (a) $Al_2S_3(s) + 6H_2O(l) \rightarrow 2Al(OH)_3(s) + 3H_2S(g)$

(b)   *Plan.* g A $\rightarrow$ mol A $\rightarrow$ mol B $\rightarrow$ g B. See Solution 3.57 (c).   *Solve.*

$6.75 \text{ g Al}_2\text{S}_3 \times \dfrac{1 \text{ mol Al}_2\text{S}_3}{150.2 \text{ g Al}_2\text{S}_3} \times \dfrac{2 \text{ mol Al(OH)}_3}{1 \text{ mol Al}_2\text{S}_3} \times \dfrac{78.00 \text{ g Al(OH)}_3}{1 \text{ mol Al(OH)}_3} = 7.01 \text{ g Al(OH)}_3$

Check. $7\left(\dfrac{2 \times 78}{150}\right) \approx 7(1) \approx 7 \text{ g Al(OH)}_3$

3.63   (a)   *Analyze.* Given: mol NaN_3. Find: mol N_2.

Plan. Use mole ratio from balanced equation.   *Solve.*

$1.50 \text{ mol NaN}_3 \times \dfrac{3 \text{ mol N}_2}{2 \text{ mol NaN}_3} = 2.25 \text{ mol N}_2$

Check. The resulting mol N_2 should be greater than mol NaN_3, (the $N_2:NaN_3$ ratio is > 1), and it is.

(b)   *Analyze.* Given: g N_2 Find: g NaN_3.

Plan. Use molar masses to get from and to grams, mol ratio to relate moles of the two substances.   *Solve.*

$10.0 \text{ g N}_2 \times \dfrac{1 \text{ mol N}_2}{28.01 \text{ g N}_2} \times \dfrac{2 \text{ mol NaN}_3}{3 \text{ mol N}_2} \times \dfrac{65.01 \text{ g NaN}_3}{1 \text{ mol NaN}_3} = 15.5 \text{ g NaN}_3$

Check. Mass relations are less intuitive than mole relations. Estimating the ratio of molar masses is sometimes useful. In this case, 65 g NaN_3/28 g $N_2 \approx 2.25$ Then, $(10 \times 2/3 \times 2.25) \approx 14$ g NaN_3. The calculated result looks reasonable.

(c)   *Analyze.* Given: vol N_2 in ft^3, density N_2 in g/L. Find: g NaN_3.

Plan. First determine how many g N_2 are in 10.0 ft^3, using the density of N_2.

Solve.

$$\frac{1.25\,g}{1\,L} \times \frac{1\,L}{1000\,cm^3} \times \frac{(2.54)^3\,cm^3}{1\,in^3} \times \frac{(12)^3\,in^3}{1\,ft^3} \times 10.0\,ft^3 = 354.0 = 354\,g\,N_2$$

$$354.0\,g\,N_2 \times \frac{1\,mol\,N_2}{28.01\,g\,N_2} \times \frac{2\,mol\,NaN_3}{3\,mol\,N_2} \times \frac{65.01\,g\,NaN_3}{1\,mol\,NaN_3} = 548\,g\,NaN_3$$

Check. 1 ft^3 ~ 28 L; 10 ft^3 ~ 280 L; 280 L $\times$ 1.25 ~ 350 g N_2

Using the ratio of molar masses from part (b), $(350 \times 2/3 \times 2.25) \approx 525$ g NaN_3

3.65   (a)   *Analyze.* Given: dimensions of Al foil. Find: mol Al.

Plan. Dimensions $\rightarrow$ vol $\xrightarrow{\text{density}}$ mass $\xrightarrow{\frac{\text{molar}}{\text{mass}}}$ mol Al

Solve. $1.00\,cm \times 1.00\,cm \times 0.550\,mm \times \frac{1\,cm}{10\,mm} = 0.0550\,cm^3\,Al$

$0.0550\,cm^3\,Al \times \frac{2.699\,g\,Al}{1\,cm^3} \times \frac{1\,mol\,Al}{26.98\,g\,Al} = 5.502 \times 10^{-3} = 5.50 \times 10^{-3}\,mol\,Al$

Check. $2.699/26.98 \approx 0.1$; $(0.055\,cm^3 \times 0.1) = 5.5 \times 10^{-3}$ mol Al

(b)   *Plan.* Write the balanced equation to get a mole ratio; change mol Al $\rightarrow$ mol $AlBr_3 \rightarrow$ g $AlBr_3$.

Solve. $2Al(s) + 3Br_2(l) \rightarrow 2AlBr_3(s)$

$5.502 \times 10^{-3}\,mol\,Al \times \frac{2\,mol\,AlBr_3}{2\,mol\,Al} \times \frac{266.69\,g\,AlBr_3}{1\,mol\,AlBr_3} = 1.467 = 1.47\,g\,AlBr_3$

Check. $(0.006 \times 1 \times 270) \approx 1.6$ g $AlBr_3$

Limiting Reactants; Theoretical Yields

3.67   (a)   The *limiting reactant* determines the maximum number of product moles resulting from a chemical reaction; any other reactant is an *excess reactant*.

(b)   The limiting reactant regulates the amount of products, because it is completely used up during the reaction; no more product can be made when one of the reactants is unavailable.

3.69   (a)   Each bicycle needs 2 wheels, 1 frame, and 1 set of handlebars. A total of 4815 wheels corresponds to 2407.5 pairs of wheels. This is more than the number of frames or handlebars. The 2255 handlebars determine that 2255 bicycles can be produced.

(b)   2305 frames – 2255 bicycles = 50 frames left over

2407.5 pairs of wheels – 2255 bicycles = 152.5 pairs of wheels left over 2(152.5) = 305 wheels left over

(c)   The handlebars are the "limiting reactant" in that they determine the number of bicycles that can be produced.

3.71   *Analyze.* Given: 1.85 mol NaOH, 1.00 mol CO_2. Find: mol Na_2CO_3.

Plan. Amounts of more than one reactant are given, so we must determine which reactant regulates (limits) product. Then apply the appropriate mole ratio from the balanced equation.

Solve. The mole ratio is $2NaOH:1CO_2$, so 1.00 mol CO_2 requires 2.00 mol NaOH for complete reaction. Less than 2.00 mol NaOH are present, so NaOH is the limiting reactant.

$$1.85 \text{ mol NaOH} \times \frac{1 \text{ mol Na}_2\text{CO}_3}{2 \text{ mol NaOH}} = 0.925 \text{ mol Na}_2\text{CO}_3 \text{ can be produced}$$

The $Na_2CO_3:CO_2$ ratio is 1:1, so 0.925 mol Na_2CO_3 produced requires 0.925 mol CO_2 consumed. (Alternately, 1.85 mol NaOH × 1 mol CO_2/2 mol NaOH = 0.925 mol CO_2 reacted). 1.00 mol CO_2 initial – 0.925 mol CO_2 reacted = 0.075 mol CO_2 remain.

Check.	$2NaOH(s)$	+	$CO_2(g)$	→	$Na_2CO_3(s)$	+	$H_2O(l)$
initial	1.85 mol		1.00 mol		0 mol		
change (reaction)	–1.85 mol		–0.925 mol		+0.925 mol		
final	0 mol		0.075 mol		0.925 mol		

Note that the "change" line (but not necessarily the "final" line) reflects the mole ratios from the balanced equation.

3.73 $3NaHCO_3(aq) + H_3C_6H_5O_7(aq) \rightarrow 3CO_2(g) + 3H_2O(l) + Na_3C_6H_5O_7(aq)$

(a)   *Analyze/Plan.* Abbreviate citric acid as H_3Cit. Follow the approach in Sample Exercise 3.19. *Solve.*

$$1.00 \text{ g NaHCO}_3 \times \frac{1 \text{ mol NaHCO}_3}{84.01 \text{ g NaHCO}_3} = 1.190 \times 10^{-2} = 1.19 \times 10^{-2} \text{ mol NaHCO}_3$$

$$1.00 \text{ g H}_3\text{C}_6\text{H}_5\text{O}_7 \times \frac{1 \text{ mol H}_3\text{Cit}}{192.1 \text{ g H}_3\text{Cit}} = 5.206 \times 10^{-3} = 5.21 \times 10^{-3} \text{ mol H}_3\text{Cit}$$

But $NaHCO_3$ and H_3Cit react in a 3:1 ratio, so 5.21×10^{-3} mol H_3Cit require $3(5.21 \times 10^{-3}) = 1.56 \times 10^{-2}$ mol $NaHCO_3$. We have only 1.19×10^{-2} mol $NaHCO_3$, so $NaHCO_3$ is the limiting reactant.

(b) $1.190 \times 10^{-2} \text{ mol NaHCO}_3 \times \dfrac{3 \text{ mol CO}_2}{3 \text{ mol NaHCO}_3} \times \dfrac{44.01 \text{ g CO}_2}{1 \text{ mol CO}_2} = 0.524 \text{ g CO}_2$

(c) 1.190×10^{-2} mol $NaHCO_3 \times \dfrac{1\,\text{mol}\,H_3Cit}{3\,\text{mol}\,NaHCO_3} = 3.968 \times 10^{-3}$

$$= 3.97 \times 10^{-3} \text{ mol } H_3Cit \text{ react}$$

5.206×10^{-3} mol $H_3Cit - 3.968 \times 10^{-3}$ mol react $= 1.238 \times 10^{-3}$

$$= 1.24 \times 10^{-3} \text{ mol } H_3Cit \text{ remain}$$

1.238×10^{-3} mol $H_3Cit \times \dfrac{192.1\,\text{g}\,H_3Cit}{\text{mol}\,H_3Cit} = 0.238$ g H_3Cit remain

3.75   *Analyze.* Given: initial g Na_2CO_3, g $AgNO_3$. Find: final g Na_2CO_3, $AgNO_3$, Ag_2CO_3, $NaNO_3$

Plan. Write balanced equation; determine limiting reactant; calculate amounts of excess reactant remaining and products, based on limiting reactant.

Solve. $2AgNO_3(aq) + Na_2CO_3(aq) \rightarrow Ag_2CO_3(s) + 2NaNO_3(aq)$

3.50 g $Na_2CO_3 \times \dfrac{1\,\text{mol}\,Na_2CO_3}{106.0\,\text{g}\,Na_2CO_3} = 0.03302 = 0.0330$ mol Na_2CO_3

5.00 g $AgNO_3 \times \dfrac{1\,\text{mol}\,AgNO_3}{169.9\,\text{g}\,AgNO_3} = 0.02943 = 0.0294$ mol $AgNO_3$

0.02943 mol $AgNO_3 \times \dfrac{1\,\text{mol}\,Na_2CO_3}{2\,\text{mol}\,AgNO_3} = 0.01471 = 0.0147$ mol Na_2CO_3 required

$AgNO_3$ is the limiting reactant and Na_2CO_3 is present in excess.

	$2AgNO_3(aq)$	+	$Na_2CO_3(aq)$	$\rightarrow$	$Ag_2CO_3(s)$	+	$2NaNO_3(aq)$
initial	0.0294 mol		0.0330 mol		0 mol		0 mol
reaction	−0.0294 mol		−0.0147 mol		+0.0147 mol		+0.0294 mol
final	0 mol		0.0183 mol		0.0147 mol		0.0294 mol

0.01830 mol $Na_2CO_3 \times 106.0$ g/mol $= 1.940 = 1.94$ g Na_2CO_3

0.01471 mol $Ag_2CO_3 \times 275.8$ g/mol $= 4.057 = 4.06$ g Ag_2CO_3

0.02943 mol $NaNO_3 \times 85.00$ g/mol $= 2.502 = 2.50$ g $NaNO_3$

Check. The initial mass of reactants was 8.50 g, and the final mass of excess reactant and products is 13.50 g; mass is conserved.

3.77   *Analyze.* Given: amounts of two reactants. Find: theoretical yield.

Plan. Determine the limiting reactant and the maximum amount of product it could produce. Then calculate % yield.   *Solve.*

(a)   30.0 g $C_6H_6 \times \dfrac{1\,\text{mol}\,C_6H_6}{78.11\,\text{g}\,C_6H_6} = 0.3841 = 0.384$ mol C_6H_6

$$65.0 \text{ g Br}_2 \times \frac{1 \text{ mol Br}_2}{159.8 \text{ g Br}_2} = 0.4068 = 0.407 \text{ mol Br}_2$$

Since C_6H_6 and Br_2 react in a 1:1 mole ratio, C_6H_6 is the limiting reactant and determines the theoretical yield.

$$0.3841 \text{ mol } C_6H_6 \times \frac{1 \text{ mol } C_6H_5Br}{1 \text{ mol } C_6H_6} \times 157.0 \text{ g } C_6H_5Br = 60.30 = 60.3 \text{ g } C_6H_5Br$$

Check. $30/78 \sim 3/8$ mol C_6H_6. $65/160 \sim 3/8$ mol Br_2. Since moles of the two reactants are similar, a precise calculation is needed to determine the limiting reactant. $3/8 \times 160 \approx 60$ g product

(b) $\% \text{ yield} = \dfrac{56.7 \text{ g } C_6H_5Br \text{ actual}}{60.3 \text{ g } C_6H_5Br \text{ theoretical}} \times 100 = 94.0\%$

3.79   *Analyze.* Given: g of two reactants, % yield. Find: g Li_3N.

Plan. Determine limiting reactant and theoretical yield. Use definition of % yield to calculate actual yield.   *Solve.*

$$5.00 \text{ g Li} \times \frac{1 \text{ mol Li}}{6.941 \text{ g Li}} = 0.7204 = 0.720 \text{ mol Li}$$

$$5.00 \text{ g } N_2 \times \frac{1 \text{ mol } N_2}{28.01 \text{ g } N_2} = 0.1785 = 0.179 \text{ mol } N_2$$

$$0.1785 \text{ mol } N_2 \times \frac{6 \text{ mol Li}}{1 \text{ mol } N_2} = 1.071 = 1.07 \text{ mol Li required}$$

Since there is less than enough Li to react exactly with 0.179 mol N_2, Li is the limiting reactant

$$0.7204 \text{ mol Li} \times \frac{2 \text{ mol } Li_3N}{6 \text{ mol Li}} \times \frac{34.83 \text{ g } Li_3N}{1 \text{ mol } Li_3N} = 8.363 = 8.36 \text{ g } Li_3N \text{ theoretical yield}$$

Check. $5/7 \approx$ mol Li; $5/(4 \times 7) \approx$ mol N_2. There are 1/4 as many mol N_2 as moles Li, but only 1/6 as many moles N_2 are required for exact reaction. N_2 is in excess and Li limits. $0.7 \times (36/3) \approx 8.4$ g Li_3N theoretical

$$\% \text{ yield} = \frac{\text{actual}}{\text{theoretical}} \times 100; \quad \frac{\% \text{ yield} \times \text{theoretical}}{100} = \text{actual yield}$$

$$\frac{88.5\%}{100} \times 8.363 \text{ g } Li_3N = 7.4013 = 7.40 \text{ g } Li_3N \text{ actual}$$

Additional Exercises

3.81   (a) $C_4H_8O_2(l) + 5O_2(g) \rightarrow 4CO_2(g) + 4H_2O(l)$

          (b) $Ni(OH)_2(s) \rightarrow NiO(s) + H_2O(g)$

          (c) $Zn(s) + Cl_2(g) \rightarrow ZnCl_2(s)$

3.83   (a) $1.25 \, \text{carat} \times \dfrac{0.200 \, \text{g}}{1 \, \text{carat}} \times \dfrac{1 \, \text{mol C}}{12.01 \, \text{g C}} = 0.020816 = 0.0208 \, \text{mol C}$

$0.020816 \, \text{mol C} \times \dfrac{6.022 \times 10^{23} \, \text{C atoms}}{1 \, \text{mol C}} = 1.25 \times 10^{22} \, \text{C atoms}$

(b) $0.500 \, \text{g C}_9\text{H}_8\text{O}_4 \times \dfrac{1 \, \text{mol C}_9\text{H}_8\text{O}_4}{180.2 \, \text{g C}_9\text{H}_8\text{O}_4} = 2.7747 \times 10^{-3} = 2.77 \times 10^{-3} \, \text{mol HC}_9\text{H}_7\text{O}_4$

$0.0027747 \, \text{mol C}_9\text{H}_8\text{O}_4 \times \dfrac{6.022 \times 10^{23} \, \text{molecules}}{1 \, \text{mol}} = 1.67 \times 10^{21} \, \text{HC}_9\text{H}_7\text{O}_4 \, \text{molecules}$

3.86   *Plan.* Assume 100 g, calculate mole ratios, empirical formula, then molecular formula from molar mass.   *Solve.*

$68.2 \, \text{g C} \times \dfrac{1 \, \text{mol C}}{12.01 \, \text{g C}} = 5.68 \, \text{mol C}; \; 5.68/0.568 \approx 10$

$6.86 \, \text{g H} \times \dfrac{1 \, \text{mol H}}{1.008 \, \text{g H}} = 6.81 \, \text{mol H}; \; 6.81/0.568 \approx 12$

$15.9 \, \text{g N} \times \dfrac{1 \, \text{mol N}}{14.01 \, \text{g N}} = 1.13 \, \text{mol N}; \; 1.13/0.568 \approx 2$

$9.08 \, \text{g O} \times \dfrac{1 \, \text{mol O}}{16.00 \, \text{g O}} = 0.568 \, \text{mol O}; \; 0.568/0.568 = 1$

The empirical formula is $C_{10}H_{12}N_2O$, FW = 176 amu (or g). Since the molar mass is 176, the empirical and molecular formula are the same, $C_{10}H_{12}N_2O$.

3.89   *Plan.* Because different sample sizes were used to analyze the different elements, calculate mass % of each element in the sample.

i.   Calculate mass % C from g CO_2.

ii.   Calculate mass % Cl from AgCl.

iii.   Get mass % H by subtraction.

iv.   Calculate mole ratios and the empirical formulas.

Solve.

i. $3.52 \, \text{g CO}_2 \times \dfrac{1 \, \text{mol CO}_2}{44.01 \, \text{g CO}_2} \times \dfrac{1 \, \text{mol C}}{1 \, \text{mol CO}_2} \times \dfrac{12.01 \, \text{g C}}{1 \, \text{mol C}} = 0.9606 = 0.961 \, \text{g C}$

$\dfrac{0.9606 \, \text{g C}}{1.50 \, \text{g sample}} \times 100 = 64.04 = 64.0\% \, \text{C}$

ii. $1.27 \, \text{g AgCl} \times \dfrac{1 \, \text{mol AgCl}}{143.3 \, \text{g AgCl}} \times \dfrac{1 \, \text{mol Cl}}{1 \, \text{mol AgCl}} \times \dfrac{35.45 \, \text{g Cl}}{1 \, \text{mol Cl}} = 0.3142 = 0.314 \, \text{g Cl}$

$\dfrac{0.3142 \, \text{g Cl}}{1.00 \, \text{g sample}} \times 100 = 31.42 = 31.4\% \, \text{Cl}$

iii.   % H = 100.0 − (64.04% C + 31.42% Cl) = 4.54 = 4.5% H

iv.   Assume 100 g sample.

$$64.04 \text{ g C} \times \frac{1 \text{ mol C}}{12.01 \text{ g C}} = 5.33 \text{ mol C}; \quad 5.33 / 0.886 = 6.02$$

$$31.42 \text{ g Cl} \times \frac{1 \text{ mol Cl}}{35.45 \text{ g Cl}} = 0.886 \text{ mol Cl}; \quad 0.886 / 0.886 = 1.00$$

$$4.54 \text{ g H} \times \frac{1 \text{ mol H}}{1.008 \text{ g H}} = 4.50 \text{ mol H}; \quad 4.50 / 0.886 = 5.08$$

The empirical formula is probably C_6H_5Cl.

The subscript for H, 5.08, is relatively far from 5.00, but C_6H_5Cl makes chemical sense. More significant figures in the mass data are required for a more accurate mole ratio.

3.92 $C_2H_5OH(l) + 3O_2(g) \rightarrow 2CO_2(g) + 3H_2O(g)$

$C_3H_8(g) + 5O_2(g) \rightarrow 3CO_2(g) + 4H_2O(g)$

$CH_3CH_2COCH_3(l) + 11/2 \, O_2(g) \rightarrow 4CO_2(g) + 4H_2O(l)$

In a combustion reaction, all H in the fuel is transformed to H_2O in the products. The reactant with most mol H/mol fuel will produce the most H_2O. C_3H_8 and $CH_3CH_2COCH_3$ (C_4H_8O) both have 8 mol H/mol fuel, so 1.5 mol of either fuel will produce the same amount of H_2O. 1.5 mol C_2H_5OH will produce less H_2O.

3.95 $2C_{57}H_{110}O_6 + 163O_2 \rightarrow 114CO_2 + 110H_2O$

molar mass of fat = 57(12.01) + 110(1.008) + 6(16.00) = 891.5

$$1.0 \text{ kg fat} \times \frac{1000 \text{ g}}{1 \text{ kg}} \times \frac{1 \text{ mol fat}}{891.5 \text{ g fat}} \times \frac{110 \text{ mol H}_2\text{O}}{2 \text{ mol fat}} \times \frac{18.02 \text{ g H}_2\text{O}}{1 \text{ mol H}_2\text{O}} \times \frac{1 \text{ kg}}{1000 \text{ g}} = 1.1 \text{ kg H}_2\text{O}$$

3.98   All of the O_2 is produced from $KClO_3$; get g $KClO_3$ from g O_2. All of the H_2O is produced from $KHCO_3$; get g $KHCO_3$ from g H_2O. The g H_2O produced also reveals the g CO_2 from the decomposition of $NaHCO_3$. The remaining CO_2 (13.2 g CO_2- g CO_2 from $NaHCO_3$) is due to K_2CO_3 and g K_2CO_3 can be derived from it.

$$4.00 \text{ g O}_2 \times \frac{1 \text{ mol O}_2}{32.00 \text{ g O}_2} \times \frac{2 \text{ mol KClO}_3}{3 \text{ mol O}_2} \times \frac{122.6 \text{ g KClO}_3}{1 \text{ mol KClO}_3} = 10.22 = 10.2 \text{ g KClO}_3$$

$$1.80 \text{ g H}_2\text{O} \times \frac{1 \text{ mol H}_2\text{O}}{18.02 \text{ g H}_2\text{O}} \times \frac{2 \text{ mol KHCO}_3}{1 \text{ mol H}_2\text{O}} \times \frac{100.1 \text{ g KHCO}_3}{1 \text{ mol KHCO}_3} = 20.00 = 20.0 \text{ g KHCO}_3$$

$$1.80 \text{ g H}_2\text{O} \times \frac{1 \text{ mol H}_2\text{O}}{18.02 \text{ g H}_2\text{O}} \times \frac{2 \text{ mol CO}_2}{1 \text{ mol H}_2\text{O}} \times \frac{44.01 \text{ g CO}_2}{1 \text{ mol CO}_2} = 8.792 = 8.79 \text{ g CO}_2 \text{ from KHCO}_3$$

13.20 g CO_2 total – 8.792 CO_2 from $KHCO_3$ = 4.408 = 4.41 g CO_2 from K_2CO_3

$$4.408 \text{ g CO}_2 \times \frac{1 \text{ mol CO}_2}{44.01 \text{ g CO}_2} \times \frac{1 \text{ mol K}_2\text{CO}_3}{1 \text{ mol CO}_2} \times \frac{138.2 \text{ g K}_2\text{CO}_3}{1 \text{ mol K}_2\text{CO}_3} = 13.84 = 13.8 \text{ g K}_2\text{CO}_3$$

100.0 g mixture – 10.22 g $KClO_3$ – 20.00 g $KHCO_3$ – 13.84 g K_2CO_3 = 56.0 g KCl

Integrative Exercises

3.101   *Plan.* Volume cube $\xrightarrow{\text{density}}$ mass $CaCO_3 \to$ moles $CaCO_3 \to$ moles $O \to O$ atoms

$\quad$ *Solve.* $(2.005)^3 \text{ in}^3 \times \dfrac{(2.54)^3 \text{ cm}^3}{1 \text{ in}^3} \times \dfrac{2.71 \text{ g CaCO}_3}{1 \text{ cm}^3} \times \dfrac{1 \text{ mol CaCO}_3}{100.1 \text{ g CaCO}_3} \times \dfrac{3 \text{ mol O}}{1 \text{ mol CaCO}_3}$

$$\times \dfrac{6.022 \times 10^{23} \text{ O atoms}}{1 \text{ mol O}} = 6.46 \times 10^{24} \text{ O atoms}$$

3.103   *Analyze.* Given: gasoline = C_8H_{18}, density = 0.69 g/mL, 20.5 mi/gal, 225 mi. Find: kg CO_2.

$\quad$ *Plan.* Write and balance the equation for the combustion of octane. Change mi $\to$ gal octane $\to$ mL $\to$ g octane. Use stoichiometry to calculate g and kg CO_2 from g octane.

$\quad$ *Solve.* $2C_8H_{18}(l) + 25O_2(g) \to 16CO_2(g) + 18H_2O(l)$

$$225 \text{ mi} \times \dfrac{1 \text{ gal}}{20.5 \text{ mi}} \times \dfrac{3.7854 \text{ L}}{1 \text{ gal}} \times \dfrac{1 \text{ mL}}{1 \times 10^{-3} \text{ L}} \times \dfrac{0.69 \text{ g octane}}{1 \text{ mL}} = 2.8667 \times 10^4 \text{ g}$$

$$= 29 \text{ kg octane}$$

$$2.8667 \times 10^4 \text{ g C}_8\text{H}_{18} \times \dfrac{1 \text{ mol C}_8\text{H}_{18}}{114.2 \text{ g C}_8\text{H}_{18}} \times \dfrac{16 \text{ mol CO}_2}{2 \text{ mol C}_8\text{H}_{18}} \times \dfrac{44.01 \text{ g CO}_2}{1 \text{ mol CO}_2} = 8.8382 \times 10^4 \text{ g}$$

$$= 88 \text{ kg CO}_2$$

$\quad$ *Check.* $\left(\dfrac{225 \times 4 \times 0.7}{20} \right) \times 10^3 = (45 \times 0.7) \times 10^3 = 30 \times 10^3 \text{ g} = 30 \text{ kg octane}$

$\dfrac{44}{114} \approx \dfrac{1}{3}$; $\dfrac{30 \text{ kg} \times 8}{3} \approx 80 \text{ kg CO}_2$

3.105   (a)$\quad$ $S(s) + O_2(g) \to SO_2(g)$; $SO_2(g) + CaO(s) \to CaSO_3(s)$

$\quad$ (b)$\quad$ $\dfrac{2000 \text{ tons coal}}{\text{day}} \times \dfrac{2000 \text{ lb}}{1 \text{ ton}} \times \dfrac{1 \text{ kg}}{2.20 \text{ lb}} \times \dfrac{1000 \text{ g}}{1 \text{ kg}} \times \dfrac{0.025 \text{ g S}}{1 \text{ g coal}} \times \dfrac{1 \text{ mol S}}{32.1 \text{ g S}}$

$$\times \dfrac{1 \text{ mol SO}_2}{1 \text{ mol S}} \times \dfrac{1 \text{ mol CaSO}_3}{1 \text{ mol SO}_2} \times \dfrac{120 \text{ g CaSO}_3}{1 \text{ mol CaSO}_3} \times \dfrac{1 \text{ kg CaSO}_3}{1000 \text{ g CaSO}_3}$$

$$= 1.7 \times 10^5 \text{ kg CaSO}_3 / \text{day}$$

$\quad$ This corresponds to about 190 tons of $CaSO_3$ per day as a waste product.

4 Aqueous Reactions and Solution Stoichiometry

Visualizing Concepts

4.1 *Analyze.* Correlate the formula of the solute with the charged spheres in the diagrams.

Plan. Determine the electrolyte properties of the solute and the relative number of cations, anions, or neutral molecules produced when the solute dissolves.

Solve. Li_2SO_4 is a strong electrolyte, a soluble ionic solid that dissociates into separate Li^+ and SO_4^{2-} when it dissolves in water. There are twice as many Li^+ cations as SO_4^{2-} anions. Diagram (c) represents the aqueous solution of a 2:1 electrolyte.

4.3 *Analyze/Plan.* Correlate the neutral molecules, cations, and anions in the diagrams with the definitions of strong, weak, and nonelectrolytes. *Solve.*

(a) AX is a nonelectrolyte, because no ions form when the molecules dissolve.

(b) AY is a weak electrolyte because a few molecules ionize when they dissolve, but most do not.

(c) AZ is a strong electrolyte because all molecules break up into ions when they dissolve.

4.5 *Analyze.* From the names and/or formulas of 3 substances determine their electrolyte and solubilities properties.

Plan. Determine whether the substance is molecular or ionic. If it is molecular, is it a weak acid or base and thus a weak electrolyte, or a nonelectrolyte? If it is ionic, is it soluble?

Solve. Glucose is a molecular compound that is neither a weak acid nor a weak base. It is a nonelectrolyte that dissolves to produce a nonconducting solution; it is solid C. NaOH is ionic; the anion is OH^-. According to Table 4.1, most hydroxides are insoluble, but NaOH is one of the soluble ones. NaOH is a strong electrolyte that dissolves to form a conducting solution; it is solid A. AgBr is ionic; the anion is Br^-. According to Table 4.1, most bromides are soluble, but AgBr is one of the insoluble ones; it is solid B.

Check. We know by elimination that AgBr is solid B, which we verified by solubility rules.

4.7 *Analyze.* Given the formulas of some ions, determine whether these ions ever form precipitates in aqueous solution. *Plan.* Use Table 4.1 to determine if the given ions can form precipitates. If not, they will always be spectator ions. *Solve.*

(a) Cl^- can form precipitates with Ag^+, Hg_2^{2+}, Pb^{2+}.

(b) NO_3^- never forms precipitates, so it is always a spectator.

(c) NH_4^+ never forms precipitates, so it is always a spectator.

(d) S^{2-} usually forms precipitates.

(e) SO_4^{2-} usually forms precipitates.

Check. NH_4^+ is a soluble exception for sulfides, phosphates, and carbonates, which usually form precipitates, so all rules indicate that it is a perpetual spectator.

4.9   *Analyze*. Given three chemical reactions, match the reaction stoichiometry to the reactant and product particles shown in the diagram. *Plan*. Determine cation:anion stoichometrics of the reactants and products, as well as spectator ions and precipitate in the products.

Solve. Anions are shown as the larger spheres in each container. The first reactant is a 1:2 strong electrolyte. The second reactant is a 2:1 strong electrolyte. The spectator ions are the anion from the first reactant and the cation from the second reactant. The precipitate is a 1:1 solid formed by the cation from the first reactant and the anion from the second reactant. Since the diagram indicates that a precipitate is formed, we can eliminate reaction (b) straight away. The first reactant in reaction (c) is a 1:1 electrolyte, so it cannot be represented by the diagram. Reaction (a) is represented by the diagram.

Check. $BaCl_2$ is a 1:2 electrolyte, Na_2SO_4 is a 2:1 electrolyte, Cl^- (anion from first reactant), and Na^+ (cation from second reactant) are spectators; $BaSO_4$ is a 1:1 precipitate.

Electrolytes

4.11   No. Electrolyte solutions conduct electricity because the dissolved ions carry charge through the solution (from one electrode to the other).

4.13   Although H_2O molecules are electrically neutral, there is an unequal distribution of electrons throughout the molecule. There are more electrons near O and fewer near H, giving the O end of the molecule a partial negative charge and the H end of the molecule a partial positive charge. Ionic compounds are composed of positively and negatively charged ions. The partially positive ends of H_2O molecules are attracted to the negative ions (anions) in the solid, while the partially negative ends are attracted to the positive ions (cations). Thus, both cations and anions in an ionic solid are surrounded and separated (dissolved) by H_2O molecules.

4.15   *Analyze/Plan*. Given the solute formula, determine the separate ions formed upon dissociation. *Solve*.

(a) $ZnCl_2(aq) \rightarrow Zn^{2+}(aq) + 2Cl^-(aq)$

(b) $HNO_3(aq) \rightarrow H^+(aq) + NO_3^-(aq)$

(c) $(NH_4)_2SO_4(aq) \rightarrow 2NH_4^+(aq) + SO_4^{2-}(aq)$

(d) $Ca(OH)_2(aq) \rightarrow Ca^{2+}(aq) + 2OH^-(aq)$

4.17   *Analyze/Plan.* Apply the definition of a weak electrolyte to $HCHO_2$.

Solve. When $HCHO_2$ dissolves in water, neutral $HCHO_2$ molecules, H^+ ions and CHO_2^- ions are all present in the solution. $HCHO_2(aq) \rightleftharpoons H^+(aq) + CHO_2^-(aq)$

Precipitation Reactions and Net Ionic Equations

4.19   *Analyze.* Given: formula of compound. Find: solubility.

Plan. Follow the guidelines in Table 4.1, in light of the anion present in the compound and notable exceptions to the "rules."   *Solve.*

(a) $NiCl_2$: soluble

(b) Ag_2S: insoluble

(c) Cs_3PO_4: soluble (Cs^+ is an alkali metal cation)

(d) $SrCO_3$: insoluble

(e) $PbSO_4$: insoluble, Pb^{2+} is an exception to soluble sulfates

4.21   *Analyze.* Given: formulas of reactants. Find: balanced equation including precipitates.

Plan. Follow the logic in Sample Exercise 4.3.

Solve. In each reaction, the precipitate is in bold type.

(a) $Na_2CO_3(aq) + 2AgNO_3(aq) \rightarrow \mathbf{Ag_2CO_3(s)} + 2NaNO_3(aq)$

(b)   No precipitate (all nitrates and most sulfates are soluble).

(c) $FeSO_4(aq) + Pb(NO_3)_2(aq) \rightarrow \mathbf{PbSO_4(s)} + Fe(NO_3)_2(aq)$

4.23   *Analyze/Plan.* Follow the logic in Sample Exercise 4.4. From the complete ionic equation, identify the ions that don't change during the reaction; these are the spectator ions. *Solve.*

(a) $2Na^+(aq) + CO_3^{2-}(aq) + Mg^{2+}(aq) + SO_4^{2-}(aq) \rightarrow MgCO_3(s) + 2Na^+(aq) + SO_4^{2-}(aq)$
       Spectators: Na^+, SO_4^{2-}

(b) $Pb^{2+}(aq) + 2NO_3^-(aq) + 2Na^+(aq) + S^{2-}(aq) \rightarrow PbS(s) + 2Na^+(aq) + 2NO_3^-(aq)$
       Spectators: Na^+, NO_3^-

(c) $6NH_4^+(aq) + 2PO_4^{3-}(aq) + 3Ca^{2+}(aq) + 6Cl^-(aq) \rightarrow Ca_3(PO_4)_2(s) + 6NH_4^+(aq) + 6Cl^-(aq)$ Spectators: NH_4^+, Cl^-

4.25   *Analyze.* Given: reactions of unknown with HBr, H_2SO_4, NaOH. Find: The unknown contains a single salt. Is K^+ or Pb^{2+} or Ba^{2+} present?

Plan. Analyze solubility guidelines for Br^-, SO_4^{2-} and OH^- and select the cation that produces a precipitate with each of the anions.

Solve. K^+ forms no precipitates with any of the anions. $BaSO_4$ is insoluble, but $BaCl_2$ and $Ba(OH)_2$ are soluble. Since the unknown forms precipitates with all three anions, it must contain Pb^{2+}.

Check. $PbBr_2$, $PbSO_4$, and $Pb(OH)_2$ are all insoluble according to Table 4.1, so our process of elimination is confirmed by the insolubility of the Pb^{2+} compounds.

4.27   *Analyze.* Given: three possible salts in an unknown solution react with $Ba(NO_3)_2$ and then NaCl. Find: Can the results identify the unknown salt? Do the three possible unknowns give distinctly different results with $Ba(NO_3)_2$ and NaCl?

Plan. Using Table 4.1, determine whether each of the possible unknowns will form a precipitate with $Ba(NO_3)_2$ and NaCl.

Solve.

Compound	$Ba(NO_3)_2$ result	NaCl result
$AgNO_3(aq)$	no ppt	AgCl ppt
$CaCl_2(aq)$	no ppt	no ppt
$Al_2(SO_4)_3$	$BaSO_4$ ppt	no ppt

This sequence of tests would definitively identify the contents of the bottle, because the results for each compound are unique.

Acid-Base Reactions

4.29   *Analyze.* Given: solute and concentration of three solutions. Find: solution with greatest concentration of solvated protons.

Plan: See Sample Exercise 4.6. Determine whether solutes are strong or weak acids or bases, or nonelectrolytes. For solutions of equal concentration, strong acids will have greatest concentration of solvated protons. Take varying concentration into consideration when evaluating the same class of solutions.

Solve. LiOH is a strong base, HI is a strong acid, CH_3OH is a molecular compound and nonelectrolyte. The strong acid HI will have the greatest concentration of solvated protons.

Check. The solution concentrations weren't needed to answer the question.

4.31   (a)   A *monoprotic acid* has one ionizable (acidic) H and a *diprotic acid* has two.

(b)   A *strong acid* is completely ionized in aqueous solution, whereas only a fraction of *weak acid* molecules are ionized.

(c)   An *acid* is an H^+ donor, a substance that increases the concentration of H^+ in aqueous solution. A *base* is an H^+ acceptor and thus increases the concentration of OH^- in aqueous solution.

4.33   As a strong acid, $HClO_4$ exists in aqueous solution almost exclusively as $H^+(aq)$ and $ClO_4^-(aq)$; there are almost no neutral $HClO_4$ molecules. As a weak acid, $HClO_2$ exists as a mixture of $H^+(aq)$, $ClO_2^-(aq)$, and $HClO_2(aq)$. There are at least as many neutral $HClO_2$ molecules as anions or cations. For equal solution concentrations, $HClO_4$ will produce a bright light like Figure 4.2(c), while $HClO_2$ will produce a dim light like Figure 4.2(b).

4.35   *Analyze.* Given: chemical formulas. Find: classify as acid, base, salt; strong, weak, or nonelectrolyte.

Plan. See Table 4.3. Ionic or molecular? Ionic, soluble: OH^-, strong base and strong electrolyte; otherwise, salt, strong electrolyte. Molecular: NH_3, weak base and weak electrolyte; H-first, acid; strong acid (Table 4.2), strong electrolyte; otherwise weak acid and weak electrolyte.   *Solve.*

(a)   HF: acid, mixture of ions and molecules (weak electrolyte)

(b) CH_3CN: none of the above, entirely molecules (nonelectrolyte)

(c) $NaClO_4$: salt, entirely ions (strong electrolyte)

(d) $Ba(OH)_2$: base, entirely ions (strong electrolyte)

4.37   *Analyze.* Given: chemical formulas. Find: electrolyte properties.

Plan. In order to classify as electrolytes, formulas must be identified as acids, bases, or salts as in Solution 4.35.   *Solve.*

(a) H_2SO_3: H first, so acid; not in Table 4.2, so weak acid; therefore, weak electrolyte

(b) C_2H_5OH: not acid, not ionic (no metal cation), contains OH group, but not as anion so not a base; therefore, nonelectrolyte

(c) NH_3: common weak base; therefore, weak electrolyte

(d) $KClO_3$: ionic compound, so strong electrolyte

(e) $Cu(NO_3)_2$: ionic compound, so strong electrolyte

4.39   *Plan.* Follow Sample Exercise 4.7.   *Solve.*

(a) $2HBr(aq) + Ca(OH)_2(aq) \rightarrow CaBr_2(aq) + 2H_2O(l)$

$H^+(aq) + OH^-(aq) \rightarrow H_2O(l)$

(b) $Cu(OH)_2(s) + 2HClO_4(aq) \rightarrow Cu(ClO_4)_2(aq) + 2H_2O(l)$

$Cu(OH)_2(s) + 2H^+(aq) \rightarrow 2H_2O(l) + Cu^{2+}(aq)$

(c) $Al(OH)_3(s) + 3HNO_3(aq) \rightarrow Al(NO_3)_3(aq) + 3H_2O(l)$

$Al(OH)_3(s) + 3H^+(aq) \rightarrow 3H_2O(l) + Al^{3+}(aq)$

4.41   *Analyze.* Given: names of reactants. Find: gaseous products.

Plan. Write correct chemical formulas for the reactants, complete and balance the metathesis reaction, and identify either H_2S or CO_2 products as gases.   *Solve.*

(a) $CdS(s) + H_2SO_4(aq) \rightarrow CdSO_4(aq) + H_2S(g)$

$CdS(s) + 2H^+(aq) \rightarrow H_2S(g) + Cd^{2+}(aq)$

(b) $MgCO_3(s) + 2HClO_4(aq) \rightarrow Mg(ClO_4)_2(aq) + H_2O(l) + CO_2(g)$

$MgCO_3(s) + 2H^+(aq) \rightarrow H_2O(l) + CO_2(g) + Mg^{2+}(aq)$

4.43   *Analyze.* Given the formulas or names of reactants, write balanced molecular and net ionic equations for the reactions.

Plan. Write correct chemical formulas for all reactants. Predict products of the neutralization reactions by exchanging ion partners. Balance the complete molecular equation, identify spectator ions by recognizing strong electrolytes, write the corresponding net ionic equation (omitting spectators).   *Solve.*

(a) $CaCO_3(s) + 2HNO_3(aq) \rightarrow Ca(NO_3)_2(aq) + H_2O(l) + CO_2(g)$

         $2H^+(aq) + CaCO_3(s) \rightarrow H_2O(l) + CO_2(g) + Ca^{2+}(aq)$

(b) $FeS(s) + 2HBr(aq) \rightarrow FeBr_2(aq) + H_2S(g)$

         $2H^+(aq) + FeS(s) \rightarrow H_2S(g) + Fe^{2+}(aq)$

Oxidation-Reduction Reactions

4.45   (a)   In terms of electron transfer, *oxidation* is the loss of electrons by a substance, and *reduction* is the gain of electrons (LEO says GER).

       (b)   Relative to oxidation numbers, when a substance is oxidized, its oxidation number increases. When a substance is reduced, its oxidation number decreases.

4.47   *Analyze.* Given the labeled periodic chart, determine which region is most readily oxidized and which is most readily reduced.

       Plan. Review the definition of oxidation and apply it to the properties of elements in the indicated regions of the chart.

       Solve. Oxidation is loss of electrons. Elements readily oxidized form positive ions; these are metals. Elements not readily oxidized tend to gain electrons and form negative ions; these are nonmetals. Elements in regions A, B, and C are metals, and their ease of oxidation is shown in Table 4.5. Metals in region A, Na, Mg, K, and Ca are most easily oxidized. Elements in region D are nonmetals and are least easily oxidized.

4.49   *Analyze.* Given the chemical formula of a substance, determine the oxidation number of a particular element in the substance.

       Plan. Follow the logic in Sample Exercise 4.8. *Solve.*

       (a)   +4   (b)   +4   (c)   +7   (d)   +1   (e)   0   (f)   –1 (O_2^{2-} is peroxide ion)

4.51   *Analyze.* Given: chemical reaction. Find: element oxidized or reduced. *Plan.* Assign oxidation numbers to all species. The element whose oxidation number becomes more positive is oxidized; the one whose oxidation number decreases is reduced.   *Solve.*

       (a) $Ni \rightarrow Ni^{2+}$, Ni is oxidized; $Cl_2 \rightarrow 2Cl^-$, Cl is reduced

       (b) $Fe^{2+} \rightarrow Fe$, Fe is reduced; $Al \rightarrow Al^{3+}$, Al is oxidized

       (c) $Cl_2 \rightarrow 2Cl$, Cl is reduced; $2I^- \rightarrow I_2$, I is oxidized

       (d) $S^{2-} \rightarrow SO_4^{2-}$(S, +6), S is oxidized; H_2O_2 (O, –1) $\rightarrow H_2O$ (O, –2); O is reduced

4.53   *Analyze.* Given: reactants. Find: balanced molecular and net ionic equations.

       Plan. Metals oxidized by H^+ form cations. Predict products by exchanging cations and balance. The anions are the spectator ions and do not appear in the net ionic equations. *Solve.*

(a) $Mn(s) + H_2SO_4(aq) \rightarrow MnSO_4(aq) + H_2(g)$; $Mn(s) + 2H^+(aq) \rightarrow Mn^{2+}(aq) + H_2(g)$

Products with the metal in a higher oxidation state are possible, depending on reaction conditions and acid concentration.

(b) $2Cr(s) + 6HBr(aq) \rightarrow 2CrBr_3(aq) + 3H_2(g)$; $2Cr(s) + 6H^+(aq) \rightarrow 2Cr^{3+}(aq) + 3H_2(g)$

(c) $Sn(s) + 2HCl(aq) \rightarrow SnCl_2(aq) + H_2(g)$; $Sn(s) + 2H^+(aq) \rightarrow Sn^{2+}(aq) + H_2(g)$

(d) $2Al(s) + 6HCHO_2(aq) \rightarrow 2Al(CHO_2)_3(aq) + 3H_2(g)$;

$2Al(s) + 6HCHO_2(aq) \rightarrow 2Al^{3+}(aq) + 6CHO_2^-(aq) + 3H_2(g)$

4.55   *Analyze.* Given: a metal and an aqueous solution. Find: balanced equation.

Plan. Use Table 4.5. If the metal is above the aqueous solution, reaction will occur; if the aqueous solution is higher, NR. If reaction occurs, predict products by exchanging cations (a metal ion or H^+), then balance the equation.   *Solve.*

(a) $Fe(s) + Cu(NO_3)_2(aq) \rightarrow Fe(NO_3)_2(aq) + Cu(s)$

(b) $Zn(s) + MgSO_4(aq) \rightarrow NR$

(c) $Sn(s) + 2HBr(aq) \rightarrow SnBr_2(aq) + H_2(g)$

(d) $H_2(g) + NiCl_2(aq) \rightarrow NR$

(e) $2Al(s) + 3CoSO_4(aq) \rightarrow Al_2(SO_4)_3(aq) + 3Co(s)$

4.57   (a)   i. $Zn(s) + Cd^{2+}(aq) \rightarrow Cd(s) + Zn^{2+}(aq)$

ii. $Cd(s) + Ni^{2+}(aq) \rightarrow Ni(s) + Cd^{2+}(aq)$

(b)   According to Table 4.5, the most active metals are most easily oxidized, and Zn is more active than Ni. Observation (i) indicates that Cd is less active than Zn; observation (ii) indicates that Cd is more active than Ni. Cd is between Zn and Ni on the activity series.

(c)   Place an iron strip in $CdCl_2(aq)$. If Cd(s) is deposited, Cd is less active than Fe; if there is no reaction, Cd is more active than Fe. Do the same test with Co if Cd is less active than Fe or with Cr if Cd is more active than Fe.

Solution Composition; Molarity

4.59   (a)   *Concentration* is an *intensive* property; it is the **ratio** of the amount of solute present in a certain quantity of solvent or solution. This ratio remains constant regardless of how much solution is present.

(b)   The term *0.50 mol HCl* defines an amount (~18 g) of the pure substance HCl. The term 0.50 *M* HCl is a ratio; it indicates that there are 0.50 mol of HCl solute in 1.0 liter of solution. This same ratio of moles solute to solution volume is present regardless of the volume of solution under consideration.

4.61   *Analyze/Plan.* Follow the logic in Sample Exercises 4.11 and 4.13.   *Solve.*

(a) $M = \dfrac{\text{mol solute}}{\text{L solution}}$; $\dfrac{0.0345 \text{ mol } NH_4Cl}{400 \text{ mL}} \times \dfrac{1000 \text{ mL}}{1 \text{ L}} = 0.0863 \ M \ NH_4Cl$

Check. $(0.035 \times 0.4) \approx 0.09 \ M$

(b) $mol = M \times L;\ \dfrac{2.20\ mol\ HNO_3}{1\ L} \times 0.0350\ L = 0.0770\ mol\ HNO_3$

Check. $(2 \times 0.035) \approx 0.07\ mol$

(c) $L = \dfrac{mol}{M};\ \dfrac{0.125\ mol\ KOH}{1.50\ mol\ KOH/L} = 0.0833\ L$ or $83.3\ mL$ of $1.50\ M\ KOH$

Check. $(0.125/1.5)$ is greater than 0.06 and less than 0.12, $\approx 0.08\ M$.

4.63   *Analyze.* Given molarity, M, and volume, L, find mass of $Na^+(aq)$ in the blood.

Plan. Calculate moles $Na^+(aq)$ using the definition of molarity: $M = \dfrac{mol}{L};\ mol = M \times L$.

Calculate mass $Na^+(aq)$ using the definition moles: $mol = g/MM/g = mol \times MM$. (MM is the symbol for molar mass in this manual.)

Solve. $\dfrac{0.135\ mol}{L} \times 5.0\ L \times \dfrac{23.0\ g\ Na^+}{mol\ Na^+} = 15.525 = 16\ g\ Na^+(aq)$

Check. Since there are more than 0.1 mol/L and we have 5.0 L, there should be more than half a mol (11.5 g) of Na^+. The calculation agrees with this estimate.

4.65   *Plan.* Proceed as in Sample Exercises 4.11 and 4.13.

$M = \dfrac{mol}{L};\ mol = \dfrac{g}{M}$ (MM is the symbol for molar mass in this manual.)

Solve.

(a) $\dfrac{0.150\ M\ KBr}{1\ L} \times 0.250\ L \times \dfrac{119.0\ g\ KBr}{1\ mol\ KBr} = 4.46\ g\ KBr$

Check. $(0.15 \times 120) \approx 18;\ 18 \times 0.25 = 18/4 \approx 4.5\ g\ KBr$

(b) $4.75\ g\ Ca(NO_3)_2 \times \dfrac{1\ mol\ Ca(NO_3)_2}{164.1\ g\ Ca(NO_3)_2} \times \dfrac{1}{0.200\ L} = 0.145\ M\ Ca(NO_3)_2$

Check. $(4.8/0.2) \approx 24;\ 24/160 = 3/20 \approx 0.15\ M\ Ca(NO_3)_2$

(c) $5.00\ g\ Na_3PO_4 \times \dfrac{1\ mol\ Na_3PO_4}{163.9\ g\ Na_3PO_4} \times \dfrac{1\ L}{1.50\ mol\ Na_3PO_4} \times \dfrac{1000\ mL}{1\ L}$
$= 20.3\ mL\ solution$

Check. $[5/(160 \times 1.5)] \approx 5/240 \approx 1/50 \approx 0.02\ L = 20\ mL$

4.67   *Analyze.* Given: formula and concentration of each solute. Find: concentration of K^+ in each solution. *Plan.* Note mol K^+/mol solute and compare concentrations or total moles. *Solve.*

(a) $KCl \rightarrow K^+ + Cl^-;\ 0.20\ M\ KCl = 0.20\ M\ K^+$

$K_2CrO_4 \rightarrow 2\ K^+ + CrO_4^{2-};\ 0.15\ M\ K_2CrO_4 = 0.30\ M\ K^+$

$K_3PO_4 \rightarrow 3\ K^+ + PO_4^{3-};\ 0.080\ M\ K_3PO_4 = 0.24\ M\ K^+$

$0.15\ M\ K_2CrO_4$ has the highest K^+ concentration.

(b) K_2CrO_4: $0.30\ M\ K^+ \times 0.0300\ L = 0.0090\ mol\ K^+$

K_3PO_4: $0.24\ M\ K^+ \times 0.0250\ L = 0.0060\ mol\ K^+$

30.0 mL of 0.15 M K_2CrO_4 has more K^+ ions.

4.69   *Analyze.* Given: formula and concentration of each solute. Find: concentration of each species in solution. *Plan.* Decide whether the solute is a strong, weak, or nonelectrolyte, which species are in solution, and concentrations.   *Solve.*

(a) $0.22\ M\ Na^+$, $0.22\ M\ OH^-$

(b) $0.16\ M\ Ca^{2+}$, $0.32\ M\ Br^-$

(c) $0.15\ M$ (CH_3OH is a molecular solute)

(d)   Mixing two solutions is, in effect, a dilution, Equation [4.35].

$M_2 = M_2V_1/V_2$, where V_2 is the total solution volume.

K^+: $\dfrac{0.15\ M \times 0.040\ L}{0.075\ L} = 0.0800 = 0.080\ M$

ClO_3^-: concentration ClO_3^- = concentration $K^+ = 0.080\ M$

SO_4^{2} : $\dfrac{0.22\ M \times 0.0350\ L}{0.075\ L} = 0.1027 = 0.10\ M\ SO_4^{2-}$

Na^+: concentration $Na^+ = 2 \times$ concentration $SO_4^{2-} = 0.21\ M$

4.71   *Analyze/Plan.* Follow the logic of Sample Exercise 4.14.

Solve.

(a) $V_1 = M_2V_2/M_1$; $\dfrac{0.250\ M\ NH_3 \times 100.0\ mL}{14.8\ M\ NH_3} = 1.689 = 1.69\ mL,\ 14.8\ M\ NH_3$

Check. $250/15 \approx 1.5$ mL

(b) $M_2 = M_1V_1/V_2$; $\dfrac{14.8\ M\ NH_3 \times 10.0\ mL}{250\ mL} = 0.592\ M\ NH_3$

Check. $150/250 \approx 0.60\ M$

4.73   (a)   *Plan/Solve.* Follow the logic in Sample Exercise 4.13. The number of moles of sucrose needed is $\dfrac{0.150\ mol}{1\ L} \times 0.125\ L = 0.01875 = 0.0188\ mol$

Weigh out $0.01875\ mol\ C_{12}H_{22}O_{11} \times \dfrac{342.3\ g\ C_{12}H_{22}O_{11}}{1\ mol\ C_{12}H_{22}O_{11}} = 6.42\ g\ C_{12}H_{22}O_{11}$

Add this amount of solid to a 125 mL volumetric flask, dissolve in a small volume of water, and add water to the mark on the neck of the flask. Agitate thoroughly to ensure total mixing.

(b)   *Plan/Solve.* Follow the logic in Sample Exercise 4.14. Calculate the moles of solute present in the final 400.0 mL of 0.100 M $C_{12}H_{22}O_{11}$ solution:

$$\text{moles } C_{12}H_{22}O_{11} = M \times L = \frac{0.100 \text{ mol } C_{12}H_{22}O_{11}}{1 \text{ L}} \times 0.4000 \text{ L} = 0.0400 \text{ mol } C_{12}H_{22}O_{11}$$

Calculate the volume of 1.50 M glucose solution that would contain 0.04000 mol $C_{12}H_{22}O_{11}$:

$$L = \text{moles}/M; 0.04000 \text{ mol } C_{12}H_{22}O_{11} \times \frac{1 \text{ L}}{1.50 \text{ mol } C_{12}H_{22}O_{11}} = 0.02667 = 0.0267 \text{ L}$$

$$0.02667 \text{ L} \times \frac{1000 \text{ mL}}{1 \text{ L}} = 26.7 \text{ mL}$$

Thoroughly rinse, clean, and fill a 50 mL buret with the 1.50 M $C_{12}H_{22}O_{11}$. Dispense 26.7 mL of this solution into a 400 mL volumetric container, add water to the mark, and mix thoroughly. (26.7 mL is a difficult volume to measure with a pipette.)

4.75   *Analyze.* Given: density of pure acetic acid, volume pure acetic acid, volume new solution. Find: molarity of new solution. *Plan.* Calculate the mass of acetic acid, $HC_2H_3O_2$, present in 20.0 mL of the pure liquid.   *Solve.*

$$20.00 \text{ mL acetic acid} \times \frac{1.049 \text{ g acetic acid}}{1 \text{ mL acetic acid}} = 20.98 \text{ g acetic acid}$$

$$20.98 \text{ g } HC_2H_3O_2 \times \frac{1 \text{ mol } HC_2H_3O_2}{60.05 \text{ g } HC_2H_3O_2} = 0.349375 = 0.3494 \text{ mol } HC_2H_3O_2$$

$$M = \text{mol}/L = \frac{0.349375 \text{ mol } HC_2H_3O_2}{0.2500 \text{ L solution}} = 1.39750 = 1.398 \ M \ HC_2H_3O_2$$

Check. $(20 \times 1) \approx 20$ g acid; $(20/60) \approx 0.33$ mol acid; $(0.33/0.25 = 0.33 \times 4) \approx 1.33 \ M$

Solution Stoichiometry; Titrations

4.77   *Analyze.* Given: volume and molarity $AgNO_3$. Find: mass NaCl.

Plan. $M \times L$ = mol $AgNO_3$ = mol Ag^+; balanced equation gives ratio mol NaCl/mol $AgNO_3$; mol NaCl $\rightarrow$ g NaCl.   *Solve.*

$$\frac{0.100 \text{ mol } AgNO_3}{1 \text{ L}} \times 0.0200 \text{ L} = 2.00 \times 10^{-3} \text{ mol } AgNO_3(aq)$$

$$AgNO_3(aq) + NaCl(aq) \rightarrow AgCl(s) + NaNO_3(aq)$$

$$\text{mol NaCl} = \text{mol } AgNO_3 = 2.00 \times 10^{-3} \text{ mol NaCl}$$

$$2.00 \times 10^{-3} \text{ mol NaCl} \times \frac{58.44 \text{ g NaCl}}{1 \text{ mol NaCl}} = 0.117 \text{ g NaCl}$$

Check. $(0.1 \times 0.02) = 0.002$ mol; $(0.002 \times 60) \approx 0.12$ g NaCl

4.79   (a)   *Analyze.* Given: M and vol base, M acid. Find: vol acid

Plan/Solve. Write the balanced equation for the reaction in question:

$$HClO_4(aq) + NaOH(aq) \rightarrow NaClO_4(aq) + H_2O(l)$$

Calculate the moles of the known substance, in this case NaOH.

$$\text{moles NaOH} = M \times L = \frac{0.0875 \text{ mol NaOH}}{1 \text{ L}} \times 0.0500 \text{ L} = 0.004375 = 0.00438 \text{ mol NaOH}$$

Apply the mole ratio (mol unknown/mol known) from the chemical equation.

$$0.004375 \text{ mol NaOH} \times \frac{1 \text{ mol HClO}_4}{1 \text{ mol NaOH}} = 0.004375 \text{ mol HClO}_4$$

Calculate the desired quantity of unknown, in this case the volume of 0.115 M $HClO_4$ solution.

$$L = \text{mol}/M; \ \ L = 0.004375 \text{ mol HClO}_4 \times \frac{1 \text{ L}}{0.115 \text{ mol HClO}_4} = 0.0380 \text{ L} = 38.0 \text{ mL}$$

Check. $(0.09 \times 0.045) = 0.0045$ mol; $(0.0045/0.11) \approx 0.040$ L ≈ 40 mL

(b)   Following the logic outlined in part (a):

$$2HCl(aq) + Mg(OH)_2(s) \rightarrow MgCl_2(aq) + 2H_2O(l)$$

$$2.87 \text{ g Mg(OH)}_2 \times \frac{1 \text{ mol Mg(OH)}_2}{58.32 \text{ g Mg(OH)}_2} = 0.049211 = 0.0492 \text{ mol Mg(OH)}_2$$

$$0.0492 \text{ mol Mg(OH)}_2 \times \frac{2 \text{ mol HCl}}{1 \text{ mol Mg(OH)}_2} = 0.0984 \text{ mol HCl}$$

$$L = \text{mol}/M = 0.09840 \text{ mol HCl} \times \frac{1 \text{ L HCl}}{0.128 \text{ mol HCl}} = 0.769 \text{ L} = 769 \text{ mL}$$

(c) $AgNO_3(aq) + KCl(aq) \rightarrow AgCl(s) + KNO_3(aq)$

$$785 \text{ mg KCl} \times \frac{1 \times 10^{-3} \text{ g}}{1 \text{ mg}} \times \frac{1 \text{ mol KCl}}{74.55 \text{ g KCl}} \times \frac{1 \text{ mol AgNO}_3}{1 \text{ mol KCl}} = 0.01053 = 0.0105 \text{ mol AgNO}_3$$

$$M = \text{mol}/L = \frac{0.01053 \text{ mol AgNO}_3}{0.0258 \text{ L}} = 0.408 \ M \text{ AgNO}_3$$

(d) $HCl(aq) + KOH(aq) \rightarrow KCl(aq) + H_2O(l)$

$$\frac{0.108 \text{ mol HCl}}{1 \text{ L}} \times 0.0453 \text{ L} \times \frac{1 \text{ mol KOH}}{1 \text{ mol HCl}} \times \frac{56.11 \text{ g KOH}}{1 \text{ mol KOH}} = 0.275 \text{ g KOH}$$

4.81   *Analyze/Plan.* See Exercise 4.79(a) for a more detailed approach.   *Solve.*

$$\frac{6.0 \text{ mol H}_2\text{SO}_4}{1 \text{ L}} \times 0.027 \text{ L} \times \frac{2 \text{ mol NaHCO}_3}{1 \text{ mol H}_2\text{SO}_4} \times \frac{84.01 \text{ g NaHCO}_3}{1 \text{ mol NaHCO}_3} = 27 \text{ g NaHCO}_3$$

4.83   *Analyze.* Given: M and vol HBr, vol Ca(OH)$_2$. Find: M Ca(OH)$_2$,

         g Ca(OH)$_2$/100 mL soln

         Plan. Write balanced equation;

$$\text{mol HBr} \xrightarrow{\frac{\text{mol}}{\text{ratio}}} \text{mol Ca(OH)}_2 \rightarrow M \text{ Ca(OH)}_2 ; \rightarrow \text{g Ca(OH)}_2 / 100 \text{ mL}$$

Solve. The neutralization reaction here is:

$$2HBr(aq) + Ca(OH)_2(aq) \rightarrow CaBr_2(aq) + 2H_2O(l)$$

$$0.0488 \text{ L HBr soln} \times \frac{5.00 \times 10^{-2} \text{ mol HBr}}{1 \text{ L soln}} \times \frac{1 \text{ mol Ca(OH)}_2}{2 \text{ mol HBr}} \times \frac{1}{0.100 \text{ L of Ca(OH)}_2}$$

$$= 1.220 \times 10^{-2} = 1.22 \times 10^{-2} \, M \text{ Ca(OH)}_2$$

From the molarity of the saturated solution, we can calculate the gram solubility of $Ca(OH)_2$ in 100 mL of H_2O.

$$0.100 \text{ L soln} \times \frac{1.220 \times 10^{-2} \text{ mol Ca(OH)}_2}{1 \text{ L soln}} \times \frac{74.10 \text{ g Ca(OH)}_2}{1 \text{ mol Ca(OH)}_2} = 0.0904 \text{ g Ca(OH)}_2 \text{ in 100 mL soln}$$

Check. $(0.05 \times 0.05 / 0.2) = 0.0125 \, M$; $(0.1 \times 0.0125 \times 64) \approx 0.085$ g/100 mL

4.85   (a) $NiSO_4(aq) + 2KOH(aq) \rightarrow Ni(OH)_2(s) + K_2SO_4(aq)$

        (b)   The precipitate is $Ni(OH)_2$.

        (c)   *Plan.* Compare mol of each reactant; mol = $M \times L$

              Solve. $0.200 \, M$ KOH $\times$ 0.1000 L KOH = 0.0200 mol KOH

              $0.150 \, M$ NiSO$_4$ $\times$ 0.2000 L KOH = 0.0300 mol NiSO$_4$

              1 mol NiSO$_4$ requires 2 mol KOH, so 0.0300 mol NiSO$_4$ requires 0.0600 mol KOH. Since only 0.0200 mol KOH is available, KOH is the limiting reactant.

        (d)   *Plan.* The amount of the limiting reactant (KOH) determines amount of product, in this case $Ni(OH)_2$.

              Solve. $0.0200 \text{ mol KOH} \times \frac{1 \text{ mol Ni(OH)}_2}{2 \text{ mol KOH}} \times \frac{92.71 \text{ g Ni(OH)}_2}{1 \text{ mol Ni(OH)}_2} = 0.927 \text{ g Ni(OH)}_2$

        (e)   *Plan/Solve.* Limiting reactant: OH^-: no excess OH^- remains in solution.

              Excess reactant: Ni^{2+}: M Ni^{2+} remaining = mol Ni^{2+} remaining/L solution

              0.0300 mol Ni^{2+} initial – 0.0100 mol Ni^{2+} reacted = 0.0200 mol Ni^{2+} remaining

              0.0200 mol Ni^{2+}/0.3000 L = $0.0667 \, M$ $Ni^{2+}(aq)$

              Spectators: SO_4^{2-}, K^+. These ions do not react, so the only change in their concentration is dilution. The final volume of the solution is 0.3000 L.

              $M_2 = M_1 V_1/V_2$: $0.200 \, M$ K^+ $\times$ 0.1000 L/0.3000 L = $0.0667 \, M$ $K^+(aq)$

              $0.150 \, M$ SO_4^{2-} $\times$ 0.2000 L/0.3000 L = $0.100 \, M$ $SO_4^{2-}(aq)$

4.87   *Analyze.* Given: mass impure $Mg(OH)_2$; M and vol **excess** HCl; M and vol NaOH.

        Find: mass % $Mg(OH)_2$ in sample. *Plan/Solve.* Write balanced equations.

        $Mg(OH)_2(s) + 2HCl(aq) \rightarrow MgCl_2(aq) + 2H_2O(l)$

        $HCl(aq) + NaOH(aq) \rightarrow NaCl(aq) + H_2O(l)$

        Calculate total moles HCl = M HCl $\times$ L HCl

$$\frac{0.2050 \text{ mol HCl}}{1 \text{ L soln}} \times 0.1000 \text{ L} = 0.02050 \text{ mol HCl total}$$

mol excess HCl = mol NaOH used = M NaOH × L NaOH

$$\frac{0.1020 \text{ mol NaOH}}{1 \text{ L soln}} \times 0.01985 \text{ L} = 0.0020247 = 0.002025 \text{ mol NaOH}$$

mol HCl reacted with $Mg(OH)_2$ = total mol HCl – excess mol HCl

0.02050 mol total – 0.0020247 mol excess = 0.0184753 = 0.01848 mol HCl reacted

(The result has 5 decimal places and 4 sig. figs.)

Use mol ratio to get mol $Mg(OH)_2$ in sample, then molar mass of $Mg(OH)_2$ to get g pure $Mg(OH)_2$.

$$0.0184753 \text{ mol HCl} \times \frac{1 \text{ mol Mg(OH)}_2}{2 \text{ mol HCl}} \times \frac{58.32 \text{ g Mg(OH)}_2}{1 \text{ mol Mg(OH)}_2} = 0.5387 \text{ Mg(OH)}_2$$

$$\text{mass \% Mg(OH)}_2 = \frac{\text{g Mg(OH)}_2}{\text{g sample}} \times 100 = \frac{0.5388 \text{ g Mg(OH)}_2}{0.5895 \text{ g sample}} \times 100 = 91.40\% \text{ Mg(OH)}_2$$

Additional Exercises

4.90   The precipitate is $CdS(s)$. $Na^+(aq)$ and $NO_3^-(aq)$ are spectator ions and remain in solution. Any excess reactant ions also remain in solution. The net ionic equation is:

$Cd^{2+}(aq) + S^{2-}(aq) \rightarrow CdS(s)$.

4.92   (a,b) Expt. 1 No reaction

Expt. 2 $2Ag^+(aq) + CrO_4^{2-}(aq) \rightarrow Ag_2CrO_4(s)$ red precipitate

Expt. 3 No reaction

Expt. 4 $2Ag^+(aq) + C_2O_4^{2-}(aq) \rightarrow Ag_2C_2O_4(s)$ white precipitate

Expt. 5 $Ca^{2+}(aq) + C_2O_4^{2-}(aq) \rightarrow CaC_2O_4(s)$ white precipitate

Expt. 6 $Ag^+(aq) + Cl^-(aq) \rightarrow AgCl(s)$ white precipitate

(c)   The silver salts of both ions are insoluble, but many silver salts are insoluble (Expt. 6). The calcium salt of CrO_4^{2-} is soluble (Expt. 3), while the calcium salt of $C_2O_4^{2-}(aq)$ is insoluble (Expt. 5). Thus, chromate salts appear more soluble than oxalate salts.

4.95 $4NH_3(g) + 5O_2(g) \rightarrow 4NO(g) + 6H_2O(g)$.
       N = –3   O = 0   N = +2   O = –2

(a)   redox reaction (b)   N is oxidized, O is reduced

$2NO(g) + O_2(g) \rightarrow 2NO_2(g)$.
N = +2   O = 0   N = +4, O = +2

(a)   redox reaction (b)   N is oxidized, O is reduced

$3NO_2(g) + H_2O(l) \rightarrow HNO_3(aq) + NO(g)$.

N = +4   N = +5   N = +2

(a)   redox reaction

(b)   N is oxidized ($NO_2 \rightarrow HNO_3$), N is reduced ($NO_2 \rightarrow NO$). A reaction where the same element is both oxidized and reduced is called disproportionation.

4.98   *Plan.* Calculate moles KBr from the two quantities of solution (mol = $M \times$ L), then new molarity (M = mol/L). KBr is nonvolatile, so no solute is lost when the solution is evaporated to reduce the total volume. *Solve.*

1.00 M KBr $\times$ 0.0350 L = 0.0350 mol KBr; 0.600 M KBr $\times$ 0.060 L = 0.0360 mol KBr

0.0350 mol KBr + 0.0360 mol KBr = 0.0710 mol KBr total

$$\frac{0.0710 \text{ mol KBr}}{0.0500 \text{ L soln}} = 1.42 \text{ } M \text{ KBr}$$

4.100   (a) $\dfrac{50 \text{ pg}}{1 \text{ mL}} \times \dfrac{1 \times 10^{-12} \text{ g}}{1 \text{ pg}} \times \dfrac{1 \times 10^3 \text{ mL}}{\text{L}} \times \dfrac{1 \text{ mol Na}}{23.0 \text{ g Na}} = 2.17 \times 10^{-9} = 2.2 \times 10^{-9} \text{ } M \text{ Na}^+$

(b) $\dfrac{2.17 \times 10^{-9} \text{ mol Na}}{1 \text{ L soln}} \times \dfrac{1 \text{ L}}{1 \times 10^3 \text{ cm}^3} \times \dfrac{6.02 \times 10^{23} \text{ Na atom}}{1 \text{ mol Na}}$

$$= 1.3 \times 10^{12} \text{ atom or Na}^+ \text{ ions/cm}^3$$

4.103   *Plan.* mol $MnO_4^- = M \times L \rightarrow$ mol ratio $\rightarrow$ mol $H_2O_2 \rightarrow M$ H_2O_2.   *Solve.*

$2MnO_4^-(aq) + 5H_2O_2(aq) + 6H^+ \rightarrow 2Mn^{2+}(aq) + 5O_2(aq) + 8H_2O(l)$

$$\frac{0.124 \text{ mol } MnO_4^-}{\text{L}} \times 0.0168 \text{ L } MnO_4^- \times \frac{5 \text{ mol } H_2O_2}{2 \text{ mol } MnO_4^-} \times \frac{1}{0.0100 \text{ L } H_2O_2}$$

$$= 0.5208 \text{ mol } H_2O_2 / \text{L} = 0.521 \text{ } M \text{ } H_2O_2$$

Integrative Exercises

4.105   *Analyze.* Given the name and mass of a solute and volume of solution, and the name of a second solute, calculate the mass of the second solute needed to make a solution of equal concentration and volume.

Plan. Using definitions of moles and molarity, find a simple ratio for calculating the mass of the second solute.

Solve. $M_1 = M_2$, so $\dfrac{\text{mol}(1)}{\text{L}} = \dfrac{\text{mol}(2)}{\text{L}}$

Since the volume is 1.00 L for both solutions, mol(1) = mol(2); $\dfrac{g(1)}{\text{MM}(1)} = \dfrac{g(2)}{\text{MM}(2)}$

(1) = sodium chlorate = $NaClO_3$, MM = 106.44

(2) = sodium chlorite = $NaClO_2$, MM = 90.44

$$\frac{1.28 \text{ g}}{106.44} = \frac{x \text{ g}}{90.44}; x = \frac{1.28(90.44)}{106.44} \text{ g} = 1.09 \text{ g NaClO}_2$$

Check. Since the molar mass of $NaClO_2$ is less than that of $NaClO_3$, a smaller mass of $NaClO_2$ will contain the same number of molecules.

4.107 $Ba^{2+}(aq) + SO_4{}^{2-}(aq) \rightarrow BaSO_4(s)$

$$0.2815 \text{ g BaSO}_4 \times \frac{137.3 \text{ g Ba}}{233.4 \text{ g BaSO}_4} = 0.16560 = 0.1656 \text{ g Ba}$$

$$\text{mass \%} = \frac{\text{g Ba}}{\text{g sample}} \times 100 = \frac{0.16560 \text{ g Ba}}{3.455 \text{ g sample}} \times 100 = 4.793 \text{ \% Ba}$$

4.110   (a) $Na_2SO_4(aq) + Pb(NO_3)_2(s) \rightarrow PbSO_4(s) + 2NaNO_3(aq)$

(b)   Calculate mol of each reactant and compare.

$$1.50 \text{ g Pb(NO}_3)_2 \times \frac{1 \text{ mol Pb(NO}_3)_2}{331.2 \text{ g Pb(NO}_3)_2} = 0.004529 = 4.53 \times 10^{-3} \text{ mol Pb(NO}_3)_2$$

$0.100 \, M \, Na_2SO_4 \times 0.125 \text{ L} = 0.0125 \text{ mol Na}_2SO_4$

Since the reactants combine in a 1:1 mol ratio, $Pb(NO_3)_2$ is the limiting reactant.

(c) $Pb(NO_3)_2$ is the limiting reactant, so no Pb^{2+} remains in solution. The remaining ions are: $SO_4{}^{2-}$ (excess reactant), Na^+ and $NO_3{}^-$ (spectators).

$SO_4{}^{2-}$: 0.0125 mol $SO_4{}^{2-}$ initial $- 0.00453$ mol $SO_4{}^{2-}$ reacted

$= 0.00797 = 0.0080$ mol $SO_4{}^{2-}$ remain

0.00797 mol $SO_4{}^{2-}$ / 0.125 L soln $= 0.064 \, M \, SO_4{}^{2-}$

Na^+:   Since the total volume of solution is the volume of $Na_2SO_4(aq)$ added, the concentration of Na^+ is unchanged.

$0.100 \, M \, Na_2SO_4 \times (2 \text{ mol Na}^+ / 1 \text{ mol Na}_2SO_4) = 0.200 \, M \, Na^+$

$NO_3{}^-$: 4.53×10^{-3} mol $Pb(NO_3)_2 \times 2$ mol $NO_3{}^-$ / 1 mol $Pb(NO_3)_2$

$= 9.06 \times 10^{-3}$ mol $NO_3{}^-$

9.06×10^{-3} mol $NO_3{}^-$ / 0.125 L $= 0.0725 \, M \, NO_3{}^-$

4.113 $Ag^+(aq) + Cl^-(aq) \rightarrow AgCl(s)$

$$\frac{0.2997 \text{ mol Ag}^+}{1 \text{ L}} \times 0.04258 \text{ L} \times \frac{1 \text{ mol Cl}^-}{1 \text{ mol Ag}^+} \times \frac{35.453 \text{ g Cl}^-}{1 \text{ mol Cl}^-} = 0.45242 = 0.4524 \text{ g Cl}^-$$

$$25.00 \text{ mL seawater} \times \frac{1.025 \text{ g}}{\text{mL}} = 25.625 = 25.63 \text{ g sea water}$$

$$\text{mass \% Cl}^- = \frac{0.45242 \text{ g Cl}^-}{25.625 \text{ g seawater}} \times 100 = 1.766\% \text{ Cl}^-$$

4.115   *Analyze.* Given 10 ppb As, find mass Na_3AsO_4 in 1.00 L of drinking water.

Plan. Use the definition of ppb to calculate g As in 1.0 L of water. Convert g As → g Na_3AsO_4 using molar masses. Assume the density of H_2O is 1.00 g/mL.

Solve. 1 billion $= 1 \times 10^9$; 1 ppb $= \dfrac{1\,\text{g solute}}{1 \times 10^9\,\text{g solution}}$

$$\frac{1\,\text{g solute}}{1 \times 10^9\,\text{g solution}} \times \frac{1\,\text{g solution}}{1\,\text{mL solution}} \times \frac{1 \times 10^3\,\text{mL}}{1\,\text{L solution}} = \frac{\text{g As}}{1 \times 10^6\,\text{L}\,H_2O}$$

$$10\,\text{ppb As} = \frac{10\,\text{g As}}{1 \times 10^6\,\text{L}\,H_2O} \times 1\,\text{L}\,H_2O = 1.0 \times 10^{-5}\,\text{g As/L}.$$

$$1.0 \times 10^{-5}\,\text{g As} \times \frac{1\,\text{mol As}}{74.92\,\text{g As}} \times \frac{1\,\text{mol}\,Na_3AsO_4}{1\,\text{mol As}} \times \frac{207.89\,\text{g}\,Na_3AsO_4}{1\,\text{mol}\,Na_3AsO_4}$$

$$= 2.8 \times 10^{-5}\,\text{g}\,Na_3AsO_4\,\text{in 1.00 L}\,H_2O$$

5 Thermochemistry

Visualizing Concepts

5.1   The book's potential energy is due to the opposition of gravity by an object of mass m at a distance d above the surface of the earth. Kinetic energy is due to the motion of the book. As the book falls, d decreases and potential energy changes into kinetic energy.

The first law states that the total energy of a system is conserved. At the instant before impact, all potential energy has been converted to kinetic energy, so the book's total kinetic energy is 85 J, assuming no transfer of energy as heat.

5.4   (a).   No. This distance traveled to the top of a mountain depends on the path taken by the hiker. Distance is a path function, not a state function.

(b)   Yes. Change in elevation depends only on the location of the base camp and the height of the mountain, not on the path to the top. Change in elevation is a state function, not a path function.

5.6   (a)   The temperature of the system and surroundings will equalize, so the temperature of the hotter system will decrease and the temperature of the colder surroundings will increase. The system loses heat by decreasing its temperature, so the sign of q is (–). The process is exothermic.

(b)   It neither volume nor pressure of the system changes, w = 0 and $\Delta E = q = \Delta H$. The change in internal energy is equal to the change in enthalpy.

The Nature of Energy

5.9   An object can possess energy by virtue of its motion or position. Kinetic energy, the energy of motion, depends on the mass of the object and its velocity. Potential energy, stored energy, depends on the position of the object relative to the body with which it interacts.

5.11   (a)   *Analyze.* Given: mass and speed of ball.   Find: kinetic energy.

Plan. Since $1\,J = 1\,kJ \cdot m^2/s^2$, convert g $\rightarrow$ kg to obtain E_k in joules.

Solve. $E_k = 1/2\,mv^2 = 1/2 \times 45\,g \times \dfrac{1\,kg}{1000\,g} \times \left(\dfrac{61\,m}{1\,s}\right)^2 = \dfrac{84\,kg \cdot m^2}{1\,s^2} = 84\,J$

Check. $1/2(45 \times 3600/1000) \approx 1/2(40 \times 4) \approx 80\,J$

(b) $83.72\,J \times \dfrac{1\,cal}{4.184\,J} = 20\,cal$

(c) As the ball hits the sand, its speed (and hence its kinetic energy) drops to zero. Most of the kinetic energy is transferred to the sand, which deforms when the ball lands. Some energy is released as heat through friction between the ball and the sand.

5.13 *Analyze.* Given: heat capacity of water $= 1\,\text{Btu/lb} \cdot °\text{F}$ Find: J/Btu

Plan. heat capacity of water $= 4.184\,\text{J/g} \cdot °\text{C}$; $\dfrac{J}{g \cdot °C} \rightarrow \dfrac{J}{lb \cdot °F} \rightarrow \dfrac{J}{Btu}$

This strategy requires changing °F to °C. Since this involves the magnitude of a degree on each scale, rather than a specific temperature, the 32 in the temperature relationship is not needed.

$100\,°\text{C} = 180\,°\text{F}; 5\,°\text{C} = 9\,°\text{F}$

Solve. $\dfrac{4.184\,\text{J}}{g \cdot °C} \times \dfrac{453.6\,\text{g}}{lb} \times \dfrac{5\,°C}{9\,°F} \times \dfrac{1\,lb \cdot °F}{1\,Btu} = 1054\,\text{J/Btu}$

5.15 The energy source of a 100 watt light bulb is electrical current from household wiring. Current passes through and heats a tungsten filament (thin wire) in the bulb. The energy is radiated in the form of heat and visible light.

The energy source for an adult person is food. When a person eats, the food undergoes a complex series of chemical reactions that release the potential energy stored in chemical bonds. Some of this energy is transferred as electrical impulses that trigger muscle action and become kinetic energy. Some is released as heat.

In both cases, the energy must travel through a network (house wiring and lamp or human body) and undergo several changes in form before it is in the correct location and form to accomplish the desired task. In both cases, the energy given off as heat is wasted; it cannot be applied to the tasks of producing light or motion.

5.17 (a) In thermodynamics, the *system* is the well-defined part of the universe whose energy changes are being studied.

(b) A *closed system* can exchange heat but not mass with its surroundings.

5.19 (a) *Work* is a force applied over a distance.

(b) The amount of work done is the magnitude of the force times the distance over which it is applied. $w = F \times d$.

5.21 (a) Gravity; work is done because the force of gravity is opposed and the pencil is lifted.

(b) Mechanical force; work is done because the force of the coiled spring is opposed as the spring is compressed over a distance.

The First Law of Thermodynamics

5.23 (a) In any chemical or physical change, energy can be neither created nor destroyed, but it can be changed in form.

(b)   The total *internal energy* (E) of a system is the sum of all the kinetic and potential energies of the system components.

(c)   The internal energy of a system increases when work is done on the system by the surroundings and/or when heat is transferred to the system from the surroundings (the system is heated).

5.25   *Analyze.* Given: heat and work. Find: magnitude and sign of ΔE.

Plan. In each case, evaluate q and w in the expression $\Delta E = q + w$. For an exothermic process, q is negative; for an endothermic process, q is positive.   *Solve.*

(a)   q is positive because the system absorbs heat and w is negative because the system does work. $\Delta E = 85$ kJ – 29 kJ = 56 kJ. The process is endothermic.

(b) $\Delta E = 1.50$ kJ – 657 J = 1.50 kJ – 0.657 kJ = 0.843 = 0.84 kJ. The process is endothermic.

(c)   q is negative because the system releases heat, and w is negative because the system does work. $\Delta E = -57.5$ kJ – 13.5 kJ = –71.0 kJ. The process is exothermic.

5.27   *Analyze.* How do the different physical situations (cases) affect the changes to heat and work of the system upon addition of 100 J of energy? *Plan.* Use the definitions of heat and work and the First Law to answer the questions.   *Solve.*

If the piston is allowed to move, case (1), the heated gas will expand and push the piston up, doing work on the surroundings. If the piston is fixed, case (2), most of the electrical energy will be manifested as an increase in heat of the system.

(a)   Since little or no work is done by the system in case (2), the gas will absorb most of the energy as heat, the case (2) gas will have the higher temperature.

(b)   In case (2), w ~ 0 and q ≈ 100 J. In case (1), a significant amount of energy will be used to do work on the surroundings (–w), but some will be absorbed as heat (+q). (The transfer of electrical energy into work is never completely efficient!)

(c) ΔE is greater for case (2), because the entire 100 J increases the internal energy of the system, rather than a part of the energy doing work on the surroundings.

5.29   (a)   A *state function* is a property of a system that depends only on the physical state (pressure, temperature, etc.) of the system, not on the route used by the system to get to the current state.

(b)   Internal energy and enthalpy **are** state functions; heat **is not** a state function.

(c)   Work **is not** a state function. The amount of work required to move from state A to state B depends on the path or series of processes used to accomplish the change.

Enthalpy

5.31   (a)   Change in enthalpy (ΔH) is usually easier to measure than change in internal energy (ΔE) because, at constant pressure, $\Delta H = q$. The heat flow associated with a process at constant pressure can easily be measured as a change in temperature. Measuring ΔE requires a means to measure both q and w.

(b)   If ΔH is negative, the enthalpy of the system decreases and the process is exothermic.

5.33   (a) $HC_2H_3O_2(l) + 2O_2(g) \rightarrow 2H_2O(l) + 2CO_2(g)$ $\Delta H = -871.7 \text{ kJ}$

        (b)   *Analyze.* How are reactants and products arranged on an enthalpy diagram?

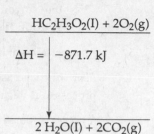

                Plan. The substances (reactants or products, collectively) with higher enthalpy are shown on the upper level, and those with lower enthalpy are shown on the lower level.

                Solve. For this reaction, ΔH is negative, so the products have lower enthalpy and are shown on the lower level; reactants are on the upper level. The arrow points in the direction of reactants to products and is labeled with the value of ΔH.

5.35   *Plan.* Consider the sign of ΔH.

        Solve. Since ΔH is negative, the reactants, $2Cl(g)$ have the higher enthalpy.

5.37   *Analyze/Plan.* Follow the strategy in Sample Exercise 5.4.   *Solve.*

        (a)   Exothermic (ΔH is negative)

        (b) $2.4 \text{ g Mg} \times \dfrac{1 \text{ mol Mg}}{24.305 \text{ g Mg}} \times \dfrac{-1204 \text{ kJ}}{2 \text{ mol Mg}} = -59 \text{ kJ heat transferred}$

                Check. The units of kJ are correct for heat. The negative sign indicates heat is evolved.

        (c) $-96.0 \text{ kJ} \times \dfrac{2 \text{ mol MgO}}{-1204 \text{ kJ}} \times \dfrac{40.30 \text{ g MgO}}{1 \text{ mol Mg}} = 6.43 \text{ g MgO produced}$

                Check. Units are correct for mass. $(100 \times 2 \times 40 / 1200) \approx (8000 / 1200) \approx 6.5 \text{ g}$

        (d) $2MgO(s) \rightarrow 2Mg(s) + O_2(g)$ $\Delta H = +1204 \text{ kJ}$

                This is the reverse of the reaction given above, so the sign of ΔH is reversed.

                $7.50 \text{ g MgO} \times \dfrac{1 \text{ mol MgO}}{40.30 \text{ g MgO}} \times \dfrac{1204 \text{ kJ}}{2 \text{ mol MgO}} = +112 \text{ kJ heat absorbed}$

                Check. The units are correct for energy. $(\sim 9000 / 80) \approx 110 \text{ kJ}$

5.39   *Analyze.* Given: balanced thermochemical equation, various quantities of substances and/or enthalpy. *Plan.* Enthalpy is an extensive property; it is "stoichiometric." Use the mole ratios implicit in the balanced thermochemical equation to solve for the desired quantity. Use molar masses to change mass to moles and vice versa where appropriate. *Solve.*

        (a) $0.200 \text{ mol AgCl} \times \dfrac{-65.5 \text{ kJ}}{1 \text{ mol AgCl}} = -13.1 \text{ kJ}$

                Check. Units are correct; sign indicates heat evolved.

(b) $2.50 \text{ g AgCl} \times \dfrac{1 \text{ mol AgCl}}{143.3 \text{ g AgCl}} \times \dfrac{-65.5 \text{ kJ}}{1 \text{ mol AgCl}} = -1.14 \text{ kJ}$

Check. Units correct; sign indicates heat evolved.

(c) $0.150 \text{ mmol AgCl} \times \dfrac{1 \times 10^{-3} \text{ mol}}{1 \text{ mmol}} \times \dfrac{+65.5 \text{ kJ}}{1 \text{ mol AgCl}} = 0.009825 \text{ kJ} = 9.83 \text{ J}$

Check. Units correct; sign of ΔH reversed; sign indicates heat is absorbed during the reverse reaction.

5.41   At constant pressure, $\Delta E = \Delta H - P\Delta V$. In order to calculate ΔE, more information about the conditions of the reaction must be known. For an ideal gas at constant pressure and temperature, $P\Delta V = RT\Delta n$. The values of either P and ΔV or T and Δn must be known to calculate ΔE from ΔH.

5.43   *Analyze/Plan.* $q = -79$ kJ (heat is given off by the system), $w = -18$ kJ (work is done by the system).   *Solve.*

$\Delta E = q + w = -79 \text{ kj} - 18 \text{ kJ} = -97 \text{ kJ}. \ \Delta H = q = -79 \text{ kJ}$ (at constant pressure).

Check. The reaction is exothermic.

5.45   *Analyze.* Given: balanced thermochemical equation. *Plan.* Follow the guidelines given in Section 5.4 for evaluating thermochemical equations.   *Solve.*

(a)   When a chemical equation is reversed, the sign of ΔH is reversed.

       $CO_2(g) + 2H_2O(l) \rightarrow CH_3OH(l) + 3/2 \, O_2(g)$ $\Delta H = 726.5$ kJ

(b)   Enthalpy is extensive. If the coefficients in the chemical equation are multiplied by 2 to obtain all integer coefficients, the enthalpy change is also multiplied by 2.

       $2CH_3OH(l) + 3O_2(g) \rightarrow 2CO_2(g) + 4H_2O(l)$ $\Delta H = 2(-726.5) \text{ kJ} = -1453 \text{ kJ}$

(c)   The exothermic forward reaction is more likely to be thermodynamically favored.

(d)   Vaporization (liquid $\rightarrow$ gas) is endothermic. If the product were $H_2O(g)$, the reaction would be more endothermic and would have a smaller negative ΔH. (Depending on temperature, the enthalpy of vaporization for 2 mol H_2O is about +88 kJ, not large enough to cause the overall reaction to be endothermic.)

Calorimetry

The specific heat of water to four significant figures, **4.184 J/g • K,** will be used in many of the following exercises; temperature units of K and °C will be used interchangeably.

5.47   (a)   J/°C or J/K. Heat capacity is the amount of heat in J required to raise the temperature of an object or a certain amount of a substance 1°C or 1 K. Since the amount is defined, units of amount are not included.

(b) $\dfrac{J}{g \cdot {}^\circ C}$ or $\dfrac{J}{g \cdot K}$ Specific heat is a particular kind of heat capacity where the amount of substance is 1 g.

(c)   To calculate heat capacity from specific heat, the **mass** of the particular piece of copper pipe must be known.

5.49   *Plan.* Manipulate the definition of specific heat to solve for the desired quantity, paying close attention to units. specific heat $= q/(m \times \Delta t)$.   *Solve.*

(a) $\dfrac{4.184\,\text{J}}{1\,\text{g}\cdot\text{K}}$ or $\dfrac{4.184\,\text{J}}{1\,\text{g}\cdot\text{°C}}$

(b) $\dfrac{4.184\,\text{J}}{1\,\text{g}\cdot\text{°C}} \times \dfrac{18.02\,\text{g H}_2\text{O}}{1\,\text{mol H}_2\text{O}} = \dfrac{75.40\,\text{J}}{\text{mol}\cdot\text{°C}}$

(c) $\dfrac{185\,\text{g H}_2\text{O} \times 4.184\,\text{J}}{1\,\text{g}\cdot\text{°C}} = 774\,\text{J/°C}$

(d) $10.00\,\text{kg H}_2\text{O} \times \dfrac{1000\,\text{g}}{1\,\text{kg}} \times \dfrac{4.184\,\text{J}}{1\,\text{g}\cdot\text{°C}} \times \dfrac{1\,\text{kJ}}{1000\,\text{J}} \times (46.2\text{°C} - 24.6\text{°C}) = 904\,\text{kJ}$

Check. $(10 \times 4 \times 20) \approx 800$ kJ; the units are correct. Note that the conversion factors for kg → g and J → kJ cancel. An equally correct form of specific heat would be kJ/kg •C°

5.51   *Analyze/Plan.* Follow the logic in Sample Exercise 5.5.   *Solve.*

$1.05\,\text{kg Fe} \times \dfrac{1000\,\text{g}}{1\,\text{kg}} \times \dfrac{0.450\,\text{J}}{\text{g}\cdot\text{K}} \times (88.5\,\text{°C} - 25.0\,\text{°C}) = 3.00 \times 10^4\,\text{J (or 30.0 kJ)}$

5.53   *Analyze.* Since the temperature of the water increases, the dissolving process is exothermic and the sign of ΔH is negative. The heat lost by the NaOH(s) dissolving equals the heat gained by the solution.

Plan/Solve. Calculate the heat gained by the solution. The temperature change is $47.4 - 23.6 = 23.8\text{°C}$. The total mass of solution is $(100.0\,\text{g H}_2\text{O} + 9.55\,\text{g NaOH}) = 109.55 = 109.6$ g.

$109.55\,\text{solution} \times \dfrac{4.184\,\text{J}}{1\,\text{g}\cdot\text{°C}} \times 23.8\text{°C} \times \dfrac{1\,\text{kJ}}{1000\,\text{J}} = 10.909 = 10.9\,\text{kJ}$

This is the amount of heat lost when 9.55 g of NaOH dissolves.

The heat loss per mole NaOH is

$\dfrac{-10.909\,\text{kJ}}{9.55\,\text{g NaOH}} \times \dfrac{40.00\,\text{g NaOH}}{1\,\text{mol NaOH}} = -45.7\,\text{kJ/mol}$ $\Delta H = q_p = -45.7\,\text{kJ/mol NaOH}$

Check. $(-11/9 \times 40) \approx -45$ kJ; the units and sign are correct.

5.55   *Analyze/Plan.* Follow the logic in Sample Exercise 5.7.   *Solve.*

$q_{bomb} = -q_{rxn}$; $\Delta T = 30.57\text{°C} - 23.44\text{°C} = 7.13\text{°C}$

$q_{bomb} = \dfrac{7.854\,\text{kJ}}{1\text{°C}} \times 7.13\text{°C} = 56.00 = 56.0\,\text{kJ}$

At constant volume, $q_v = \Delta E$. ΔE and ΔH are very similar.

$$\Delta H_{rxn} \approx \Delta E_{rxn} = q_{rxn} = -q_{bomb} = \frac{-56.0\,kJ}{2.20\,g\,C_6H_4O_2} = -25.454 = -25.5\,kJ/g\ C_6H_4O_2$$

$$\Delta H_{rxn} = \frac{-25.454\,kJ}{1\,g\,C_6H_4O_2} \times \frac{108.1\,g\,C_6H_4O_2}{1\,mol\,C_6H_4O_2} = -2.75 \times 10^3\,kJ/mol\,C_6H_4O_2$$

5.57   *Analyze.* Given: specific heat and mass of glucose, ΔT for calorimeter. Find: heat capacity, C, of calorimeter. *Plan.* All heat from the combustion raises the temperature of the calorimeter. Calculate heat from combustion of glucose, divide by ΔT for calorimeter to get kJ/°C.   *Solve.*

   (a) $C_{total} = 2.500\,g\ glucose \times \dfrac{15.57\,kJ}{1\,g\ glucose} \times \dfrac{1}{2.70^\circ C} = 14.42 = 14.4\,kJ/^{\circ}C$

   (b)   Qualitatively, assuming the same exact initial conditions in the calorimeter, twice as much glucose produces twice as much heat, which raises the calorimeter temperature by twice as many °C. Quantitatively,

   $$5.000\,g\ glucose \times \frac{15.57\,kJ}{1\,g\ glucose} \times \frac{1^{\circ}C}{14.42\,kJ} = 5.40^{\circ}C$$

   Check. Units are correct. ΔT is twice as large as in part (a). The result has 3 sig figs, because the heat capacity of the calorimeter is known to 3 sig figs.

Hess's Law

5.59   Hess's Law is a consequence of the fact that enthalpy is a state function. Since ΔH is independent of path, we can describe a process by any series of steps that add up to the overall process and ΔH for the process is the sum of the ΔH values for the steps.

5.61   *Analyze/Plan.* Follow the logic in Sample Exercise 5.8. Manipulate the equations so that "unwanted" substances can be canceled from reactants and products. Adjust the corresponding sign and magnitude of ΔH.   *Solve.*

$$
\begin{array}{ll}
P_4O_5(s) \rightarrow P_4(s) + 3O_2(g) & \Delta H = 1640.1\,kJ \\
\underline{P_4(s) + 5O_2(g) \rightarrow P_4O_{10}(s)} & \underline{\Delta H = -2940.1\,kJ} \\
P_4O_6(s) + 2O_2(g) \rightarrow P_4O_{10}(s) & \Delta H = -1300.0\,kJ
\end{array}
$$

   Check. We have obtained the desired reaction.

5.63   *Analyze/Plan.* Follow the logic in Sample Exercise 5.9. Manipulate the equations so that "unwanted" substances can be canceled from reactants and products. Adjust the corresponding sign and magnitude of ΔH.   *Solve.*

$$
\begin{array}{ll}
C_2H_4(g) \rightarrow 2H_2(g) + 2C(s) & \Delta H = -52.3\,kJ \\
2C(s) + 4F_2(g) \rightarrow 2CF_4(g) & \Delta H = 2(-680\,kJ) \\
\underline{2H_2(g) + 2F_2(g) \rightarrow 4HF(g)} & \underline{\Delta H = 2(-537\,kJ)} \\
C_2H_4(g) + 6F_2(g) \rightarrow 2CF_4(g) + 4HF(g) & \Delta H = -2.49 \times 10^3\,kJ
\end{array}
$$

   Check. We have obtained the desired reaction.

Enthalpies of Formation

5.65 (a) *Standard conditions* for enthalpy changes are usually $P = 1$ atm and $T = 298$ K. For the purpose of comparison, standard enthalpy changes, $\Delta H°$, are tabulated for reactions at these conditions.

(b) *Enthalpy of formation*, ΔH_f, is the enthalpy change that occurs when a compound is formed from its component elements.

(c) Standard enthalpy of formation, $\Delta H_f°$, is the enthalpy change that accompanies formation of one mole of a substance from elements in their standard states.

5.67 (a) $1/2\,N_2(g) + 3/2\,H_2(g) \rightarrow NH_3(g)$ $\qquad\qquad$ $\Delta H_f° = -46.19\,kJ$

(b) $1/8\,S_8(s) + O_2(g) \rightarrow SO_2(g)$ $\qquad\qquad$ $\Delta H_f° = -296.9\,kJ$

(c) $Rb(s) + 1/2\,Cl_2(g) + 3/2\,O_2(g) \rightarrow RbClO_3(s)$ $\qquad$ $\Delta H_f° = -392.4\,kJ$

(d) $N_2(g) + 2H_2(g) + 3/2\,O_2(g) \rightarrow NH_4NO_3(s)\ \Delta H_f° = -365.6\,kJ$

5.69 *Plan.* $\Delta H_{rxn}° = \Sigma n \Delta H_f°$ (products) $- \Sigma n \Delta H_f°$ (reactants). Be careful with coefficients, states, and signs. *Solve.*

$\Delta H_{rxn}° = \Delta H_f°\,Al_2O_3(s) + 2\Delta H_f°\ Fe(s) - \Delta H_f°\,Fe_2O_3 - 2\Delta H_f°\,Al(s)$

$\Delta H_{rxn}° = (-1669.8\,kJ) + 2(0) - (-822.16\,kJ) - 2(0) = -847.6\,kJ$

5.71 *Plan.* $\Delta H_{rxn}° = \Sigma n \Delta H_f°$ (products) $- \Sigma n \Delta H_f°$ (reactants). Be careful with coefficients, states and signs. *Solve.*

(a) $\Delta H_{rxn}° = 2\Delta H_f°\,SO_3(g) - 2\Delta H_f°\,SO_2(g) - \Delta H_f°\,O_2(g)$

$\qquad = 2(-395.2\,kJ) - 2(-296.9\,kJ) - 0 = -196.6\,kJ$

(b) $\Delta H_{rxn}° = \Delta H_f°\,MgO(s) + \Delta H_f°\,H_2O(l) - \Delta H_f°\,Mg(OH)_2(s)$

$\qquad = -601.8\,kJ + (-285.83\,kJ) - (-924.7\,kJ) = 37.1\,kJ$

(c) $\Delta H_{rxn}° = \Delta H_f°\,CCl_4(l) + 4\Delta H_f°\,HCl(g) - \Delta H_f°\,CH_4(g) - 4\Delta H_f°\,Cl_2(g)$

$\qquad = -139.3\,kJ + 4(-92.30\,kJ) - (-74.8\,kJ) - 4(0) = -433.7\,kJ$

(d) $\Delta H_{rxn}° = \Delta H_f°\,SiO_2(s) + 4\Delta H_f°\,HCl(g) - \Delta H_f°\,SiCl_4(g) - 2\Delta H_f°\,H_2O(l)$

$\qquad = -910.9\,kJ + 4(-92.30\,kJ) - (-640.1\,kJ) - 2(-285.83\,kJ) = -68.3\,kJ$

5.73 *Analyze.* Given: combustion reaction, enthalpy of combustion, enthalpies of formation for most reactants and products. Find: enthalpy of formation for acetone. *Plan.* Rearrange the expression for enthalpy of reaction to calculate the desired enthalpy of formation. *Solve.*

$$\Delta H_{rxn}^{\circ} = 3\Delta H_f^{\circ}\, CO_2(g) + 3\Delta H_f^{\circ}\, H_2O(l) - \Delta H_f^{\circ}\, C_3H_6O(l)$$

$$-1790\, kJ = 3(-393.5\, kJ) + 3(-285.83\, kJ) - \Delta H_f^{\circ}\, C_3H_6O(l)$$

$$\Delta H_f^{\circ}\, C_3H_6O(l) = -248\, kJ$$

5.75   *Plan.* Use Hess's Law to arrange the given reactions so the overall sum is the formation
        reaction for $Mg(OH)_2(s)$. Adjust the corresponding ΔH values and calculate ΔH_f° for
        $Mg(OH)_2(s)$.   *Solve.*

$Mg(s) + 1/2\, O_2(g) \rightarrow MgO(s)$	$\Delta H^{\circ} = 1/2(-1203.6\, kJ)$
$MgO(s) + H_2O(l) \rightarrow Mg(OH)_2(s)$	$\Delta H^{\circ} = -(37.1\, kJ)$
$H_2(g) + 1/2\, O_2(g) \rightarrow H_2O(l)$	$\Delta H^{\circ} = 1/2(-571.7\, kJ)$

$$Mg(s) + O_2(g) + H_2(g) \rightarrow Mg(OH)_2(s) \qquad \Delta H_f^{\circ} = -924.8\, kJ$$

Check. The overall reaction is correct.

5.77   (a) $C_8H_{18}(l) + 25/2\, O_2(g) \rightarrow 8CO_2(g) + 9H_2O(g) \qquad \Delta H^{\circ} = -5069\, kJ$

        (b) $8C(s, gr) + 9H_2(g) \rightarrow C_8H_{18}(l) \qquad\qquad\qquad \Delta H_f^{\circ} = ?$

        (c)   *Plan.* Follow the logic in Solution 5.73.   *Solve.*

$$\Delta H_{rxn}^{\circ} = 8\Delta H_f^{\circ}\, CO_2(g) + 9\Delta H_f^{\circ}\, H_2O(g) - \Delta H_f^{\circ}\, C_8H_{18}(l) - 25/2\, \Delta H_f^{\circ}\, O_2(g)$$

$$-5069\, kJ = 8(-393.5\, kJ) + 9(-241.82\, kJ) - \Delta H_f^{\circ}\, C_8H_{18}(l) - 25/2(0)$$

$$\Delta H_f^{\circ}\, C_8H_{18}(l) = 8(-393.5\, kJ) + 9(-241.82\, kJ) + 5069\, kJ = -255\, kJ$$

Foods and Fuels

5.79   (a)   *Fuel value* is the amount of heat produced when 1 gram of a substance (fuel) is
               combusted.

        (b)   The fuel value of fats is 9 kcal/g and of carbohydrates is 4 kcal/g. Therefore, 5 g
               of fat produce 45 kcal, while 9 g of carbohydrates produce 36 kcal; 5 g of fat are a
               greater energy source.

5.81   *Plan.* Calculate the Cal (kcal due to each nutritional component of the Campbell's®
        soup, then sum.   *Solve.*

$$9\, g\ carbohydrates \times \frac{17\, kJ}{1\, g\ carbohydrate} = 153\ or\ 2 \times 10^2\, kJ$$

$$1\, g\ protein \times \frac{17\, kJ}{1\, g\ protein} = 17\ or\ 0.2 \times 10^2\, kJ$$

$$7\, g\ fat \times \frac{38\, kJ}{1\, g\ fat} = 266\ or\ 3 \times 10^2\, kJ$$

total energy = 153 kJ + 17 kJ + 266 kJ = 436 or 4×10^2 kJ

$$436 \text{ kJ} \times \frac{1 \text{ kcal}}{4.184 \text{ kJ}} \times \frac{1 \text{ Cal}}{1 \text{ kcal}} = 104 \text{ or } 1 \times 10^2 \text{ Cal/serving}$$

Check. 100 Cal/serving is a reasonable result; units are correct. The data and the result have 1 sig fig.

5.83   *Plan.* g $\rightarrow$ mol $\rightarrow$ kJ $\rightarrow$ Cal   *Solve.*

$$16.0 \text{ g } C_6H_{12}O_6 \times \frac{1 \text{ mol } C_6H_{15}O_6}{180.2 \text{ g } C_6H_{12}O_6} \times \frac{2812 \text{ kJ}}{\text{mol } C_6H_{12}O_6} \times \frac{1 \text{ Cal}}{4.184 \text{ kJ}} = 59.7 \text{ Cal}$$

Check. 60 Cal is a reasonable result for most of the food value in an apple.

5.85   *Plan.* Use enthalpies of formation to calculate molar heat (enthalpy) of combustion using Hess's Law. Use molar mass to calculate heat of combustion per kg of hydrocarbon. *Solve.*

Propyne: $C_3H_4(g) + 4O_2(g) \rightarrow 3CO_2(g) + 2H_2(g)$

(a) $\Delta H = 3(-393.5 \text{ kJ}) + 2(-241.82 \text{ kJ}) - (185.4 \text{ kJ}) - 4(0) = -1849.5 = -1850 \text{ kJ/mol}$ C_3H_4

(b) $\dfrac{-1849.5 \text{ kJ}}{1 \text{ mol } C_3H_4} \times \dfrac{1 \text{ mol } C_3H_4}{40.065 \text{ g } C_3H_4} \times \dfrac{1000 \text{ g } C_3H_4}{1 \text{ kg } C_3H_4} = -4.616 \times 10^4 \text{ kJ/kg } C_3H_4$

Propylene: $C_3H_6(g) + 9/2 \, O_2(g) \rightarrow 3CO_2(g) + 3H_2O(g)$

(a) $\Delta H = 3(-393.5 \text{ kJ}) + 3(-241.82 \text{ kJ}) - (20.4 \text{ kJ}) - 9/2(0) = -1926.4 = -1926 \text{ kJ/mol}$ C_3H_6

(b) $\dfrac{-1926.4 \text{ kJ}}{1 \text{ mol } C_3H_6} \times \dfrac{1 \text{ mol } C_3H_6}{42.080 \text{ g } C_3H_6} \times \dfrac{1000 \text{ g } C_3H_6}{1 \text{ kg } C_3H_6} = -4.578 \times 10^4 \text{ kJ/kg } C_3H_6$

Propane: $C_3H_8(g) + 5O_2(g) \rightarrow 3CO_2(g) + 4H_2O(g)$

(a) $\Delta H = 3(-393.5 \text{ kJ}) + 4(-241.82 \text{ kJ}) - (-103.8 \text{ kJ}) - 5(0) = -2044.0 = -2044 \text{ kJ/mol}$ C_3H_8

(b) $\dfrac{-2044.0 \text{ kJ}}{1 \text{ mol } C_3H_8} \times \dfrac{1 \text{ mol } C_3H_8}{44.096 \text{ g } C_3H_8} \times \dfrac{1000 \text{ g } C_3H_8}{1 \text{ kg } C_3H_8} = -4.635 \times 10^4 \text{ kJ/kg } C_3H_8$

(c)   These three substances yield nearly identical quantities of heat per unit mass, but propane is marginally higher than the other two.

Additional Exercises

5.87   (a)   mi/hr $\rightarrow$ m/s

$$1050 \, \frac{\text{mi}}{\text{hr}} \times \frac{1.6093 \text{ km}}{1 \text{ mi}} \times \frac{1000 \text{ m}}{1 \text{ km}} \times \frac{1 \text{ hr}}{3600 \text{ s}} = 469.38 = 469.4 \text{ m/s}$$

(b)   Find the mass of one N_2 molecule in kg.

$$\frac{28.0134 \text{ g } N_2}{1 \text{ mol}} \times \frac{1 \text{ mol}}{6.022 \times 10^{23} \text{ molecules}} \times \frac{1 \text{ kg}}{1000 \text{ g}} = 4.6518 \times 10^{-26}$$

$$= 4.652 \times 10^{-26} \text{ kg}$$

$$E_k = 1/2 \, mv^2 = 1/2 \times 4.6518 \times 10^{-26} \, kg \times (469.38 \, m/s)^2$$

$$= 5.1244 \times 10^{-21} \, \frac{kg \bullet m^2}{s^2} = 5.124 \times 10^{-21} \, J$$

(c) $\dfrac{5.1244 \times 10^{21} \, J}{molecule} \times \dfrac{6.022 \times 10^{23} \, molecules}{1 \, mol} = 3086 \, J/mol = 3.086 \, kJ/mol$

5.90   Like the combustion of $H_2(g)$ and $O_2(g)$ described in Section 5.4, the reaction that inflates airbags is spontaneous after initiation. Spontaneous reactions are usually exothermic, $-\Delta H$. The airbag reaction occurs at constant atmospheric pressure, $\Delta H = q_p$; both are likely to be large and negative. When the bag inflates, work is done by the system on the surroundings, so the sign of w is negative.

5.93 $\Delta E = q + w = +38.95 \, kJ - 2.47 \, kJ = +36.48 \, kJ$

$\Delta H = q_p = +38.95 \, kJ$

5.95   Find the heat capacity of 1.7×10^3 gal H_2O.

$$C_{H_2O} = 1.7 \times 10^3 \, gal \, H_2O \times \frac{4 \, qt}{1 \, gal} \times \frac{1 \, L}{1.057 \, qt} \times \frac{1 \times 10^3 \, cm^3}{1 \, L} \times \frac{1 \, g}{1 \, cm^3} \times \frac{4.184 \, J}{1 \, g \bullet {}^\circ C}$$

$$= 2.692 \times 10^7 \, J/{}^\circ C = 2.7 \times 10^4 \, kJ/{}^\circ C; \text{ then,}$$

$$\frac{2.692 \times 10^7 \, J}{1 \, {}^\circ C} \times \frac{1 {}^\circ C \bullet g}{0.85 \, J} \times \frac{1 \, kg}{1 \times 10^3 \, g} \times \frac{1 \, brick}{1.8 \, kg} = 1.8 \times 10^4 \text{ or } 18,000 \text{ bricks}$$

Check. $(1.7 \times \sim 16 \times 10^6)/(\sim 1.6 \times 10^3) \approx 17 \times 10^3$ bricks; the units are correct.

5.98   *Plan.* Use the heat capacity of H_2O to calculate the energy required to heat the water. Use the enthalpy of combustion of CH_4 to calculate the amount of CH_4 needed to provide this amount of energy, assuming 100% transfer. Assume $H_2O(l)$ is the product of combustion at standard conditions.

$$1.00 \, kg \, H_2O \times \frac{1000 \, g}{1 \, kg} \times \frac{4.184 \, J}{g \bullet {}^\circ C} \times (90.0 {}^\circ C - 25.0 {}^\circ C) = 2.720 \times 10^5 \, J = 272 \, kJ \text{ required}$$

$$CH_4(g) + 2O_2(g) \rightarrow CO_2(g) + 2H_2O(l)$$

$$\Delta H^\circ_{rxn} = \Delta H^\circ_f \, CO_2(g) + 2\Delta H^\circ_f \, H_2O(g) - \Delta H^\circ_f \, CH_4(g) - \Delta H^\circ_f \, O_2(g)$$

$$= -393.5 \, kJ + 2(-285.83 \, kJ) - (-74.8 \, kJ) - (0) = -890.4 \, kJ$$

$$2.720 \times 10^2 \, kJ \times \frac{1 \, mol \, CH_4}{-890.4 \, kJ} \times \frac{16.04 \, g \, CH_4}{mol \, CH_4} = 4/899 = 4.90 \, g \, CH_4$$

5.101   (a) $3C_2H_2(g) \rightarrow C_6H_6(l)$

            $\Delta H^\circ_{rxn} = \Delta H^\circ_f \, C_6H_6(l) - 3\Delta H^\circ_f \, C_2H_2(g) = 49.0 \, kJ - 3(226.7 \, kJ) = -631.1 \, kJ$

     (b)   Since the reaction is exothermic (ΔH is negative), the product, 1 mole of $C_6H_6(l)$, has less enthalpy than the reactants, 3 moles of $C_2H_2(g)$.

(c)   The fuel value of a substance is the amount of heat (kJ) produced when 1 gram of the substance is burned. Calculate the molar heat of combustion (kJ/mol) and use this to find kJ/g of fuel.

$$C_2H_2(g) + 5/2\, O_2(g) \rightarrow 2CO_2(g) + H_2O(l)$$

$$\Delta H^{\circ}_{rxn} = 2\Delta H^{\circ}_f\, CO_2(g) + \Delta H^{\circ}_f\, H_2O(l) - \Delta H^{\circ}_f\, C_2H_2(g) - 5/2\, \Delta H^{\circ}_f\, O_2(g)$$

$$= 2(-393.5\ kJ) + (-285.83\ kJ) - 226.7\ kJ - 5/2\ (0) = -1299.5\ kJ/mol\ C_2H_2$$

$$\frac{-1299.5\ kJ}{1\ mol\ C_2H_2} \times \frac{1\ mol\ C_2H_2}{26.036\ g\ C_2H_2} = 49.912 = 50\ kJ/g$$

$$C_6H_6(g) + 15/2\, O_2(g) \rightarrow 6CO_2(g) + 3H_2O(l)$$

$$\Delta H^{\circ}_{rxn} = 6\Delta H^{\circ}_f\, CO_2(g) + 3\Delta H^{\circ}_f\, H_2O(l) - \Delta H^{\circ}_f\, C_6H_6(l) - 15/2\, \Delta H^{\circ}_f\, O_2(g)$$

$$= 6(-393.5\ kJ) + 3(-285.83\ kJ) - 49.0\ kJ - 15/2\ (0) = -3267.5\ kJ/mol\ C_6H_6$$

$$\frac{-3267.5\ kJ}{1\ mol\ C_6H_6} \times \frac{1\ mol\ C_6H_6}{78.114\ g\ C_6H_6} = 41.830 = 42\ kJ/g$$

5.105 $\Delta E_p = m\, g\, \Delta h$. Be careful with units. $1\ J = 1\ kg \bullet m^2/s^2$

$$200\ lb \times \frac{1\ kg}{2.205\ lb} \times \frac{9.81\ m}{s^2} \times \frac{45\ ft}{time} \times \frac{1\ yd}{3\ ft} \times \frac{1\ m}{1.0936\ yd} \times 20\ times$$

$$= 2.441 \times 10^5\ kg \bullet m^2/s^2 = 2.441 \times 10^5\ J = 2.4 \times 10^2\ kJ$$

$1\ Cal = 1\ kcal = 4.184\ kJ$

$$2.441 \times 10^2\ kJ \times \frac{1\ Cal}{4.184\ kJ} = 58.34 = 58\ Cal$$

If all work is used to increase the man's potential energy, 20 rounds of stair-climbing will not compensate for one extra order of 245 Cal fries. In fact, more than 58 Cal of work will be required to climb the stairs, because some energy is required to move limbs and some energy will be lost as heat (see Solution 5.92).

Integrative Exercises

5.108   (a) $CH_4(g) + 2O_2(g) \rightarrow CO_2(g) + 2H_2O(l)$

$$\Delta H^{\circ} = \Delta H^{\circ}_f\, CO_2(g) + 2\Delta H^{\circ}_f\, H_2O(l) - \Delta H^{\circ}_f\, CH_4(g) - 2\Delta H^{\circ}_f\, O_2(g)$$

$$= -393.5\ kJ + 2(-285.83\ kJ) - (-74.8\ kJ) - 2(0) = -890.36 = -890.4\ kJ/mol\ CH_4$$

$$\frac{-890.36\ kJ}{mol\ CH_4} \times \frac{1000\ J}{1\ kJ} \times \frac{1\ mol}{6.022 \times 10^{23}\ molecules\ CH_4} = 1.4785 \times 10^{-18}$$

$$= 1.479 \times 10^{18}\ J/molecule$$

(b) $1 \text{eV} = 96.485 \text{ kJ/mol}$

$$8 \text{ keV} \times \frac{1000 \text{ eV}}{1 \text{ keV}} \times \frac{96.485 \text{ kJ}}{\text{eV} \cdot \text{mol}} \times \frac{1 \text{ mol}}{6.022 \times 10^{23}} \times \frac{1000 \text{ J}}{\text{kJ}} = 1.282 \times 10^{-15} = 1 \times 10^{-15} \text{ J/X-ray}$$

The X-ray has approximately 1000 times more energy than is produced by the combustion of 1 molecule of $CH_4(g)$.

5.111   (a) $\Delta H^\circ = \Delta H_f^\circ \text{ NaNO}_3(aq) + \Delta H_f^\circ \text{ H}_2\text{O}(l) - \Delta H_f^\circ \text{ HNO}_3(aq) - \Delta H_f^\circ \text{ NaOH}(aq)$

$\Delta H^\circ = -446.2 \text{ kJ} - 285.83 \text{ kJ} - (-206.6 \text{ kJ}) - (-469.6 \text{ kJ}) = -55.8 \text{ kJ}$

$\Delta H^\circ = \Delta H_f^\circ \text{ NaCl}(aq) + \Delta H_f^\circ \text{ H}_2\text{O}(l) - \Delta H_f^\circ \text{ HCl}(aq) - \Delta H_f^\circ \text{ NaOH}(aq)$

$\Delta H^\circ = -407.1 \text{ kJ} - 285.83 \text{ kJ} - (-167.2 \text{ kJ}) - (-469.6 \text{ kJ}) = -56.1 \text{ kJ}$

$\Delta H^\circ = \Delta H_f^\circ \text{ NH}_3(aq) + \Delta H_f^\circ \text{ Na}^+(aq) + \Delta H_f^\circ \text{ H}_2\text{O}(l) - \Delta H_f^\circ \text{ NH}_4^+(aq) - \Delta H_f^\circ \text{ NaOH}(aq)$

$= -80.29 \text{ kJ} - 240.1 \text{ kJ} - 285.83 \text{ kJ} - (-132.5 \text{ kJ}) - (-469.6 \text{ kJ}) = -4.1 \text{ kJ}$

(b) $H^+(aq) + OH^-(aq) \rightarrow H_2O(l)$ is the net ionic equation for the first two reactions.

$NH_4^+(aq) + OH^-(aq) \rightarrow NH_3(aq) + H_2O(l)$

(c)   The ΔH° values for the first two reactions are nearly identical, –55.9 kJ and –56.2 kJ. The spectator ions by definition do not change during the course of a reaction, so ΔH° is the enthalpy change for the net ionic equation. Since the first two reactions have the same net ionic equation, it is not surprising that they have the same ΔH°.

(d)   Strong acids are more likely than weak acids to donate H^+. The neutralization of the two strong acids is energetically favorable, while the third reaction is not. $NH_4^+(aq)$ is probably a weak acid.

6 Electronic Structure of Atoms

Electronic Structure of Atoms

6.2   (a)   No, visible radiation has wavelengths of 4×10^{-7} to 7×10^{-7} m, much shorter than 1 cm.

       (b)   Energy and wavelength are inversely proportional. Photons of the longer 1 cm radiation have less energy than visible photons.

       (c)   Radiation of 1 cm, or 1×10^{-2} m, is in the microwave region. The appliance is probably a microwave oven.

6.5   (a) $n = 1, n = 4$ (b) $n = 1, n = 2$

       (c)   Wavelength and energy are inversely proportional; the smaller the energy, the longer the wavelength. In order of increasing wavelength (and decreasing energy): (iii) $n = 2$ to $n = 4$ < (iv) $n = 3$ to $n = 1$ < (ii) $n = 3$ to $n = 2$ < (i) $n = 1$ to $n = 2$

6.8   (a)   In the left-most box, the two electrons cannot have the same spin. The *Pauli principle* states that no two electrons can have the same set of quantum numbers. Since the first three quantum numbers describe an orbital, the fourth quantum number, m_s, must have different values for two electrons in the same orbital; their "spins" must be opposite.

       (b)   Flip one of the arrows in the left-most box, so that one points up and the other down.

       (c)   Group 6A. The drawing shows three boxes or orbitals at the same energy, so it must represent p orbitals. Since some of these p orbitals are partially filled, they must be the valence orbitals of the element. Elements with four valence electrons in their p orbitals belong to group 6A.

Radiant Energy

6.9   (a)   Meters (m)   (b)   1/seconds (s^{-1})   (c)   meters/second ($m \cdot s^{-1}$ or m/s)

6.11   (a)   True.

       (b)   False. The frequency of radiation decreases as the wavelength increases.

       (c)   False. Ultraviolet light has shorter wavelengths than visible light. [See Solution 6.10(b).]

       (d)   False. Electromagnetic radiation and sound waves travel at different speeds.

6.13   *Analyze/Plan.* Use the electromagnetic spectrum in Figure 6.4 to determine the wavelength of each type of radiation; put them in order from shortest to longest wavelength.   *Solve.*

Wavelength of X-rays < ultraviolet < green light < red light < infrared < radio waves

Check. These types of radiation should read from left to right on Figure 6.4

6.15   *Analyze/Plan.* These questions involve relationships between wavelength, frequency, and the speed of light. Manipulate the equation $v = c/\lambda$ to obtain the desired quantities, paying attention to units.   *Solve.*

(a) $v = c/\lambda$; $\dfrac{2.998 \times 10^8 \text{ m}}{\text{s}} \times \dfrac{1}{955 \text{ }\mu\text{m}} \times \dfrac{1 \text{ }\mu\text{m}}{1 \times 10^{-6} \text{ m}} = 3.14 \times 10^{11} \text{ s}^{-1}$

(b) $\lambda = c/v$; $\dfrac{2.998 \times 10^8 \text{ m}}{\text{s}} \times \dfrac{1 \text{ s}}{5.50 \times 10^{14}} = 5.45 \times 10^{-7} \text{ m}$ (545 nm)

(c)   The radiation in (b) is in the visible region and is "visible" to humans. The ~1×10^{-3} m radiation in (a) is in the microwave region near the infrared edge and is not visible.

(d) $50.0 \text{ }\mu\text{s} \times \dfrac{1 \text{ s}}{1 \times 10^6 \text{ }\mu\text{s}} \times \dfrac{2.998 \times 10^8 \text{ m}}{\text{s}} = 1.50 \times 10^4 \text{ m}$

Check. Confirm that powers of 10 make sense and units are correct.

6.17   *Analyze/Plan.* $v = c/\lambda$; change nm → m.

Solve. $v = c/\lambda$; $\dfrac{2.998 \times 10^8 \text{ m}}{1 \text{ s}} \times \dfrac{1}{436 \text{ nm}} \times \dfrac{1 \text{ nm}}{1 \times 10^{-9} \text{ m}} = 6.88 \times 10^{14} \text{ s}^{-1}$

The color is blue.

Check. $(3000 \times 10^5 / 500 \times 10^{-9}) = 6 \times 10^{14} \text{ s}^{-1}$; units are correct.

Quantized Energy and Photons

6.19   (a)   Quantization means that energy can only be absorbed or emitted in specific amounts or multiples of these amounts. This minimum amount of energy is called a quantum and is equal to a constant times the frequency of the radiation absorbed or emitted. $E = hv$.

(b)   In everyday activities, we deal with macroscopic objects such as our bodies or our cars, which gain and lose total amounts of energy much larger than a single quantum, hv. The gain or loss of the relatively minuscule quantum of energy is unnoticed.

6.21   *Analyze/Plan.* These questions deal with the relationships between energy, wavelength, and frequency. Use the relationships $E = hv = hc/\lambda$ to calculate the desired quantities. Pay attention to units.   *Solve.*

(a) $E = h\nu = hc/\lambda = 6.626 \times 10^{-34} \text{ J} \cdot \text{s} \times \dfrac{2.998 \times 10^{8} \text{ m}}{1 \text{ s}} \times \dfrac{1}{438 \text{ nm}} \times \dfrac{1 \text{ nm}}{1 \times 10^{-9} \text{ m}}$

$$= 4.54 \times 10^{-19} \text{ J}$$

(b) $E = h\nu = 6.626 \times 10^{-34} \text{ J} \cdot \text{s} \times \dfrac{6.75 \times 10^{12}}{1 \text{ s}} = 4.47 \times 10^{-21} \text{ J}$

(c) $\lambda = hc/E = 6.626 \times 10^{-34} \text{ J} \cdot \text{s} \times \dfrac{2.998 \times 10^{8} \text{ m}}{1 \text{ s}} \times \dfrac{1}{2.87 \times 10^{-18} \text{ J}} = 6.92 \times 10^{-8} \text{ m}$

$$= 69.2 \text{ nm}$$

This radiation is in the ultraviolet region.

Check. Units are correct and powers of 10 are reasonable.

6.23 *Analyze/Plan.* Use $E = hc/\lambda$; pay close attention to units. *Solve.*

(a) $E = hc/\lambda = 6.626 \times 10^{-34} \text{ J} \cdot \text{s} \times \dfrac{2.998 \times 10^{8} \text{ m}}{1 \text{ s}} \times \dfrac{1}{3.3 \text{ μm}} \times \dfrac{1 \text{ μm}}{1 \times 10^{-6} \text{ m}}$

$$= 6.0 \times 10^{-20} \text{ J}$$

$E = hc/\lambda = 6.626 \times 10^{-34} \text{ J} \cdot \text{s} \times \dfrac{2.998 \times 10^{8} \text{ m}}{1 \text{ s}} \times \dfrac{1}{0.154 \text{ nm}} \times \dfrac{1 \text{ nm}}{1 \times 10^{-9} \text{ m}}$

$$= 1.29 \times 10^{-15} \text{ J}$$

Check. $(6.6 \times 3/3.3) \times (10^{-34} \times 10^{8}/10^{-6}) \approx 6 \times 10^{-20} \text{ J}$

$(6.6 \times 3/0.15) \times (10^{-34} \times 10^{8}/10^{-9}) \approx 120 \times 10^{-17} \approx 1.2 \times 10^{-15} \text{ J}$

The results are reasonable. We expect the longer wavelength 3.3 μm radiation to have the lower energy.

(b) The 3.3 μm photon is in the infrared and the 0.154 nm (1.54×10^{-10} m) photon is in the X-ray region; the X-ray photon has the greater energy.

6.25 *Analyze/Plan.* Use $E = hc/\lambda$ to calculate J/photon; Avogadro's number to calculate J/mol; photon/J [the result from part (a)] to calculate photons in 1.00 mJ. Pay attention to units.

Solve.

(a) $E_{photon} = hc/\lambda = \dfrac{6.626 \times 10^{-34} \text{ J} \cdot \text{s}}{325 \times 10^{-9} \text{ m}} \times \dfrac{2.998 \times 10^{8} \text{ m}}{\text{s}} = 6.1122 \times 10^{-19}$

$$= 6.11 \times 10^{-19} \quad \text{J/photon}$$

(b) $\dfrac{6.1122 \times 10^{-19} \text{ J}}{1 \text{ photon}} \times \dfrac{6.022 \times 10^{23} \text{ photons}}{1 \text{ mol}} = 3.68 \times 10^{5} \text{ J/mol} = 368 \text{ kJ/mol}$

(c) $\dfrac{1 \text{ photon}}{6.1122 \times 10^{-19} \text{ J}} \times 1.00 \text{ mJ} \times \dfrac{1 \times 10^{-3}}{1 \text{ mJ}} = 1.64 \times 10^{15} \text{ photons}$

Check. Powers of 10 (orders of magnitude) and units are correct.

6.27 *Analyze/Plan.* $E = hc/\lambda$ gives J/photon. Use this result with J/s (given) to calculate photons/s. *Solve.*

(a)   The $\sim 1 \times 10^{-6}$ m radiation is infrared but very near the visible edge.

(b) $E_{photon} = hc/\lambda = \dfrac{6.626 \times 10^{-34} \text{ J}\bullet\text{s}}{987 \times 10^{-9} \text{ m}} \times \dfrac{2.998 \times 10^8 \text{ m}}{1 \text{ s}} = 2.0126 \times 10^{-19}$

$$= 2.01 \times 10^{-19} \quad \text{J/photon}$$

$$\dfrac{0.52 \text{ J}}{32 \text{ s}} \times \dfrac{1 \text{ photon}}{2.0126 \times 10^{-19} \text{ J}} = 8.1 \times 10^{16} \text{ photons/s}$$

Check. $(7 \times 3/1000) \times (10^{-34} \times 10^8 / 10^{-9}) \approx 21 \times 10^{-20} \approx 2.1 \times 10^{-19}$ J/photon

$(0.5/30/2) \times (1/10^{-19}) = 0.008 \times 10^{19} = 8 \times 10^{16}$ photons/s

Units are correct; powers of 10 are reasonable.

6.29   *Analyze/Plan.* Use $E = h\nu$ and $\nu = c/\lambda$. Calculate the desired characteristics of the photons.

Compare E_{min} and E_{120} to calculate maximum kinetic energy of the emitted electron. *Solve.*

(a) $E = h\nu = 6.626 \times 10^{-34} \text{ J}\bullet\text{s} \times 1.09 \times 10^{15} \text{ s}^{-1} = 7.22 \times 10^{-19}$ J

(b) $\lambda = c/\nu = \dfrac{2.998 \times 10^8 \text{ m}}{1 \text{ s}} \times \dfrac{1 \text{ s}}{1.09 \times 10^{15}} = 2.75 \times 10^{-7} \text{ m} = 275$ nm

(c) $E_{120} = hc/\lambda = 6.626 \times 10^{-34} \text{ J}\bullet\text{s} \times \dfrac{2.998 \times 10^8 \text{ m}}{1 \text{ s}} \times \dfrac{1}{120 \text{ nm}} \times \dfrac{1 \text{ nm}}{1 \times 10^{-9} \text{ m}}$

$$= 1.655 \times 10^{-18} = 1.66 \times 10^{-18} \quad \text{J}$$

The excess energy of the 120 nm photon is converted into the kinetic energy of the emitted electron.

$E_k = E_{120} - E_{min} = 16.55 \times 10^{-19} \text{ J} - 7.22 \times 10^{-19} \text{ J} = 9.3 \times 10^{-19}$ J/electron

Check. E_{120} must be greater than E_{min} in order for the photon to impart kinetic energy to the emitted electron. Our calculations are consistent with this requirement.

Bohr's Model; Matter Waves

6.31   When applied to atoms, the notion of quantized energies means that only certain energies can be gained or lost, only certain values of ΔE are allowed. The allowed values of ΔE are represented by the lines in the emission spectra of excited atoms.

6.33   *Analyze/Plan.* An isolated electron is assigned an energy of zero; the closer the electron comes to the nucleus, the more negative its energy. Thus, as an electron moves closer to the nucleus, the energy of the electron decreases and the excess energy is emitted. Conversely, as an electron moves further from the nucleus, the energy of the electron increases and energy must be absorbed.   *Solve.*

(a)   As the principle quantum number decreases, the electron moves toward the nucleus and energy is emitted.

(b)   An increase in the radius of the orbit means the electron moves away from the nucleus; energy is absorbed.

(c)   An isolated electron is assigned an energy of zero. As the electron moves to the n = 3 state closer to the H^+ nucleus, its energy becomes more negative (decreases) and energy is emitted.

6.35   *Analyze/Plan.* Equation 6.5: $E = (-2.18 \times 10^{-18} \text{ J})(1/n^2)$.   *Solve.*

$E_2 = -2.18 \times 10^{-18} \text{ J}/(2)^2 = -5.45 \times 10^{-19} \text{ J}$

$E_6 = -2.18 \times 10^{-18} \text{ J}/(6)^2 = -6.0556 \times 10^{-20} = -0.606 \times 10^{-19} \text{ J}$

$\Delta E = E_6 - E_2 = (-0.606 \times 10^{-19} \text{ J}) - (-5.45 \times 10^{-19} \text{ J}) = 4.844 \times 10^{-19} \text{ J} = 4.84 \times 10^{-19} \text{ J}$

$\lambda = hc/\Delta E = \dfrac{6.626 \times 10^{-34} \text{ J} \cdot \text{s}}{4.844 \times 10^{-9} \text{ J}} \times \dfrac{2.998 \times 10^8 \text{ m}}{\text{s}} = 4.10 \times 10^{-7} \text{ m} = 410 \text{ nm}$

The visible range is 400–700 nm, so this line is visible; the observed color is violet.

Check. We expect E_6 to be a more positive (or less negative) than E_2, and it is. ΔE is positive, which indicates emission. The orders of magnitude make sense and units are correct.

6.37   (a)   Only lines with $n_f = 2$ represent ΔE values and wavelengths that lie in the visible portion of the spectrum. Lines with $n_f = 1$ have larger ΔE values and shorter wavelengths that lie in the ultraviolet. Lines with $n_f > 2$ have smaller ΔE values and lie in the lower energy longer wavelength regions of the electromagnetic spectrum.

(b)   *Analyze/Plan.* Use Equation 6.7 to calculate ΔE, then $\lambda = hc/\Delta E$.   *Solve.*

$n_i = 3, n_f = 2; \quad \Delta E = -2.18 \times 10^{-18} \text{ J} \left[\dfrac{1}{n_f^2} - \dfrac{1}{n_i^2} \right] = -2.18 \times 10^{-18} \text{ J} (1/4 - 1/9)$

$\lambda = hc/E = \dfrac{6.626 \times 10^{-34} \text{ J} \cdot \text{s} \times 2.998 \times 10^8 \text{ m/s}}{-2.18 \times 10^{-18} \text{ J} (1/4 - 1/9)} = 6.56 \times 10^{-7} \text{ m}$

This is the red line at 656 nm.

$n_i = 4, n_f = 2; \quad \lambda = hc/E = \dfrac{6.626 \times 10^{-34} \text{ J} \cdot \text{s} \times 2.998 \times 10^8 \text{ m/s}}{-2.18 \times 10^{-18} \text{ J} (1/4 - 1/16)} = 4.86 \times 10^{-7} \text{ m}$

This is the blue line at 486 nm.

$n_i = 5, n_f = 2; \quad \lambda = hc/E = \dfrac{6.626 \times 10^{-34} \text{ J} \cdot \text{s} \times 2.998 \times 10^8 \text{ m/s}}{-2.18 \times 10^{-18} \text{ J} (1/4 - 1/25)} = 4.34 \times 10^{-7} \text{ m}$

This is the violet line at 434 nm.

Check. The calculated wavelengths correspond well to three lines in the H emission spectrum in Figure 6.12, so the results are sensible.

6.39   (a) $93.8 \text{ nm} \times \dfrac{1 \times 10^{-9} \text{ m}}{1 \text{ nm}} = 9.38 \times 10^{-8} \text{ m};$ this line is in the ultraviolet region.

(b)   *Analyze/Plan.* Only lines with $n_f = 1$ have a large enough ΔE to lie in the ultraviolet region (see Solutions 6.37 and 6.38). Solve Equation 6.7 for n_i, recalling that ΔE is negative for emission.   *Solve.*

$$\frac{-hc}{\lambda} = -2.18 \times 10^{-18} \ J\left[\frac{1}{n_f^2} - \frac{1}{n_i^2}\right]; \quad \frac{hc}{\lambda(2.18 \times 10^{-18} \ J)} = \left[1 - \frac{1}{n_i^2}\right]$$

$$\frac{1}{n_i^2} = \left[\frac{hc}{\lambda(2.18 \times 10^{-18} \ J)} - 1\right]; \frac{1}{n_i^2} = \left[1 - \frac{hc}{\lambda(2.18 \times 10^{-18} \ J)}\right]$$

$$n_i^2 = \left[1 - \frac{hc}{\lambda(2.18 \times 10^{-18} \ J)}\right]^{-1}; n_i = \left[1 - \frac{hc}{\lambda(2.18 \times 10^{-18} \ J)}\right]^{-1/2}$$

$$n_i = \left(1 - \frac{6.626 \times 10^{-34} \ J \bullet s \times 2.998 \times 10^8 \ m/s}{9.38 \times 10^{-8} \ m \times 2.18 \times 10^{-18} \ J}\right)^{-1/2} = 6 \ (n \text{ values must be integers})$$

$$n_i = 6, \ n_f = 1$$

Check. From Solution 6.38, we know that $n_i > 4$ for $\lambda = 93.8$ nm. The calculated result is close to 6, so the answer is reasonable.

6.41   *Analyze/Plan.* $\lambda = \dfrac{h}{mv}; 1 \ J = \dfrac{1 \ kg \bullet m^2}{s^2}$; Change mass to kg and velocity to m/s in each case.   *Solve.*

(a) $\dfrac{50 \ km}{1 \ hr} \times \dfrac{1000 \ m}{1 \ km} \times \dfrac{1 \ hr}{60 \ min} \times \dfrac{1 \ min}{60 \ s} = 13.89 = 14 \ m/s$

$$\lambda = \frac{6.626 \times 10^{-34} \ kg \bullet m^2 \bullet s}{1 \ s^2} \times \frac{1}{85 \ kg} \times \frac{1 \ s}{13.89 \ m} = 5.6 \times 10^{-37} \ m$$

(b) $10.0 \ g \times \dfrac{1 \ kg}{1000 \ g} = 0.0100 \ kg$

$$\lambda = \frac{6.626 \times 10^{-34} \ kg \bullet m^2 \bullet s}{1 \ s^2} \times \frac{1}{0.0100 \ kg} \times \frac{1 \ s}{250 \ m} = 2.65 \times 10^{-34} \ m$$

(c)   We need to calculate the mass of a single Li atom in kg.

$$\frac{6.94 \ g \ Li}{1 \ mol \ Li} \times \frac{1 \ kg}{1000 \ g} \times \frac{1 \ mol}{6.022 \times 10^{23} \ Li \ atoms} = 1.152 \times 10^{-26} = 1.15 \times 10^{-26} \ kg$$

$$\lambda = \frac{6.626 \times 10^{-34} \ kg \bullet m^2 \bullet s}{1 \ s^2} \times \frac{1}{1.152 \times 10^{-26} \ kg} \times \frac{1 \ s}{2.5 \times 10^5 \ m} = 2.3 \times 10^{-13} \ m$$

6.43   *Analyze/Plan.* Use $v = h/m\lambda$; change wavelength to meters and mass of neutron (back inside cover) to kg.   *Solve.*

$$\lambda = 0.955 \text{ Å} \times \frac{1 \times 10^{-10} \text{ m}}{1 \text{ Å}} = 0.955 \times 10^{-10} \text{ m; m} = 1.6749 \times 10^{-27} \text{ kg}$$

$$v = \frac{6.626 \times 10^{-34} \text{ kg} \cdot \text{m}^2 \cdot \text{s}}{1 \text{ s}^2} \times \frac{1}{1.6749 \times 10^{-27} \text{ kg}} \times \frac{1}{0.955 \times 10^{-10} \text{ m}} = 4.14 \times 10^3 \text{ m/s}$$

Check. $(6.6/1.6/1) \times (10^{-34}/10^{-27}/10^{-10}) \approx 4 \times 10^3$ m/s

6.45   *Analyze/Plan.* Use $\Delta x \geq h/4\pi$ m Δv, paying attention to appropriate units. Note that the uncertainty in speed of the particle (Δv) is important, rather than the speed itself. *Solve.*

(a) $m = 1.50 \text{ mg} \times \dfrac{1 \text{ g}}{1000 \text{ mg}} \times \dfrac{1 \text{ kg}}{1000 \text{ g}} = 1.50 \times 10^{-6}$ kg; $\Delta v = 0.01$ m/s

$$\Delta x \geq = \frac{6.626 \times 10^{-34} \text{ J} \cdot \text{s}}{4\pi(1.50 \times 10^{-6} \text{ kg})(0.01 \text{ m/s})} \geq 3.52 \times 10^{-27} = 4 \times 10^{-27} \text{ m}$$

(b) $m = 1.673 \times 10^{-24}$ g $= 1.673 \times 10^{-27}$ kg; $\Delta v = 0.01 \times 10^4$ m/s

$$\Delta x \geq = \frac{6.626 \times 10^{-34} \text{ J} \cdot \text{s}}{4\pi(1.673 \times 10^{-27} \text{ kg})(0.01 \times 10^4 \text{ m/s})} \geq 3 \times 10^{-10} \text{ m}$$

Check. The more massive particle in (a) has a much smaller uncertainty in position.

Quantum Mechanics and Atomic Orbitals

6.47   (a)   The uncertainty principle states that there is a limit to how precisely we can simultaneously know the position and momentum (related to energy) of an electron. The Bohr model states that electrons move about the nucleus in precisely circular orbits of known radius; each permitted orbit has an allowed energy associated with it. Thus, according to the Bohr model, we can know the exact distance of an electron from the nucleus and its energy. This violates the uncertainty principle.

(b)   deBroglie stated that electrons demonstrate the properties of both particles and waves, that each particle has a wave associated with it. A wave function is the mathematical description of the matter wave of an electron.

(c)   Although we cannot predict the exact location of an electron in an allowed energy state, we can determine the likelihood or probability of finding an electron at a particular position (or within a particular volume). This statistical knowledge of electron location is called the *probability density* or electron density and is a function of ψ^2, the square of the wave function ψ.

6.49   (a)   The possible values of l are (n – 1) to 0. $n = 4$, $l = 3, 2, 1, 0$

(b)   The possible values of m_l are $-l$ to $+l$. $l = 2$, $m_l = -2, -1, 0, 1, 2$

6.51   (a)   3p: $n = 3$, $l = 1$ (b)   2s: $n = 2$, $l = 0$

(c)   4f: $n = 4$, $l = 3$ (d)   5d: $n = 5$, $l = 2$

6.53   Impossible: (a) 1p, only $l = 0$ is possible for $n = 1$; (d) 2d, for $n = 2$, $l = 1$ or 0, but not 2

6.55

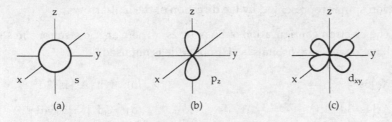

        (a)   (b)   (c)

6.57   (a)   The 1s and 2s orbitals of a hydrogen atom have the same overall spherical shape. The 2s orbital has a larger radial extension and one node, while the 1s orbital has continuous electron density. Since the 2s orbital is "larger," there is greater probability of finding an electron further from the nucleus in the 2s orbital.

        (b)   A single 2p orbital is directional in that its electron density is concentrated along one of the three Cartesian axes of the atom. The $d_{x^2-y^2}$ orbital has electron density along both the x- and y-axes, while the p_x orbital has density only along the x-axis.

        (c)   The average distance of an electron from the nucleus in a 3s orbital is greater than for an electron in a 2s orbital. In general, for the same kind of orbital, the larger the n value, the greater the average distance of an electron from the nucleus of the atom.

        (d)   1s < 2p < 3d < 4f < 6s. In the hydrogen atom, orbitals with the same n value are degenerate and energy increases with increasing n value. Thus, the order of increasing energy is given above.

Many-Electron Atoms and Electron Configurations

6.59   (a)   In the hydrogen atom, orbitals with the same principle quantum number, n, have the same energy; they are degenerate.

        (b)   In a many-electron atom, for a given n-value, orbital energy increases with increasing l-value: s < p < d < f.

6.61   (a)   +1/2, –1/2

        (b)   Electrons with opposite spins are affected differently by a strong inhomogeneous magnetic field. An apparatus with a strong, inhomogeneous magnetic field, similar to the diagram in Figure 6.26, can be used to distinguish electrons with opposite spins.

        (c)   The Pauli exclusion principle states that no two electrons can have the same four quantum numbers. Two electrons in a 1s orbital have the same, n, l, and m_l values. They must have different m_s values.

6.63   *Analyze/Plan.* Each subshell has an l-value associated with it. For a particular l-value, permissible m_l-values are $-l$ to $+l$. Each m_l-value represents an orbital, which can hold two electrons.   *Solve.*

        (a)   6   (b)   10   (c)   2   (d)   14

6.65   (a)   Each box represents an orbital.

        (b)   Electron spin is represented by the direction of the half-arrows.

        (c)   No. The electron configuration of Be is $1s^2 2s^2$. There are no electrons in subshells that have degenerate orbitals, so Hund's rule is not used.

6.67   (a)   Cs: $[Xe]6s^1$ (b)   Ni: $[Ar]4s^2 3d^8$

        (c)   Se: $[Ar]4s^2 3d^{10}4p^4$ (d)   Cd: $[Kr]5s^2 4d^{10}$

        (e)   Ac: $[Rn]7s^2 6d^1$ (f)   Pb: $[Xe]6s^2 4f^{14}5d^{10}6p^2$

6.69   Mt: $[Rn]7s^2 5f^{14}6d^7$

6.71   (a)   Mg   (b)   Al   (c)   Cr   (d)   Te

6.73   (a)   The fifth electron would fill the 2p subshell (same n-value as 2s) before the 3s.

        (b)   The Ne core has filled 2s and 2p subshells. Either the core is [He] or the outer electron configuration should be $3s^2 3p^3$.

        (c)   The 3p subshell would fill before the 3d because it has the lower l-value and the same n-value.

Additional Exercises

6.75   (a) $\lambda_A = 1.6 \times 10^{-7}\,m\,/\,4.5 = 3.56 \times 10^{-8} = 3.6 \times 10^{-8}\,m$

            $\lambda_B = 1.6 \times 10^{-7}\,m\,/\,2 = 8.0 \times 10^{-8}\,m$

        (b) $\nu = c/\lambda; \quad \nu_A = \dfrac{2.998 \times 10^8\,m}{1\,s} \times \dfrac{1}{3.56 \times 10^{-8}\,m} = 8.4 \times 10^{15}\,s^{-1}$

            $\nu_B = \dfrac{2.998 \times 10^8\,m}{1\,s} \times \dfrac{1}{8.0 \times 10^{-8}\,m} = 3.7 \times 10^{15}\,s^{-1}$

        (c)   A: ultraviolet, B: ultraviolet

6.77   All electromagnetic radiation travels at the same speed, 2.998×10^8 m/s. Change miles to meters and seconds to some appropriate unit of time.

        $746 \times 10^6\,mi \times \dfrac{1.6093\,km}{1\,mi} \times \dfrac{1000\,m}{1\,km} \times \dfrac{1\,s}{2.998 \times 10^8\,m} \times \dfrac{1\,min}{60\,s} = 66.7\,min$

6.80 $E/photon = hc/\lambda$

        $E = \dfrac{6.626 \times 10^{-34}\,J \cdot s \times 2.998 \times 10^8\,m/s}{455 \times 10^{-9}\,m} \times \dfrac{6.022 \times 10^{23}\,photons}{mol} = 2.63 \times 10^5\,J/mol$

                                              $= 263\,kJ/mol$

6.84 (a) Lines with $n_f = 1$ lie in the ultraviolet (see Solution 6.30) and with $n_f = 2$ lie in the visible (see Solution 6.29). Lines with $n_f = 3$ will have smaller ΔE and longer wavelengths and lie in the infrared.

  (b) Use Equation 6.7 to calculate ΔE, then $\lambda = hc/\Delta E$.

$$n_i = 4, n_f = 3; \Delta E = -2.18 \times 10^{-18}\ J \left[\frac{1}{n_f^2} - \frac{1}{n_i^2} \right] = -2.18 \times 10^{-18}\ J\ (1/9 - 1/16)$$

$$\lambda = hc/E = \frac{6.626 \times 10^{-34}\ J \bullet s \times 2.998 \times 10^8\ m/s}{-2.18 \times 10^{-18}\ (1/9 - 1/16)} = 1.87 \times 10^{-6}\ m$$

$$n_i = 5, n_f = 3; \lambda = hc/E = \frac{6.626 \times 10^{-34}\ J \bullet s \times 2.998 \times 10^8\ m/s}{-2.18 \times 10^{-18}\ (1/9 - 1/25)} = 1.28 \times 10^{-6}\ m$$

$$n_i = 6, n_f = 3; \lambda = hc/E = \frac{6.626 \times 10^{-34}\ J \bullet s \times 2.998 \times 10^8\ m/s}{-2.18 \times 10^{-18}\ (1/9 - 1/36)} = 1.09 \times 10^{-6}\ m$$

  These three wavelengths are all greater than 1 μm or 1×10^{-6} m. They are in the infrared, close to the visible edge (0.7×10^{-6} m).

6.87 $\lambda = h/mv; v = h/m\lambda. \lambda = 0.711\,\text{Å} \times \dfrac{1 \times 10^{-10}\ m}{1\,\text{Å}} = 7.11 \times 10^{-11}\ m; m_e = 9.1094 \times 10^{-31}\ kg$

$$v = \frac{6.626 \times 10^{-34}\ J \bullet s}{9.1094 \times 10^{-31}\ kg \times 7.11 \times 10^{-11}\ m} \times \frac{1\,kg \bullet m^2/s^2}{1\,J} = 1.02 \times 10^7\ m/s$$

6.90 (a) l (b) n and l (c) m (d) m_l

6.96 If m_s had three allowed values instead of two, each orbital would hold three electrons instead of two. Assuming that the same orbitals are available (that there is no change in the n, l, and m_l values), the number of elements in each of the first four rows would be:

  1st row: 1 orbital $\times 3 =$ 3 elements

  2nd row: 4 orbitals $\times 3 = 12$ elements

  3rd row: 4 orbitals $\times 3 = 12$ elements

  4th row: 9 orbitals $\times 3 = 27$ elements

  The s-block would be 3 columns wide, the p-block 9 columns wide and the d-block 15 columns wide.

Integrative Exercises

6.99 We know the wavelength of microwave radiation, the volume of coffee to be heated, and the desired temperature change. Assume the density and heat capacity of coffee are the same as pure water. We need to calculate: (i) the total energy required to heat the coffee and (ii) the energy of a single photon in order to find (iii) the number of photons required.

(i)   From Chapter 5, the heat capacity of liquid water is 4.184 J/g°C.

To find the mass of 200 mL of coffee at 23°C, use the density of water given in Appendix B.

$$200 \text{ mL} \times \frac{0.997 \text{ g}}{1 \text{ mL}} = 199.4 = 199 \text{ g coffee}$$

$$\frac{4.184 \text{ J}}{1 \text{ g} \,°\text{C}} \times 199.4 \text{ g} \times (60°\text{C} - 23°\text{C}) = 3.087 \times 10^4 \text{ J} = 31 \text{ kJ}$$

(ii) $E = hc/\lambda = 6.626 \times 10^{-34} \text{ J} \bullet \text{s} \times \dfrac{2.998 \times 10^8 \text{ m}}{1 \text{ s}} \times \dfrac{1}{0.112 \text{ m}} = \dfrac{1.77 \times 10^{-24} \text{ J}}{1 \text{ photon}}$

(iii) $3.087 \times 10^4 \text{ J} \times \dfrac{1 \text{ photon}}{1.774 \times 10^{-24} \text{ J}} = 1.7 \times 10^{28}$ photons

(The answer has 2 sig figs because the temperature change, 43°C, has 2 sig figs.)

6.103   (a)   Bohr's theory was based on the Rutherford "nuclear" model of the atom. That is, Bohr theory assumed a dense positive charge at the center of the atom and a diffuse negative charge (electrons) surrounding it. Bohr's theory then specified the nature of the diffuse negative charge. The prevailing theory before the nuclear model was Thomson's plum pudding or watermelon model, with discrete electrons scattered about a diffuse positive charge cloud. Bohr's theory could not have been based on the Thomson model of the atom.

(b)   DeBroglie's hypothesis is that electrons exhibit both particle and wave properties. Thomson's conclusion that electrons have mass is a particle property, while the nature of cathode rays is a wave property. De Broglie's hypothesis actually rationalizes these two seemingly contradictory observations about the properties of electrons.

7 Periodic Properties of the Elements

Visualizing Concepts

7.2   The billiard ball has a definite "hard" boundary, while an atom has no definitive edge. According to the quantum mechanical model, we cannot know the exact location of an electron, only the probability of finding it at a particular radius. Thus, there is no exact edge of an atom, only decreasing probability density as atomic radius increases.

Billiard balls can be used to model nonbonding interactions, such as Ne atoms colliding, because there is no penetration of electron clouds. In a bonding interaction, electron clouds do penetrate and the bonding atomic radius is smaller than the nonbonding radius.

7.5

$$A(g) \rightarrow A^+(g) + e^-$$ ionization energy of A

$$A(g) + e^- \rightarrow A^-(g)$$ electron affinity of A

$$A(g) + A(g) \rightarrow A^+(g) + A^-(g)$$ ionization energy of A + electron affinity of A

The energy change for the reaction is the ionization energy of A plus the electron affinity of A.

Periodic Table; Effective Nuclear Charge

7.7   Mendeleev insisted that elements with similar chemical and physical properties be placed within a family or column of the table. Since many elements were as yet undiscovered, Mendeleev left blanks. He predicted properties for the "blanks" based on properties of other elements in the family and on either side.

7.9   (a)   *Effective nuclear charge*, Z_{eff}, is a representation of the average electrical field experienced by a single electron. It is the average environment created by the nucleus and the other electrons in the molecule, expressed as a net positive charge at the nucleus. It is approximately the nuclear charge, Z, minus the number of core electrons.

        (b)   Going from left to right across a period, nuclear charge increases while the number of electrons in the core is constant. This results in an increase in Z_{eff}.

7.11   (a) $Z_{eff} = Z - S$; K: $[Ar]4s^1$; Z = 19. In the [Ar] core, there are 18 electrons. If these are 100% effective at screening, S = 18. $Z_{eff} = 19 - 18 = 1$.

        (b)   The valence electron in K is a 4s electron. All s electrons have a finite probability of being close to the nucleus and inside the core (Figure 7.3). For this reason, shielding of s electrons is never perfect and calculated values of Z_{eff} reflect this.

7.13 Krypton has a larger nuclear charge (Z = 36) than argon (Z = 18). The shielding of electrons in the $n = 3$ shell by the 1s and 2s core electrons in the two atoms is approximately equal, so the $n = 3$ electrons in Kr experience a greater effective nuclear charge and are thus situated closer to the nucleus.

Atomic and Ionic Radii

7.15 Atomic radii are determined by distances between atoms (interatomic distances) in various situations. Bonding radii are calculated from the internuclear separation of two atoms joined by a chemical bond. Nonbonding radii are calculated from the internuclear separation between two gaseous atoms that collide and move apart, but do not bond.

7.17 The atomic (*metallic*) radius of W is the interatomic W-W distance divided by 2, 2.74 Å/2= 1.37 Å.

7.19 From atomic radii, As–I = 1.19 Å + 1.33 Å = 2.52 Å. This is very close to the experimental value of 2.55 Å.

7.21 (a) Atomic radii decrease moving from left to right across a row and (b) increase from top to bottom within a group.

(c) F < S < P <As. The order is unambiguous according to the trends of increasing atomic radius moving down a column and to the left in a row of the table.

7.23 *Plan*. Locate each element on the periodic charge and use trends in radii to predict their order. *Solve*.

(a) Be < Mg < Ca   (b) Br < Ge < Ga   (c) Si < Al < Tl

7.25 (a) Electrostatic repulsions are reduced by removing an electron from a neutral atom, Z_{eff} increases, and the cation is smaller.

(b) The additional electrostatic repulsion produced by adding an electron to a neutral atom causes the electron cloud to expand, so that the radius of the anion is larger than the radius of the neutral atom.

(c) Going down a column, the n value of the valence electrons increases and they are farther from the nucleus. Valence electrons also experience greater shielding by core electrons. The greater radial extent of the valence electrons outweighs the increase in Z, and the size of particles with like charge increases.

7.27 The size of the red sphere decreases on reaction, so it loses one or more electrons and becomes a cation. Metals lose electrons when reacting with nonmetals, so the red sphere represents a metal. The size of the blue sphere increases on reaction, so it gains one or more electrons and becomes an anion. Nonmetals gain electrons when reacting with metals, so the blue sphere represents a nonmetal.

7.29 (a) An isoelectronic series is a group of atoms or ions that have the same number of electrons, and thus the same electron configuration.

(b) (i) N^{3-}: Ne   (ii) Ba^{2+}: Xe   (iii) Se^{2-}: Kr   (iv) Bi^{3+}: Hg

7.31 (a) Since the electron configurations of the ions in an isoelectronic series are the same, shielding effects do not vary for the different particles. As Z increases, Z_{eff} increases, the valence electrons are more strongly attracted to the nucleus and the size of the particle decreases.

(b) Because F^-, Ne and Na^+ have the same electron configuration, the 2p electron in the particle with the largest Z experiences the largest effective nuclear charge. A 2p electron in Na^+ experiences the greatest effective nuclear charge.

7.33 *Plan.* Use relative location on periodic chart and trends in ionic radii to establish the order.

Solve. (a) $Se < Se^{2-} < Te^{2-}$ (b) $Co^{3+} < Fe^{3+} < Fe^{2+}$ (c) $Ti^{4+} < Sc^{3+} < Ca$ (d) $Be^{2+} < Na^+ < Ne$

Ionization Energies; Electron Affinities

7.35 $B(g) \rightarrow B^+(g) + 1e^-$; $B^+(g) \rightarrow B^{2+}(g) + 1e^-$; $B^{2+}(g) \rightarrow B^{3+}(g) + 1e^-$

7.37 (a) According to Coulomb's law, the energy of an electron in an atom is negative, because of the electrostatic attraction of the electron for the nucleus. In order to overcome this attraction, remove the electron and increase its energy; energy must be added to the atom. Ionization energy, ΔE for this process, is positive, regardless of the magnitude of Z or the quantum numbers of the electron.

(b) F has a greater first ionization energy than O, because F has a greater Z_{eff} and the outer electrons in both elements are approximately the same distance from the nucleus.

(c) The second ionization energy of an element is greater than the first because Z_{eff} is larger for the +1 cation than the neutral atom; more energy is required to overcome the larger Z_{eff}.

7.39 (a) In general, the smaller the atom, the larger its first ionization energy.

(b) According to Figure 7.11, He has the largest and Cs the smallest first ionization energy of the nonradioactive elements.

7.41 *Plan.* Use periodic trends in first ionization energy.   *Solve.*

(a) Ar   (b) Be   (c) Co   (d) S   (e) Te

7.43 *Plan.* Follow the logic of Sample Exercise 7.7.   *Solve.*

(a) Si^{2+}: $[Ne]3s^2$ (b) Bi^{3+}: $[Xe]6s^24f^{14}5d^{10}$

(c) Te^{2-}: $[Kr]5s^24d^{10}5p^6$ or $[Xe]$ (d) V^{3+}: $[Ar]3d^2$

(e) Hg^{2+}: $[Xe]4f^{14}5d^{10}$ (f) Ni^{2+}: $[Ar]3d^8$

7.45 *Plan.* Follow the logic in Sample Exercise 7.7. Construct a mental box diagram for the outer electrons to determine how many are unpaired.   *Solve.*

(a) Mn^{2+}: $[Ar]3d^5$, 5 unpaired electrons   (b) Si^{2-}: $[Ne]3s^23p^4$, 2 unpaired electrons

7.47   Ionization energy : $Se(g)$ $\rightarrow$ $Se^+(g) + 1e^-$

$$[Ar]4s^2 3d^{10} 4p^4 \qquad [Ar]4s^2 3d^{10} 4p^3$$

Electron affinity : $Se(g) + 1e^- \rightarrow$ $Se^-(g)$

$$[Ar]4s^2 3d^{10} 4p^4 \qquad [Ar]4s^2 3d^{10} 4p^5$$

7.49 $Li + 1e^- \rightarrow Li^-$; $Be + 1e^- \rightarrow Be^-$

$[He]2s^1 \qquad [He]2s^2 \quad [He]2s^2 \qquad\qquad [He]2s^2 2p^1$

Adding an electron to Li completes the 2s subshell. The added electron experiences essentially the same effective nuclear charge as the other valence electron, except for the repulsion of pairing electrons in an orbital. There is an overall stabilization; ΔE is negative.

An extra electron in Be would occupy the higher energy 2p subshell. This electron is shielded from the full nuclear charge by the 2s electrons and does not experience a stabilization in energy; ΔE is positive.

7.51   Ionization energy of F^-: $F^-(g) \rightarrow F(g) + 1e^-$

Electron affinity of F: $F(g) + 1e^- \rightarrow F^-(g)$

The two processes are the reverse of each other. The energies are equal in magnitude but opposite in sign. $I_1 (F^-) = -E (F)$

Properties of Metals and Nonmetals

7.53   The smaller the first ionization energy of an element, the greater the metallic character of that element.

7.55   *Analyze/Plan.* Metallic character increases moving down a family and to the left in a period. Use these trends to select the element with greater metallic character.   *Solve.*

(a)   Li   (b)   Na   (c)   Sn   (d)   Al

7.57   *Analyze/Plan.* Ionic compounds are formed by combining a metal and a nonmetal; molecular compounds are formed by two or more nonmetals. *Solve.*

Ionic: MgO, Li_2O, Y_2O_3; molecular: SO_2, P_2O_5, N_2O, XeO_3

7.59   (a)   When dissolved in water, an "acidic oxide" produces an acidic (pH < 7) solution. A "basic oxide" dissolved in water produces a basic (pH > 7) solution.

(b)   Oxides of nonmetals are acidic. Example: $SO_3(g) + H_2O(l) \rightarrow H_2SO_4(aq)$. Oxides of metals are basic. Example: CaO (quick lime). $CaO(s) + H_2O(l) \rightarrow Ca(OH)_2(aq)$.

7.61   *Analyze/Plan.* Cl_2O_7 is a molecular compound formed by two nonmetallic elements. More specifically, it is a nonmetallic oxide and acidic.   *Solve.*

(a)   Dichlorineseptoxide

(b)   Elemental chlorine and oxygen are diatomic gases.
$$2Cl_2(g) + 7O_2(g) \rightarrow 2Cl_2O_7(l)$$

(c)   Most nonmetallic oxides we have seen, such as CO_2 and SO_3, are gases. However, oxides with more atoms, such as $P_2O_3(l)$ and $P_2O_5(s)$, exist in other states. A boiling point of 81°C is not totally unexpected for a large molecule like Cl_2O_7.

(d) Cl_2O_7 is an acidic oxide, so it will be more reactive to base, OH^-.

$$Cl_2O_7(l) + 2OH^-(aq) \rightarrow 2ClO_4^-(aq) + H_2O(l)$$

7.63   (a) $BaO(s) + H_2O(l) \rightarrow Ba(OH)_2(aq)$

        (b) $FeO(s) + 2HClO_4(aq) \rightarrow Fe(ClO_4)_2(aq) + H_2O(l)$

        (c) $SO_3(g) + H_2O(l) \rightarrow H_2SO_4(aq)$

        (d) $CO_2(g) + 2NaOH(aq) \rightarrow Na_2CO_3(aq) + H_2O(l)$

Group Trends in Metals and Nonmetals

7.65

	Na	Mg
(a)	$[Ne]3s^1$	$[Ne]3s^2$
(b)	+1	+2
(c)	+496 kJ/mol	+738 kJ/mol
(d)	very reactive	reacts with steam, but not $H_2O(l)$
(e)	1.54 Å	1.30 Å

(b)   When forming ions, both adopt the stable configuration of Ne, but Na loses one electron and Mg two electrons to achieve this configuration.

(c),(e) The nuclear charge of Mg ($Z = 12$) is greater than that of Na, so it requires more energy to remove a valence electron with the same n value from Mg than Na. It also means that the 2s electrons of Mg are held closer to the nucleus, so the atomic radius (e) is smaller than that of Na.

(d)   Mg is less reactive because it has a filled subshell and it has a higher ionization energy.

7.67   (a)   Ca and Mg are both metals; they tend to lose electrons and form cations when they react. Ca is more reactive because it has a lower ionization energy than Mg. The Ca valence electrons in the 4s orbital are less tightly held because they are farther from the nucleus than the 3s valence electrons of Mg.

        (b)   K and Ca are both metals; they tend to lose electrons and form cations when they react. K is more reactive because it has a lower ionization energy. The 4s valence electron in K is less tightly held because it experiences a smaller nuclear charge ($Z = 19$ for K versus $Z = 20$ for Ca) with similar shielding effects than the 4s valence electrons of Ca.

7.69   (a) $2K(s) + Cl_2(g) \rightarrow 2KCl(s)$

        (b) $SrO(s) + H_2O(l) \rightarrow Sr(OH)_2(aq)$

(c) $4Li(s) + O_2(g) \rightarrow 2Li_2O(s)$

(d) $2Na(s) + S(l) \rightarrow Na_2S(s)$

7.71   H: $1s^1$; Li: $[He]2s^1$; F: $[He]2s^22p^5$. Like Li, H has only one valence electron, and its most common oxidation number is +1, which both H and Li adopt after losing the single valence electron. Like F, H needs only one electron to adopt the stable electron configuration of the nearest noble gas. Both H and F can exist in the –1 oxidation state, when they have gained an electron to complete their valence shells.

7.73

	F	**Cl**
(a)	$[He]2s^22p^5$	$[Ne]3s^23p^5$
(b)	–1	–1
(c)	1681 kJ/mol	1251 kJ/mol
(d)	reacts exothermically to form HF	reacts slowly to form HCl
(e)	–328 kJ/mol	–349 kJ/mol
(f)	0.71 Å	0.99 Å

(b)   F and Cl are in the same group, have the same valence electron configuration and common ionic charge.

(c),(f) The $n = 2$ valence electrons in F are closer to the nucleus and more tightly held than the $n = 3$ valence electrons in Cl. Therefore, the ionization energy of F is greater, and the atomic radius is smaller.

(d)   In its reaction with H_2O, F is reduced; it gains an electron. Although the electron affinity, a gas phase single atom property, of F is less negative than that of Cl, the tendency of F to hold its own electrons (high ionization energy) coupled with a relatively large exothermic electron affinity makes it extremely susceptible to reduction and chemical bond formation. Cl is unreactive to water because it is less susceptible to reduction.

(e)   While F has approximately the same Z_{eff} as Cl, its small atomic radius gives rise to large repulsions when an extra electron is added, so the overall electron affinity of F is smaller (less exothermic) than that of Cl.

(f)   The $n = 2$ valence electrons in F are closer to the nucleus so the atomic radius is smaller than that of Cl.

7.75   Under ambient conditions, the Group 8A elements are all gases that are extremely unreactive, owing to their stable core electron configurations. Thus, the name "inert gases" seemed appropriate.

In the 1960s, scientists discovered that Xe, which has the lowest ionization energy of the nonradioactive noble gases, would react with substances having a strong tendency to remove electrons, such as PtF_6 or F_2. Thus, the term "inert" no longer described all the Group 8A elements. (Kr also reacts with F_2, but reactions of Ar, Ne, and He are as yet unknown.)

7.77   (a) $2O_3(g) \rightarrow 3O_2(g)$

          (b) $Xe(g) + F_2(g) \rightarrow XeF_2(g)$

                    $Xe(g) + 2F_2(g) \rightarrow XeF_4(s)$

                    $Xe(g) + 3F_2(g) \rightarrow XeF_6(s)$

          (c) $S(s) + H_2(g) \rightarrow H_2S(g)$

          (d) $2F_2(g) + 2H_2O(l) \rightarrow 4HF(aq) + O_2(g)$

Additional Exercises

7.79   Up to Z = 82, there are three instances where atomic weights are reversed relative to atomic numbers: Ar and K; Co and Ni; Te and I.

        In each case, the most abundant isotope of the element with the larger atomic number (Z) has one more proton, but fewer neutrons than the element with the smaller atomic number. The smaller number of neutrons causes the element with the larger Z to have a smaller than expected atomic weight.

7.81   (a)   4s

          (b)   To a first approximation, s and p valence electrons do not shield each other, so we expect the 4s and 4p electrons in As to experience a similar Z_{eff}. However, since s electrons have a finite probability of being very close to the nucleus (Figure 7.4), they experience less shielding than p electrons with the same n-value. Since $Z_{eff} = Z - S$ and Z is the same for all electrons in As, if S is smaller for 4s than 4p, Z_{eff} will be greater for 4s electrons and they will have a lower energy.

7.85   Atomic size (bonding atomic radius) is strongly correlated to Z_{eff}, which is determined by Z and S. Moving across the representative elements, electrons added to ns or np valence orbitals do not effectively screen each other. The increase in Z is not accompanied by a similar increase in S; Z_{eff} increases and atomic size decreases. Moving across the transition elements, electrons are added to (n–1)d orbitals and become part of the core electrons, which do significantly screen the ns valence electrons. The increase in Z is accompanied by a larger increase in S for the ns valence electrons; Z_{eff} increases more slowly and atomic size decreases more slowly.

7.88   Y: $[Kr]5s^24d^1$, Z = 39   Zr: $[Kr]5s^24d^2$, Z = 40

        La: $[Xe]6s^25d^1$, Z = 57   Hf: $[Xe]6s^24f^{14}5d^2$, Z = 72

        The completed 4f subshell in Hf leads to a much larger change in Z going from Zr to Hf (72 – 40 = 32) than in going from Y to La (57 – 39 = 18). The 4f electrons in Hf do not completely shield the valence electrons, so there is also a larger increase in Z_{eff}. This larger increase in Z_{eff} going from Zr to Hf leads to a smaller increase in atomic radius than in going from Y to La.

7.91   C: $1s^22s^22p^2$. I_1 through I_4 represent loss of the 2p and 2s electrons in the outer shell of the atom. The values of I_1–I_4 increase as expected. The nuclear charge is constant, but removing each electron reduces repulsive interactions between the remaining electrons, so effective nuclear charge increases and ionization energy increases. I_5 and I_6 represent loss of the 1s core electrons. These 1s electrons are much closer to the nucleus and experience the full nuclear charge (they are not shielded), so the values of I_5 and I_6 are significantly greater than I_1–I_4. I_6 is larger than I_5 because all repulsive interactions have been eliminated.

7.94   O: $[He]2s^2 2p^4$

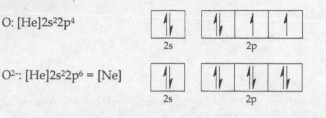

O^{2-}: $[He]2s^2 2p^6 = [Ne]$

O^{3-}: $[Ne]3s^1$ The third electron would be added to the 3s orbital, which is farther from the nucleus and more strongly shielded by the [Ne] core. The overall attraction of this 3s electron for the O nucleus is not large enough for O^{3-} to be a stable particle.

7.96   (a)   The group 2B metals have complete $(n–1)$d subshells. An additional electron would occupy an np subshell and be substantially shielded by both ns and $(n–1)$d electrons. Overall this is not a lower energy state than the neutral atom and a free electron.

(b)   Valence electrons in Group 1B elements experience a relatively large effective nuclear charge due to the buildup in Z with the filling of the $(n–1)$d subshell. Thus, the electron affinities are large and negative. Group 1B elements are exceptions to the usual electron filling order and have the generic electron configuration $n s^1 (n–1) d^{10}$. The additional electron would complete the ns subshell and experience repulsion with the other ns electron. Going down the group, size of the ns subshell increases and repulsion effects decrease. That is, effective nuclear charge is greater going down the group because it is less diminished by repulsion, and electron affinities become more negative.

7.99 $O_2 < Br_2 < K < Mg$. O_2 and Br_2 are (nonpolar) nonmetals. We expect O_2, with the much lower molar mass, to have the lower melting point. This is confirmed by data in Tables 7.6 and 7.7. K and Mg are metallic solids (all metals are solids), with higher melting points than the two nonmetals. Since alkaline earth metals (Mg) are typically harder, more dense and higher melting than alkali metals (K), we expect Mg to have the highest melting point of the group. This is confirmed by data in Tables 7.4 and 7.5.

7.103   Ionic "inorganic" halogen compounds are formed when a metal with low ionization energy and small negative electron affinity combines with a halogen with large ionization energy and large negative electron affinity. That is, it is relatively easy to remove an electron from a metal, and there is only a small energy payback if a metal gains an electron. The opposite is true of a halogen; it is hard to remove an electron and there is a large energy advantage if a halogen gains an electron. Thus, the metal "gives up" an electron to the halogen and an ionic compound is formed. Carbon, on the other hand, is much closer in ionization energy and electron affinity to the halogens. Carbon has a much greater tendency than a metal to keep its own electrons and at least some attraction for the electrons of other elements. Thus, compounds of carbon and the halogens are molecular, rather than ionic.

Integrative Exercises

7.105 (a) Li: $[He]2s^1$. Assume that the [He] core is 100% effective at shielding the 2s valence electron $Z_{eff} = Z - S \approx 3 - 2 = 1+$.

(b) The first ionization energy represents loss of the 2s electron.

ΔE = energy of free electron (n = ∞) – energy of electron in ground state (n = 2)

$\Delta E = I_1 = [-2.18 \times 10^{-18} J (Z^2/\infty^2)] - [-2/18 \times 10^{-18} J (Z^2/2^2)]$

$\Delta E = I_1 = 0 + 2.18 \times 10^{-18} J (Z^2/2^2)$

For Li, which is not a one-electron particle, let $Z = Z_{eff}$.

$\Delta E \approx 2.18 \times 10^{-18} J (+1^2/4) \approx 5.45 \times 10^{-19} J/atom$

(c) Change the result from part (b) to kJ/mol so it can be compared to the value in

Table 7.4. $5.45 \times 10^{-19} \dfrac{J}{atom} \times \dfrac{6.022 \times 10^{23} \ atom}{mol} \times \dfrac{1 \ kJ}{1000 \ J} = 328 \ kJ/mol$

The value in Table 7.4 is 520 kJ/mol. This means that our estimate for Z_{eff} was a lower limit, that the [He] core electrons do not perfectly shield the 2s electron from the nuclear charge.

(d) From Table 7.4, $I_1 = 520$ kJ/mol.

$\dfrac{520 \ kJ}{mol} \times \dfrac{1000 \ J}{kJ} \times \dfrac{1 \ mol}{6.022 \times 10^{23} \ atoms} = 8.6350 \times 10^{-19} \ J/atom$

Use the relationship for I_1 and Z_{eff} developed in part (b).

$Z_{eff}^2 = \dfrac{4(8.6350 \times 10^{-19} \ J)}{2.18 \times 10^{-18} \ J} = 1.5844 = 1.58; Z_{eff} = 1.26$

This value, $Z_{eff} = 1.26$, based on the experimental ionization energy, is greater than our estimate from part (a), which is consistent with the explanation in part (c).

7.108 (a) Mg_3N_2

(b) $Mg_3N_2(s) + 3H_2O(l) \rightarrow 3MgO(s) + 2NH_3(g)$

The driving force is the production of $NH_3(g)$.

(c) After the second heating, all the Mg is converted to MgO.

Calculate the initial mass Mg.

$$0.486 \text{ g MgO} \times \frac{24.305 \text{ g Mg}}{40.305 \text{ g MgO}} = 0.293 \text{ g Mg}$$

$x = $ g Mg converted to MgO; $y = $ g Mg converted to Mg_3N_2; $x = 0.293 - y$

$$\text{g MgO} = x\left(\frac{40.305 \text{ g MgO}}{24.305 \text{ g Mg}}\right); \text{g Mg}_3\text{N}_2 = y\left(\frac{100.929 \text{ g Mg}_3\text{N}_2}{72.915 \text{ g Mg}}\right)$$

$\text{g MgO} + \text{g Mg}_3\text{N}_2 = 0.470$

$$(0.293 - y)\left(\frac{40.305}{24.305}\right) + y\left(\frac{100.929}{72.915}\right) = 0.470$$

$$(0.293 - y)(1.6583) + y(1.3842) = 0.470$$

$$-1.6583\, y + 1.3842\, y = 0.470 - 0.48588$$

$$-0.2741\, y = -0.016$$

$$y = 0.05794 = 0.058 \text{ g Mg in Mg}_3\text{N}_2$$

$$\text{g Mg}_3\text{N}_2 = 0.05794 \text{ g Mg} \times \frac{100.929 \text{ g Mg}_3\text{N}_2}{72.915 \text{ g Mg}} = 0.0802 = 0.080 \text{ g Mg}_3\text{N}_2$$

$$\text{mass \% Mg}_3\text{N}_2 = \frac{0.0802 \text{ g Mg}_3\text{N}_2}{0.470 \text{ g (MgO} + \text{Mg}_3\text{N}_2)} \times 100 = 17\%$$

(The final mass % has two sig figs because the mass of Mg obtained from solving simultaneous equations has two sig figs.)

(d) $3Mg(s) + 2NH_3(g) \rightarrow Mg_3N_2(s) + 3H_2(g)$

$$6.3 \text{ g Mg} \times \frac{1 \text{ mol Mg}}{24.305 \text{ g Mg}} = 0.2592 = 0.26 \text{ mol Mg}$$

$$2.57 \text{ g NH}_3 \times \frac{1 \text{ mol NH}_3}{17.031 \text{ g NH}_3} = 0.1509 = 0.16 \text{ mol NH}_3$$

$$0.2592 \text{ mol Mg} \times \frac{2 \text{ mol NH}_3}{3 \text{ mol Mg}} = 0.1728 = 0.17 \text{ mol NH}_3$$

0.26 mol Mg requires more than the available NH_3 so NH_3 is the limiting reactant.

$$0.1509 \text{ mol NH}_3 \times \frac{3 \text{ mol H}_2}{2 \text{ mol NH}_3} \times \frac{2.016 \text{ g H}_2}{\text{mol H}_2} = 0.4563 = 0.46 \text{ g H}_2$$

(e) $\Delta H^\circ_{rxn} = \Delta H^\circ_f \text{ Mg}_3\text{N}_2(s) + 3\Delta H^\circ_f \text{ H}_2(g) - 3\Delta H^\circ_f \text{ Mg}(s) - 2\Delta H^\circ_f \text{ NH}_3(g)$

$= -461.08 \text{ kJ} + 0 - 3(0) - 2(-46.19) = -368.70 \text{ kJ}$

8 Basic Concepts of Chemical Bonding

Visualizing Concepts

8.2   (a)   No. Oppositely charged ions combine (via Coulombic attraction) to form ionic compounds. A_1 and A_2 have like charges and repel each other.

       (b) A_1Z_1 has the largest lattice energy. Lattice energy increases as ionic charge increases and decreases as inter-ionic distance increases. Since the magnitude of all charges is 1, this factor doesn't vary among possible ion combinations. A_1 and Z_1 have the smallest radii, which leads to the smallest inter-ionic distance and the largest lattice energy.

       (c) A_2Z_2 has the smallest lattice energy, because it has the largest inter-ion distance.

8.5   *Analyze/Plan.* Since there are no unshared pairs in the molecule, we use single bonds to H to complete the octet of each C atom. For the same pair of bonded atoms, the greater the bond order, the shorter and stronger the bond.   *Solve.*

       (a)   Moving from left to right along the molecule, the first C needs two H atoms, the second needs one, the third needs none, and the fourth needs one. The complete molecule is:

$$H-\overset{\overset{\displaystyle H}{|}}{C}\overset{1}{=}\overset{\overset{\displaystyle H}{|}}{C}\overset{2}{-}C\overset{3}{\equiv}C-H$$

       (b)   In order of increasing bond length: $3 < 1 < 2$

       (c)   In order of increasing bond enthalpy (strength): $2 < 1 < 3$

Lewis Symbols

8.7   (a)   Valence electrons are those that take part in chemical bonding, those in the outermost electron shell of the atom. This usually means the electrons beyond the core noble-gas configuration of the atom, although it is sometimes only the outer shell electrons.

       (b) $N: [He]\underset{\text{Valence electrons}}{\underbrace{2s^2 2p^3}}$ A nitrogen atom has 5 valence electrons.

       (c) $\underset{[Ne]}{\underbrace{1s^2 2s^2 2p^6}}\;\underset{\text{valence electrons}}{\underbrace{3s^2 3p^2}}$ The atom (Si) has 4 valence electrons.

8.9   P: $1s^2 2s^2 2p^6 3s^2 3p^3$. A 3s electron is a valence electron; a 2s (or 1s) electron is a non-valence electron. The 3s valence electron is involved in chemical bonding, while the 2s or 1s non-valence electron is not.

8.11   (a) $\cdot \dot{A}l \cdot$ (b) $: \dot{\ddot{B}r} :$ (c) $: \ddot{\ddot{A}r} :$ (d) $\cdot \dot{S}r$

Ionic Bonding

8.13 $\overset{\frown}{Mg} \cdot \; + \; \cdot \ddot{\underset{\cdot\cdot}{O}} : \; \longrightarrow \; Mg^{2+} + \left[: \ddot{\underset{\cdot\cdot}{O}} : \right]^{2-}$

8.15   (a) AlF_3 (b) K_2S (c) Y_2O_3 (d) Mg_3N_2

8.17   (a) Sr^{2+}: [Kr], noble-gas configuration

         (b) Ti^{2+}: $[Ar]3d^2$

         (c) Se^{2-}: $[Ar]4s^2 3d^{10} 4p^6$ = [Kr], noble-gas configuration

         (d) Ni^{2+}: $[Ar]3d^8$

         (e) Br^-: $[Ar]4s^2 3d^{10} 4p^6$ = [Kr], noble-gas configuration

         (f) Mn^{3+}: $[Ar]3d^4$

8.19   (a)   *Lattice energy* is the energy required to totally separate one mole of solid ionic compound into its gaseous ions.

         (b)   The magnitude of the lattice energy depends on the magnitudes of the charges of the two ions, their radii, and the arrangement of ions in the lattice. The main factor is the charges, because the radii of ions do not vary over a wide range.

8.21   KF, 808 kJ/mol; CaO, 3414 kJ/mol; ScN, 7547 kJ/mol

         The sizes of the ions vary as follows: $Sc^{3+} < Ca^{2+} < K^+$ and $F^- < O^{2-} < N^{3-}$. Therefore, the inter-ionic distances are similar. According to Coulomb's law for compounds with similar ionic separations, the lattice energies should be related as the product of the charges of the ions. The lattice energies above are approximately related as (1)(1): (2)(2): (3)(3) or 1:4:9. Slight variations are due to the small differences in ionic separations.

8.23   Since the ionic charges are the same in the two compounds, the K–Br and Cs–Cl separations must be approximately equal. Since the radii are related as $Cs^+ > K^+$ and $Br^- > Cl^-$, the difference between Cs^+ and K^+ must be approximately equal to the difference between Br^- and Cl^-. This is somewhat surprising, since K^+ and Cs^+ are two rows apart and Cl^- and Br^- are only one row apart.

8.25   Equation 8.4 predicts that as the oppositely charged ions approach each other, the energy of interaction will be large and negative. This more than compensates for the energy required to form Ca^{2+} and O^{2-} from the neutral atoms (see Figure 8.4 for the formation of NaCl).

8.27 $RbCl(s) \rightarrow Rb^+(g) + Cl^-(g)$ ΔH (lattice energy) = ?

         By analogy to NaCl, Figure 8.4, the lattice energy is

         $$\Delta H_{latt} = -\Delta H_f^{\circ} \, RbCl(s) + \Delta H_f^{\circ} \, Rb(g) + \Delta H_f^{\circ} \, Cl(g) + I_1 \, (Rb) + E \, (Cl)$$
         $$= -(-430.5 \, kJ) + 85.8 \, kJ + 121.7 \, kJ + 403 \, kJ + (-349 \, kJ) = +692 \, kJ$$

This value is smaller than that for NaCl (+788 kJ) because Rb^+ has a larger ionic radius than Na^+. This means that the value of d in the denominator of Equation 8.4 is larger for RbCl, and the potential energy of the electrostatic attraction is smaller.

Covalent Bonding, Electronegativity, and Bond Polarity

8.29 (a) A *covalent bond* is the bond formed when two atoms share one or more pairs of electrons.

    (b) Any simple compound whose component atoms are nonmetals, such as H_2, SO_2, and CCl_4, are molecular and have covalent bonds between atoms.

    (c) Covalent because it is a gas even below room temperature.

8.31 *Analyze/Plan.* Follow the logic in Sample Exercise 8.3.   *Solve.*

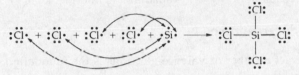

    Check. Each pair of shared electrons in $SiCl_4$ is shown as a line; each atom is surrounded by an octet of electrons.

8.33 (a) $:\ddot{O}\!=\!\underset{..}{O}:$

    (b) A double bond is required because there are not enough electrons to satisfy the octet rule with single bonds and unshared pairs.

    (c) The greater the number of shared electron pairs between two atoms, the shorter the distance between the atoms. If O_2 has a double bond, the O–O distance will be shorter than the O–O single bond distance.

8.35 (a) Electronegativity is the ability of an atom in a molecule (a bonded atom) to attract electrons to itself.

    (b) The range of electronegativities on the Pauling scale is 0.7–4.0.

    (c) Fluorine, F, is the most electronegative element.

    (d) Cesium, Cs, is the least electronegative element that is not radioactive.

8.37 *Plan.* Electronegativity increases going up and to the right in the periodic table. *Solve.*

    (a) Br   (b) C   (c) P   (d) Be

    Check. The electronegativity values in Figure 8.6 confirm these selections.

8.39 The bonds in (a), (c) and (d) are polar because the atoms involved differ in electronegativity. The more electronegative element in each polar bond is:

    (a) F   (c) O   (d) I

8.41 *Analyze/Plan.* Q is the charge at either end of the dipole. $Q = \mu/r$. From Table 8.3, the values for HF are $\mu = 1.82$ D and $r = 0.92$ Å. Change Å to m and use the definition of the Debye and the charge of an electron to calculate the charge in units of *e*.   *Solve.*

$$Q = \frac{\mu}{r} = \frac{1.82\,D}{0.92\,\text{Å}} \times \frac{1\,\text{Å}}{1 \times 10^{-10}\,m} \times \frac{3.34 \times 10^{-30}\,C \bullet m}{1\,D} \times \frac{1\,e}{1.60 \times 10^{-19}\,C} = 0.41e$$

Check. The calculated charge on H and F is 0.41 *e*. This can be thought of as the amount of charge "transferred" from H to F. This value is consistent with our idea that HF is a polar covalent molecule; the bonding electron pair is unequally shared, but not totally transferred from H to F.

8.43   *Analyze/Plan.* Generally, compounds formed by a metal and a nonmetal are described as ionic, while compounds formed from two or more nonmetals are covalent.   *Solve.*

(a) MnO_2, ionic   (b) P_2S_3, covalent

(c)   CoO, ionic   (d)   copper(l) sulfide, ionic

(e)   chlorine trifluoride, covalent   (f)   vanadium(V) fluoride, ionic

Lewis Structures; Resonance Structures

8.45   *Analyze.* Counting the **correct number of valence electrons** is the foundation of every Lewis structure.   *Plan/Solve.*

(a)   Count valence electrons: $4 + (4 \times 1) = 8$ e⁻, 4 e⁻ pairs. Follow the procedure in Sample Exercise 8.6.

$$\begin{array}{c} H \\ | \\ H\!-\!Si\!-\!H \\ | \\ H \end{array}$$

(b)   Valence electrons: $4 + 6 = 10$ e⁻, 5 e⁻ pairs

$$:C \equiv O:$$

(c)   Valence electrons: $[6 + (2 \times 7)] = 20$ e⁻, 10 e⁻ pairs

$$:\!\ddot{F}\!-\!\ddot{S}\!-\!\ddot{F}\!:$$

i.   Place the S atom in the middle and connect each F atom with a single bond; this requires 2 e⁻ pairs.

ii.   Complete the octets of the F atoms with nonbonded pairs of electrons; this requires an additional 6 e⁻ pairs.

iii.   The remaining 2 e⁻ pairs complete the octet of the central S atom.

(d)   32 valence e⁻, 16 e⁻ pairs

$$\begin{array}{c} :\!\ddot{O}\!: \\ | \\ :\!\ddot{O}\!-\!S\!-\!\ddot{O}\!-\!H \\ | \\ :\!\ddot{O}\!: \\ | \\ H \end{array}$$

(Choose the Lewis structure that obeys the octet rule, Section 8.7.)

(e)   Follow Sample Exercise 8.8. 20 valence e⁻, 10 e⁻ pairs

$$\left[:\!\ddot{O}\!-\!\ddot{Cl}\!-\!\ddot{O}\!:\right]^{-}$$

(f)   14 valence e⁻, 7 e⁻ pairs

$$H—\overset{\displaystyle|}{\underset{\displaystyle H}{N}}—\ddot{\underset{\displaystyle\cdot\cdot}{O}}—H$$

Check. In each molecule, bonding e⁻ pairs are shown as lines, and each atom is surrounded by an octet of electrons (duet for H).

8.47   (a)   *Formal charge* is the charge on each atom in a molecule, assuming all atoms have the same electronegativity.

       (b)   Formal charges are not actual charges. They assume perfect covalency, one extreme for the possible electron distribution in a molecule.

       (c)   The other extreme is represented by oxidation numbers, which assume that the more electronegative element holds all electrons in a bond. The true electron distribution is some composite of the two extremes.

8.49   *Analyze/Plan.* Draw the correct Lewis structure: count valence electrons in each atom, total valence electrons and electron pairs in the molecule or ion; connect bonded atoms with a line, place the remaining e⁻ pairs as needed, in nonbonded pairs or multiple bonds, so that each atom is surrounded by an octet (or duet for H). Calculate formal charges: assign electrons to individual atoms [nonbonding e⁻ + 1/2 (bonding e⁻)]; formal charge = valence electrons – assigned electrons. Assign oxidation numbers, assuming that the more electronegative element holds all electrons in a bond.

       Solve. Formal charges are shown near the atoms, oxidation numbers (ox. #) are listed below the structures.

(a)   10 e⁻, 5 e⁻ pairs

$$\left[:N\equiv O:\right]^{+}$$
$$\;\;\;\;0\;\;\;\;+1$$

ox. #: N, +3; O, –2

(b)   32 valence e⁻, 16 e⁻ pairs

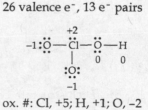

ox #: P, +5; Cl, –1; O, –2

(c)   32 valence e⁻, 16 e⁻ pairs

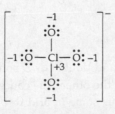

ox. #: Cl, +7; O, –2

(d)   26 valence e⁻, 13 e⁻ pairs

$$-1:\ddot{O}—\overset{+2}{\underset{\displaystyle|}{Cl}}—\ddot{O}—H$$
$$\;\;\;\;\;\;\;\;\;\;\;\;\;:\ddot{O}:\;\;\;\;\;\;0\;\;\;\;\;0$$
$$\;\;\;\;\;\;\;\;\;\;\;\;\;\;\;-1$$

ox. #: Cl, +5; H, +1; O, –2

Check. Each atom is surrounded by an octet (or duet) and the sum of the formal charges and oxidation numbers is the charge on the particle.

8.51   (a)   *Plan.* Count valence electrons, draw all possible correct Lewis structures, taking note of alternate placements for multiple bonds.   *Solve.*

18 e⁻, 9 e⁻ pairs

$$\left[\ddot{O}=\ddot{N}-\ddot{O}:\right]^{-} \longleftrightarrow \left[:\ddot{O}-\ddot{N}=\ddot{O}\right]^{-}$$

Check. The octet rule is satisfied.

(b)   *Plan.* Isoelectronic species have the same number of valence electrons and the same electron configuration.   *Solve.*

A single O atom has 6 valence electrons, so the neutral ozone molecule O_3 is isoelectronic with NO_2^-.

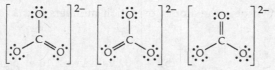

Check. The octet rule is satisfied.

(c)   Since each N–O bond has partial double bond character, the N–O bond length in NO_2^- should be shorter than in species with formal N–O single bonds.

8.53   *Plan/Solve.* The Lewis structures are as follows:

5 e⁻ pairs   8 e⁻ pairs

:C≡O: $\ddot{O}=C=\ddot{O}$

12 e⁻ pairs

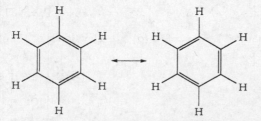

The more pairs of electrons shared by two atoms, the shorter the bond between the atoms. The average number of electron pairs shared by C and O in the three species is 3 for CO, 2 for CO_2, and 1.33 for CO_3^{2-}. This is also the order of increasing bond length: $CO < CO_2 < CO_3^{2-}$.

8.55   (a)   Two equally valid Lewis structures can be drawn for benzene.

Each structure consists of alternating single and double C–C bonds; a particular bond is single in one structure and double in the other. The concept of resonance dictates that the true description of bonding is some hybrid or blend of the two Lewis structures. The most obvious blend of these two resonance structures is a molecule with six equivalent C–C bonds, each with some but not total double-bond character. If the molecule has six equivalent C–C bonds, the lengths of these bonds should be equal.

(b)   The resonance model described in (a) has six equivalent C–C bonds, each with some double bond character. That is, more than one pair but less than two pairs of electrons is involved in each C–C bond. This model predicts a uniform C–C bond length that is shorter than a single bond but longer than a double bond.

Exceptions to the Octet Rule

8.57   (a)   The octet rule states that atoms will gain, lose, or share electrons until they are surrounded by eight valence electrons.

(b)   The octet rule applies to the individual ions in an ionic compound. That is, the cation has lost electrons to achieve an octet and the anion has gained electrons to achieve an octet. For example, in $MgCl_2$, Mg loses 2 e^- to become Mg^{2+} with the electron configuration of Ne. Each Cl atom gains one electron to form Cl^- with the electron configuration of Ar.

8.59   The most common exceptions to the octet rule are molecules with more than eight electrons around one or more atoms, usually the central atom. Examples: SF_6, PF_5

8.61   (a)   26 e^-, 13 e^- pairs

$$\left[:\overset{..}{\underset{..}{O}} - \overset{..}{\underset{|}{S}} - \overset{..}{\underset{..}{O}}: \right]^{2-}$$
$$\underset{\qquad :\overset{}{\underset{..}{O}}:}{}$$

Other resonance structures with one, two, or three double bonds can be drawn. While a structure with three double bonds minimizes formal charges, all structures with double bonds violate the octet rule. Theoretical calculations show that the single best Lewis structure is the one that doesn't violate the octet rule. Such a structure is shown above.

(b)   6 e^-, 3 e^- pairs

H—Al—H
    |
    H

6 electrons around Al; impossible to satisfy octet rule with only 6 valence electrons.

(c)   16 e^-, 8 e^- pairs

$$\left[:N\equiv N - \overset{..}{\underset{..}{N}}: \right]^- \longleftrightarrow \left[:\overset{..}{\underset{..}{N}} - N \equiv N: \right]^- \longleftrightarrow \left[:\overset{..}{N} = N = \overset{..}{\underset{..}{N}}: \right]^-$$

3 resonance structures; all obey octet rule.

(d)   20 e^-, 10 e^- pairs

      :C̈l:
       |
:C̈l—C—H
      |
      H

Obeys octet rule.

(e)   40 e⁻, 20 e⁻ pairs

:F:
|
:F—Sb⟨F:
|
:F:

(with additional F atoms bonded to Sb)

Does not obey octet rule; 10 e⁻ around central Sb

8.63   (a)   16 e⁻, 8 e⁻ pairs

:Cl—Be—Cl:

This structure violates the octet rule; Be has only 4 e⁻ around it.

(b)   Cl=Be=Cl ⟷ :Cl—Be≡Cl: ⟷ :Cl≡Be—Cl:

(c)   The formal charges on each of the atoms in the four resonance structures are:

:Cl—Be—Cl:	Cl=Be=Cl	:Cl—Be≡Cl:	:Cl≡Be—Cl:
0 0 0	+1 −2 +1	0 −2 +2	+2 −2 0

Formal charges are minimized on the structure that violates the octet rule; this form is probably most important. Note that this is a different conclusion than for molecules that have resonance structures with expanded octets that minimize formal charge.

Bond Enthalpies

8.65   *Analyze.* Given: structural formulas. Find: enthalpy of reaction.

Plan. Count the number and kinds of bonds that are broken and formed by the reaction. Use bond enthalpies from Table 8.4 and Equation 8.12 to calculate the overall enthalpy of reaction, ΔH.   *Solve.*

(a) $\Delta H = 2D(O\text{-}H) + D(O\text{-}O) + 4D(C\text{-}H) + D(C=C)$

           $-2D(O\text{-}H) - 2D(O\text{-}C) - 4D(C\text{-}H) - D(C\text{-}C)$

$\Delta H = D(O\text{-}O) + D(C=C) - 2D(O\text{-}C) - D(C\text{-}C)$

     $= 146 + 614 - 2(358) - 348 = -304$ kJ

(b) $\Delta H = 5D(C\text{-}H) + D(C \equiv N) + D(C=C) - 5D(C\text{-}H) - D(C \equiv N) - 2D(C\text{-}C)$

     $= D(C=C) - 2D(C\text{-}C) = 614 - 2(348) = -82$ kJ

(c) $\Delta H = 6D(N\text{-}Cl) - 3D(Cl\text{-}Cl) - D(N \equiv N)$

     $= 6(200) - 3(242) - 941 = -467$ kJ

8.67   *Plan.* Draw structural formulas so bonds can be visualized. Then use Table 8.4 and Equation 8.12.   *Solve.*

(a)   2 H—C(H)(H)—H + O=O ⟶ 2 H—C(H)(H)—O—H

$$\Delta H = 8D(C-H) + D(O=O) - 6D(C-H) - 2D(C-O) - 2D(O-H)$$

$$= 2D(C-H) + D(O=O) - 2D(C-O) - 2D(O-H)$$

$$= 2(413) + (495) - 2(358) - 2(463) = -321 \text{ kJ}$$

(b)   H–H   +   Br–Br $\rightarrow$ 2 H–Br

$$\Delta H = D(H-H) + D(Br-Br) - 2D(H-Br)$$

$$= (436) + (193) - 2(366) = -103 \text{ kJ}$$

(c)   2 H–O–O–H $\rightarrow$ 2 H–O–H + O = O

$$\Delta H = 4D(O-H) + 2D(O-O) - 4D(O-H) - D(O=O)$$

$$\Delta H = 2D(O-O) - D(O=O) = 2(146) - (495) = -203 \text{ kJ}$$

8.69   *Plan.* Draw structural formulas so bonds can be visualized. Then use Table 8.4 and Equation 8.12. *Solve.*

(a)   :N≡N: + 3 H— H   ⟶   2 H—N̈—H
                                         |
                                         H

$$\Delta H = D(N \equiv N) + 3D(H-H) - 6(N-H) = 941 \text{ kJ} + 3(436 \text{ kJ}) - 6(391 \text{ kJ})$$

$$= -97 \text{ kJ}/2 \text{ mol } NH_3; \text{ exothermic}$$

(b)   Plan. Use Eq. 5.31 to calculate ΔH_{rxn} from ΔH_f° values.

$$\Delta H_{rxn}^{\circ} = \Sigma n \, \Delta H_f^{\circ} \text{ (products)} - \Sigma n \, \Delta H_f^{\circ} \text{ (reactants)}. \quad \Delta H_f^{\circ} \, NH_3(g) = -46.19 \text{ kJ}.$$
Solve.

$$\Delta H_{rxn}^{\circ} = 2 \, \Delta H_f^{\circ} \, NH_3(g) - 3 \, \Delta H_f^{\circ} \, H_2(g) - \Delta H_f^{\circ} \, N_2(g)$$

$$\Delta H_{rxn}^{\circ} = 2(-46.19) - 3(0) - 0 = -92.38 \text{ kJ}/2 \text{ mol } NH_3$$

The ΔH calculated from bond enthalpies is slightly more exothermic (more negative) than that obtained using ΔH_f° values.

8.71   The average Ti–Cl bond enthalpy is just the average of the four values listed. 430 kJ/mol.

Additional Exercises

8.73   Six nonradioactive elements in the periodic table have Lewis symbols with single dots. Yes, they are in the same family, assuming H is placed with the alkali metals, as it is on the inside cover of the text. This is because the Lewis symbol represents the number of valence electrons of an element, and all elements in the same family have the same number of valence electrons. By definition of a family, all elements with the same Lewis symbol must be in the same family.

8.75   (a)

	Compound	Lattice Energy (kJ)			Compound	Lattice Energy (kJ)	
106 kJ	NaCl	788	56 kJ	104 kJ	LiCl	834	55 kJ
	NaBr	732			**LiBr**	**779**	
	Na I	682			Li I	730	

The difference in lattice energy between LiCl and LiI is 104 kJ. The difference between NaCl and NaI is 106 kJ; the difference between NaCl and NaBr is 56 kJ, or 53% of the difference between NaCl and NaI. Applying this relationship to the Li salts, 0.53(104 kJ) = 55 kJ difference between LiCl and LiBr. The approximate lattice energy of LiBr is (834 – 55) kJ = 779 kJ.

(b)

Compound	Lattice Energy (kJ)		Compound	Lattice Energy (kJ)	
NaCl	788	56 kJ	CsCl	657	30 kJ
106 kJ NaBr	732		57 kJ CsBr	**627**	
Na I	682		Cs I	600	

By analogy to the Na salts, the difference between lattice energies of CsCl and CsBr should be approximately 53% of the difference between CsCl and CsI. The lattice energy of CsBr is approximately 627 kJ.

(c)

Compound	Lattice Energy (kJ)		Compound	Lattice Energy (kJ)	
MgO	3795	381 kJ	$MgCl_2$	2326	131 kJ
578 kJ CaO	3414		199 kJ $CaCl_2$	**2195**	
SrO	3217		$SrCl_2$	2127	

By analogy to the oxides, the difference between the lattice energies of $MgCl_2$ and $CaCl_2$ should be approximately 66% of the difference between $MgCl_2$ and $SrCl_2$. That is, 0.66(199 kJ) = 131 kJ. The lattice energy of $CaCl_2$ is approximately (2326 – 131) kJ = 2195 kJ.

8.78 (a) A polar molecule has a measurable dipole moment; its centers of positive and negative charge do not coincide. A nonpolar molecule has a zero net dipole moment; its centers of positive and negative charge do coincide.

(b) Yes. If X and Y have different electronegativities, they have different attractions for the electrons in the molecule. The electron density around the more electronegative atom will be greater, producing a charge separation or dipole in the molecule.

(c) $\mu = Qr$. The dipole moment, μ, is the product of the magnitude of the separated charges, Q, and the distance between them, r.

8.81 (a) 12 + 3 + 15 = 30 valence e^-, 15 e^- pairs.

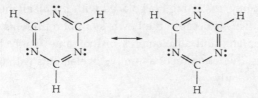

Structures with H bound to N and nonbonded electron pairs on C can be drawn, but the structures above minimize formal charges on the atoms.

(b) The resonance structures indicate that triazine will have six equal C–N bond lengths, intermediate between C–N single and C–N double bond lengths. (See

Solutions 8.55 and 8.56.) From Table 8.5, an average C–N length is 1.43 Å, a C=N length is 1.38 Å. The average of these two lengths is 1.405 Å. The C–N bond length in triazine should be in the range 1.40–1.41 Å.

8.84   Formal charge (FC) = # valence e⁻ – (# nonbonding e⁻ + 1/2 # bonding e⁻)

    (a)   18 e⁻, 9 e⁻ pairs

$$:\ddot{O}—\ddot{O}=\ddot{O} \longleftrightarrow \ddot{O}=\ddot{O}—\ddot{O}:$$

    FC for the central O = 6 – [2 + 1/2 (6)] = +1

    (b)   48 e⁻, 24 e⁻ pairs

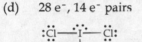   FC for P = 5 – [0 + 1/2 (12)] = –1

    The three nonbonded pairs on each F have been omitted.

    (c)   17 e⁻; 8 e⁻ pairs, 1 odd e⁻

$$\ddot{O}=\dot{N}—\ddot{O}: \longleftrightarrow :\ddot{O}—\dot{N}=\ddot{O}$$

    The odd electron is probably on N because it is less electronegative than O. Assuming the odd electron is on N, FC for N = 5 – [1 + 1/2 (6)] = +1. If the odd electron is on O, FC for N = 5 – [2 + 1/2 (6)] = 0.

    (d)   28 e⁻, 14 e⁻ pairs   (e)   32 e⁻, 16 e⁻ pairs

$$:\ddot{Cl}—\ddot{I}—\ddot{Cl}:$$
$$\ \ \ \ \ \ |$$
$$\ \ \ :\ddot{Cl}:$$

    FC for I = 7 – [4 + 1/2 (6)] = 0

$$:\ddot{O}:$$
$$\ \ |$$
$$:\ddot{O}—\ddot{O}—\ddot{O}—H$$
$$\ \ |$$
$$:\ddot{O}:$$

    FC for Cl = 7 – [0 + 1/2 (8)] = +3

8.87   ΔH = 8D(C–H) – D(C–C) – 6D(C–H) – D(H–H)

        = 2D(C–H) – D(C–C) – D(H–H)

        = 2(413) – 348 – 436 = +42 kJ

    ΔH = 8D(C–H) + 1/2 D(O=O) – D(C–C) – 6D(C–H) – 2D(O–H)

    (a) 2D(C–H) + 1/2 D(O=O) – D(C–C) – 2D(O–H)

        = 2(413) + 1/2 (495) – 348 – 2(463) = –200 kJ

The fundamental difference in the two reactions is the formation of 1 mol of H–H bonds versus the formation of 2 mol of O–H bonds. The latter is much more exothermic, so the reaction involving oxygen is more exothermic.

8.90   (a) $C_3H_6N_6O_6$ 12 + 6 + 30 + 36 = 84 e⁻, 42 e⁻ pairs

           42 e⁻ pairs – 24 shared e⁻ pairs 18 unshared (lone) e⁻ pairs

           Use unshared pairs to complete octets on terminal O atoms (15 unshared pairs) and ring N atoms (3 unshared pairs).

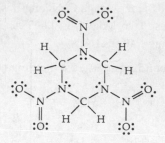

(b)   No C=N bonds in the 6-membered ring are possible, because all C octets are complete with 4 bonds to other atoms. N=N are possible, as shown below. There are 8 possibilities involving some combination of N–N and N=N groups [1 with 0 N=N, 3 with 1 N=N, 3 with 2N=N, 1 with 3N=N]. A resonance structure with 1 N=N is shown below.

Each terminal O=N–O group has two possible placements for the N=O. This generates 8 structures with 0 N=N groups (and 3 O = N–O groups), 4 with 1 N=N and 2 O=N–O, 2 with 2 N=N and 1 O=N–O, and 1 with 3 N=N and no O=N–O. This sums to a total of 15 resonance structures (that I can visualize). Can you find others?

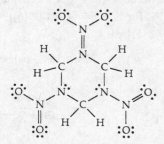

(c) $C_3H_6N_6O_6(s) \rightarrow 3CO(g) + 3N_2(g) + 3H_2O(g)$

(d)   The molecule contains N=O, N=N, C–H, C–N, N–O, and N–N bonds. According to Table 8.4, N–N bonds have the smallest bond enthalpy and are weakest.

(e)   Calculate the enthalpy of decomposition for the resonance structure drawn in part (a).

$\Delta H = 3D(N=O) + 3D(N–O) + 3D(N–N) + 6D(N–C) + 6D(C–H)$

$\quad – 3D(C\equiv O) – 3D(N\equiv N) – 6D(O–H)$

$\quad = 3(607) + 3(201) + 3(163) + 6(293) + 6(413) – 3(1072) – 3(941) – 6(463)$

$\quad = –1668 \text{ kJ/mol } C_3H_6N_6O_6$

$$5.0 \text{ g } C_3H_6N_6O_6 \times \frac{1 \text{ mol } C_3H_6N_6O_6}{222.1 \text{ g } C_3H_6N_6O_6} \times \frac{–1668 \text{ kJ}}{\text{mol } C_3H_6N_6O_6} = 37.55 = 38 \text{ kJ}$$

While exchanging N=O and N–O bonds has no effect on the enthalpy calculation, structures with N=N and 2 N–O do have different enthalpy of decomposition. For the resonance structure with 3 N=N and 6 N–O bonds instead of 3 N–N,

3 N–O and 3 N=O, ΔH = –2121 kJ/mol. The actual enthalpy of decomposition is probably somewhere between –1668 and –2121 kJ/mol. The enthalpy charge for the decomposition of 5.0 g RDX is then in the range 38–48 kJ.

8.93   (a)   S–N ≈ 1.77 Å (sum of the bonding atomic radii from Figure 7.6).

        (b)   S–O ≈ 1.75 Å (the sum of the bonding atomic radii from Figure 7.6.) Alternatively, half of the S–S distance in S_8 (1.02) plus half of the O–O distance from Table 8.5 (0.74) is 1.76 Å.

        (c)   Owing to the resonance structures for SO_2, we assume that the S–O bond in SO_2 is intermediate between a double and single bond, so the distance of 1.43 Å should be significantly shorter than an S–O single bond distance, 1.75 Å.

        (d)   54 e^-, 27 e^- pair

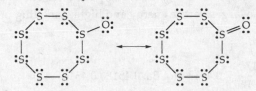

The observed S–O bond distance, 1.48 Å, is similar to that in SO_2, 1.43 Å, which can be described by resonance structures showing both single and double S–O bonds. Thus, S_8O must have resonance structures with both single and double S–O bonds. The structure with the S–O bond has 5 e pairs about this S atom. To the extent that this resonance form contributes to the true structure, the S atom bound to O has more than an octet of electrons around it.

Integrative Exercises

8.94   (a) Ti^{2+} : $[Ar]3d^2$; Ca : $[Ar]4s^2$. Yes. The two valence electrons in Ti^{2+} and Ca are in different principle quantum levels and different subshells.

        (b)   According to the Aufbau Principle, valence electrons will occupy the lowest energy empty orbital. Thus, in Ca the 4s is lower in energy than the 3d, while in Ti^{2+}, the 3d is lower in energy than the 4s.

        (c)   Since there is only one 4s orbital, the two valence electrons in Ca are paired. There are five degenerate 3d orbitals, so the two valence electrons in Ti^{2+} are unpaired. Ca has no unpaired electrons, Ti^{2+} has two.

8.97   (a)   Assume 100 g.

            A: 87.7 g In/114.82   = 0.764 mol In; 0.764/0.384 ≈ 2

                12.3 g S/32.07   = 0.384 mol S; 0.384/0.384 = 1

            B: 78.2 g In/114.82   = 0.681 mol In; 0.681/0.68 0 ≈ 1

                21.8 g S/32.07   = 0.680 mol S; 0.680/0.680 = 1

            C: 70.5 g In/114.82   = 0.614 mol In; 0.614/0.614 = 1

                29.5 g S/32.07   = 0.920 mol S; 0.920/0.614 = 1.5

            A: In_2S; B: InS; C: In_2S_3

(b)   A: In(I); B: In(II); C: In(III)

(c)   In(I) : $[Kr]5s^2 4d^{10}$; In(II) : $[Kr]5s^1 4d^{10}$; In(III) : $[Kr]4d^{10}$
       None of these is a noble-gas configuration.

(d)   The ionic radius of In^{3+} in compound C will be smallest. Removing successive electrons from an atom reduces electron repulsion, increases the effective nuclear charge experienced by the valence electrons and decreases the ionic radius. The higher the charge on a cation, the smaller the radius.

(e)   Lattice energy is directly related to the charge on the ions and inversely related to the interionic distance. Only the charge and size of the In varies in the three compounds. In(I) in compound A has the smallest charge and the largest ionic radius, so compound A has the smallest lattice energy and the lowest melting point. In(III) in compound C has the greatest charge and the smallest ionic radius, so compound C has the largest lattice energy and highest melting point.

8.100   (a)   Assume 100 g.

$$62.04 \text{ g Ba} \times \frac{1 \text{ mol}}{137.33 \text{ g Ba}} = 0.4518 \text{ mol Ba}; 0.4518 / 0.4518 = 1.0$$

$$37.96 \text{ g N} \times \frac{1 \text{ mol}}{14.007 \text{ g N}} = 2.710 \text{ mol N}; 2.710 / 0.4518 = 6.0$$

The empirical formula is BaN_6. Ba has an ionic charge of 2+, so there must be two 1– azide ions to balance the charge. The formula of each azide ion is N_3^-.

(b)   16 e^-, 8 e^- pairs

$$\left[:\ddot{N}=N=\ddot{N}:\right]^- \longleftrightarrow \left[:N\equiv N-\ddot{N}:\right]^- \longleftrightarrow \left[:\ddot{N}-N\equiv N:\right]^-$$

$$\begin{array}{ccc} -1 \quad +1 \quad -1 & \quad 0 \quad +1 \quad -2 & \quad -2 \quad +1 \quad 0 \end{array}$$

(c)   The left structure minimizes formal charges and is probably the main contributor.

(d)   The two N–N bond lengths will be equal. The two minor contributors would individually cause unequal N–N distances, but collectively they contribute equally to the lengthening and shortening of each bond. The N–N distance will be approximately 1.24 Å, the average N=N distance.

8.104   (a)

$$\begin{array}{llll}
HF(g) \rightarrow H(g) + F(g) & D(H-F) & 567 \text{ kJ} \\
H(g) \rightarrow H^+(g) + 1 e^- & I(H) & 1312 \text{ kJ} \\
F(g) + 1 e^- \rightarrow F^-(g) & E(F) & -328 \text{ kJ} \\
\hline
HF(g) \rightarrow H^+(g) + F^-(g) & \Delta H & 1551 \text{ kJ}
\end{array}$$

(b) $\Delta H = D(H-Cl) + I(H) + E(Cl)$

       $\Delta H = 431 \text{ kJ} + 1312 \text{ kJ} + (-349) \text{ kJ} = 1394 \text{ kJ}$

(c) $\Delta H = D(H-Br) + I(H) + E(Br)$

       $\Delta H = 366 \text{ kJ} + 1312 \text{ kJ} + (-325) \text{ kJ} = 1353 \text{ kJ}$

9 Molecular Geometry and Bonding Theories

Visualizing Concepts

9.3 (a) Square pyramidal

(b) Yes, there is one nonbonding electron domain on A. If there were only five bonding domains, the shape would be trigonal bipyramidal. With five bonding and one nonbonding electron domains, the molecule has octahedral domain geometry.

(c) Yes. If the B atoms are halogens, each will have three nonbonding electron pairs; there are five bonding pairs, and A has one nonbonded pair, for a total of $[5(3) + 5 + 1] = 21$ e^- pairs and 42 electrons in the Lewis structure. If the five halogens contribute 35 e^-, A must contribute seven valence electrons. A is also a halogen.

9.6 (a) 90° angles are characteristic of hybrids with a d atomic orbital contribution. This pair of orbitals could be sp^3d or sp^3d^2.

(b) Angles of 109.5° are characteristic of sp^3 hybrid orbitals only.

(c) Angles of 120° can be formed by sp^2 hybrids or sp^3d hybrids.

9.9 *Analyze/Plan.* σ molecular orbitals (MOs) are symmetric about the internuclear axis, π MOs are not. Bonding MOs have most of their electron density in the area between the nuclei, antibonding MOs have a node between the nuclei.

(a) (i) The shape of the molecular orbital (MO) indicates that it is formed by two s atomic orbitals (electron density at each nucleus).

(ii) σ-type MO (symmetric about the internuclear axis, s orbitals can produce only σ overlap).

(iii) antibonding MO (node between nuclei)

(b) (i) Shape indicates this MO is formed by two p atomic orbitals overlapping end-to-end (node near each nucleus).

(ii) σ-type MO (symmetric about internuclear axis)

(iii) bonding MO (concentration of electron density between nuclei)

(c) (i) Shape indicates this MO is formed by two p atomic orbitals overlapping side-to-side (node near each nucleus).

(ii) π-type MO (not symmetric about internuclear axis, side-to-side overlap)

(iii) antibonding MO (node between nuclei)

Molecular Shapes; the VSEPR Model

9.11   (a)   Yes. The bond angle specifies the shape and the bond length tells the size.

(b)   Yes. This description means that the three terminal atoms point toward the corners of an equilateral triangle and the central atom is in the plane of this triangle. Only 120° bond angles are possible in this arrangement.

9.13   (a)   An *electron domain* is a region in a molecule where electrons are most likely to be found.

(b)   Each balloon in Figure 9.5 occupies a volume of space. The best arrangement is one where each balloon has its "own" space, where they are as far apart as possible and repulsions are minimized. Electron domains are negatively charged regions, so they also adopt an arrangement where repulsions are minimized.

9.15   *Analyze/Plan.* See Table 9.1.   *Solve.*

(a)   trigonal planar   (b)   tetrahedral

(c)   trigonal bipyramidal   (d)   octahedral

9.17   The electron-domain geometry indicated by VSEPR describes the arrangement of all bonding and nonbonding electron domains. The molecular geometry describes just the atomic positions. H_2O has the Lewis structure given below; there are four electron domains around oxygen so the electron-domain geometry is tetrahedral, but the molecular geometry of the three atoms is bent.

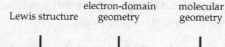

Lewis structure   electron-domain   molecular geometry
                      geometry

9.19   *Analyze/Plan.* See Tables 9.2 and 9.3.   *Solve.*

Lewis structure   electron-domain   molecular
                      geometry   geometry

(a)   tetrahedral   tetrahedral

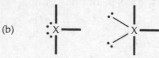

(b)   trigonal bipyramidal   T-shaped

(c)   octahedral   square pyramidal

9.21   *Analyze/Plan.* Follow the logic in Sample Exercise 9.1.

Solve. bent (b), linear (l), octahedral (oh), seesaw (ss) square pyramidal (sp),
square planar (spl), tetrahedral (td), trigonal bipyramidal (tbp), trigonal planar (tr),
trigonal pyramidal (tp), T-shaped (T)

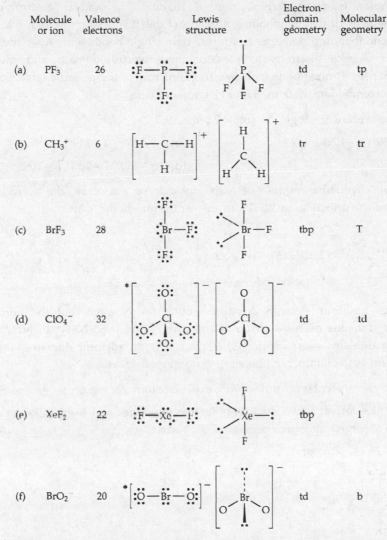

Molecule or ion	Valence electrons	Lewis structure	Electron-domain geometry	Molecular geometry
(a) PF_3	26		td	tp
(b) CH_3^+	6		tr	tr
(c) BrF_3	28		tbp	T
(d) ClO_4^-	32		td	td
(e) XeF_2	22		tbp	l
(f) BrO_2^-	20		td	b

*More than one resonance structure is possible. All equivalent resonance structures
predict the same molecular geometry.

9.23   *Analyze/Plan* Work backwards from molecular geometry, using Tables 9.2 and 9.3.
        Solve.

(a)   Electron-domain geometries: i, trigonal planar; ii, tetrahedral; iii, trigonal
        bipyramidal

(b)   nonbonding electron domains: i, 0; ii, 1; iii, 2

(c)   N and P. Shape ii has three bonding and one nonbonding electron domains. Li
        and Al would form ionic compounds with F, so there would be no nonbonding

electron domains. Assuming that F always has three nonbonding domains, BF_3 and ClF_3 would have the wrong number of nonbonding domains to produce shape ii.

(d)   Cl (also Br and I, since they have seven valence electrons). This T-shaped molecular geometry arises from a trigonal bipyramidal electron-domain geometry with two nonbonding domains (Table 9.3). Assuming each F atom has three nonbonding domains and forms only single bonds with A, A must have seven valence electrons to produce these electron-domain and molecular geometries. It must be in or below the third row of the periodic table, so that it can accommodate more than four electron domains.

9.25   *Analyze/Plan.* Follow the logic in Sample Exercise 9.3.   *Solve.*

(a)   1 – 109°, 2 – 109°   (b)   3 – 109°, 4 – 109°

(c)   5 – 180°   (d)   6 – 120°, 7 – 109°, 8 – 109°

9.27   *Analyze/Plan.* Given the formula of each molecule or ion, draw the correct Lewis structure and use principles of VSEPR to answer the question.   *Solve.*

$$\left[H-\overset{..}{\underset{..}{N}}-H \right]^{-} \quad\quad H-\overset{..}{N}-H \quad\quad \left[H-\overset{\overset{\displaystyle H}{|}}{\underset{\underset{\displaystyle H}{|}}{N}}-H \right]^{+}$$
$$\qquad\qquad\qquad\qquad\qquad |$$
$$\qquad\qquad\qquad\qquad\quad H$$

Each species has four electron domains around the N atom, but the number of nonbonding domains decreases from two to zero, going from NH_2^- to NH_4^+. Since nonbonding domains exert greater repulsive forces on adjacent domains, the bond angles expand as the number of nonbonding domains decreases.

9.29   *Analyze.* Given: molecular formulas. Find: explain features of molecular geometries.

Plan. Draw the correct Lewis structures for the molecules and use VSEPR to predict and explain observed molecular geometry.   *Solve.*

(a) BrF_4^- 36 e^-, 18 e^- pr

$$\left[\begin{array}{c} :\overset{..}{F}: \\ | \\ :\overset{..}{F}-\overset{..}{Br}-\overset{..}{F}: \\ | \\ :\overset{..}{F}: \end{array} \right]^{-}$$

6 e^- pairs around Br, octahedral e^- domain geometry,
square planar molecular geometry

BF_4^- 32 e^-, 16 e^- pr

$$\left[\begin{array}{c} :\overset{..}{F}: \\ | \\ :\overset{..}{F}-B-\overset{..}{F}: \\ | \\ :\overset{..}{F}: \end{array} \right]^{-}$$

4 e^- pairs around B, tetrahedral e^- domain geometry,
tetrahedral molecular geometry

The fundamental feature that determines molecular geometry is the number of electron domains around the central atom, and the number of these that are bonding domains. Although BrF_4^- and BF_4^- are both of the form AX_4^-, the central atoms and thus the number of valence electrons in the two ions are different. This leads to different numbers of e^- domains about the two central atoms. Even though both ions have four bonding electron domains, the six total domains around Br require octahedral domain geometry and square planar molecular geometry, while the four total domains about B lead to tetrahedral domain and molecular geometry.

(b) CF_4 32 e^-, 16 e^- pr

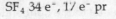

4 e^- domains around C, tetrahedral e^- domain geometry, tetrahedral molecular geometry

SF_4 34 e^-, 17 e^- pr

5 e^- domains around S, trigonal bipyramidal e^- domain geometry, seesaw molecular geometry

CF_4 will have bond angles closest to the value predicted by VSEPR, because there are no nonbonding e^- domains around C. The four bonding domains in CF_4 are equivalent and lead to the balance of repulsions implicit in VSEPR theory. In SF_4, one of the e^- domains is nonbonding. A nonbonding domain is surely not equivalent to a bonding domain; we expect it to be more diffuse. That is, nonbonding domains will occupy more space, "push back" the bonding domains, and lead to bond angles that are nonideal.

Polarity of Polyatomic Molecules

9.31   See Sample Exercise 9.4(b) for the correct resonance structures and analysis of S–O bond dipoles. According to the electron density model, the net dipole moment vector points along the O–S–O angle bisector with the negative end pointing away from S. the magnitude of this vector is 1.63 D.

9.33   (a)   In Exercise 9.23, molecules ii and iii will have nonzero dipole moments. Molecule i has no nonbonding electron pairs on A, and the 3 A–F dipoles are oriented so that the sum of their vectors is zero (the bond dipoles cancel). Molecules ii and iii have nonbonding electron pairs on A and their bond dipoles do not cancel. A nonbonding electron pair (or pairs) on a central atom guarantees at least a small molecular dipole moment, because no bond dipole exactly cancels a nonbonding pair.

(b) AF_4 molecules will have a zero dipole moment if there are no nonbonding electron pairs on the central atom and the 4 A–F bond dipoles are arranged (symmetrically) so that they cancel. Therefore, in Exercise 9.24, molecules i and ii have zero dipole moments and are nonpolar.

9.35   *Analyze/Plan.* Given molecular formulas, draw correct Lewis structures, determine molecular structure and polarity. *Solve.*

(a)   Nonpolar, in a symmetrical tetrahedral structure (Figure 9.1) the bond dipoles cancel.

(b)   Polar, there is an unequal charge distribution due to the nonbonded electron pair on N.

(c)   Polar, there is an unequal charge distribution due to the nonbonded electron pair on S.

(d)   Nonpolar, the bond dipoles and the nonbonded electron pairs cancel.

(e)   Polar, the C–H and C–Br bond dipoles are not equal and do not cancel.

(f)   Nonpolar, in a symmetrical trigonal planar structure, the bond dipoles cancel.

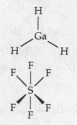

9.37

polar	nonpolar	polar

All three isomers are planar. The molecules on the left and right are polar because the C–Cl bond dipoles do not point in opposite directions. In the middle isomer, the C–Cl bonds and dipoles are pointing in opposite directions (as are the C–H bonds), the molecule is nonpolar and has a measured dipole moment of zero.

Orbital Overlap; Hybrid Orbitals

9.39   (a)   *Orbital overlap* occurs when a valence atomic orbital on one atom shares the same region of space with a valence atomic orbital on an adjacent atom.

(b)   In valence bond theory, overlap of orbitals allows the two electrons in a chemical bond to mutually occupy the space between the bonded nuclei.

(c)   Valence bond theory is a combination of the atomic orbital concept with the Lewis model of electron pair bonding.

9.41   (a)   4 valence e^-, 2 e^- pairs

$H-Mg-H$

2 bonding e^- domains, linear e^- domain and molecular geometry

(b)   The Mg atom has two electrons in its 2s orbital; these electrons are paired. Without promotion, Mg has no unpaired electrons available to form bonds with H.

(c)   The linear electron domain geometry in MgH_2 requires sp hybridization.

(d)

9.43   *Analyze/Plan.* Given electron domain geometry, list the appropriate orbital hybridization and associated bond angles; refer to Table 9.4.   *Solve.*

(a)   sp – 180°   (b) sp^3 – 109°   (c) sp^2 – 120°

(d) sp^3d^2 – 90° and 180°   (e) sp^3d – 90°, 120° and 180°

9.45   (a)   B: $[He]2s^2 2p^1$

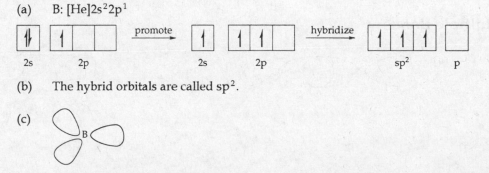

(b)   The hybrid orbitals are called sp^2.

(c)

(d)   A single 2p orbital is unhybridized. It lies perpendicular to the trigonal plane of the sp^2 hybrid orbitals.

9.47   *Analyze/Plan.* Given the molecular (or ionic) formula, draw the correct Lewis structure and determine the electron domain geometry, which determines hybridization. *Solve.*

(a)   24 e⁻, 12 e⁻ pairs

3 e⁻ pairs around B trigonal, planar e⁻ domain geometry, sp² hybrid orbitals

(b)   32 e⁻, 16 e⁻ pairs

4 e⁻ domains around Al, tetrahedral e⁻ domain geometry, sp³ hybrid orbitals

(c)   16 e⁻, 8 e⁻ pairs

$$\overset{..}{\underset{..}{S}} = C = \overset{..}{\underset{..}{S}}$$

2 e⁻ domains around C, linear e⁻ domain geometry, sp hybrid orbitals

(d)   22 e⁻, 11 e⁻ pairs

$$\overset{..}{\underset{..}{F}} - \overset{.}{\underset{.}{Kr}} - \overset{..}{\underset{..}{F}}$$

5 e⁻ pairs around Kr, trigonal bipyramidal e⁻ domain geometry, sp³d hybrid orbitals

(e)   48 e⁻, 24 e⁻ pairs

6 e⁻ pairs around P, octahedral e⁻ domain geometry, sp³d² orbitals

Multiple Bonds

9.49   (a)   (b)

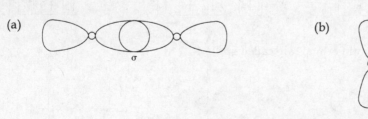

(c)   A σ bond is generally stronger than a π bond, because there is more extensive orbital overlap.

9.51   (a)

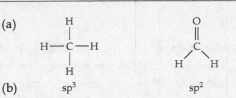

        (b) sp^3 sp^2

        (c)   The C atom in CH_4 is sp^3 hybridized; there are no unhybridized p orbitals available for the π overlap required by multiple bonds. In CH_2O, the C atom is sp^2 hybridized, with one p atomic orbital available to form the π overlap in the C=O double bond.

9.53   *Analyze/Plan.* Single bonds are σ bonds, double bonds consist of 1 σ and 1 π bond. Each bond is formed by a pair of valence electrons.   *Solve.*

        (a) C_3H_6 has $3(4) + 6(1) = 18$ valence electrons

        (b)   8 pairs or 16 total valence electrons form σ bonds

        (c)   1 pair or 2 total valence electrons form π bonds

        (d)   no valence electrons are nonbonding

        (e)   The left and central C atoms are sp^2 hybridized; the right C atom is sp^3 hybridized.

9.55   *Analyze/Plan.* Given the correct Lewis structure, analyze the electron domain geometry at each central atom. This determines the hybridization and bond angles at that atom. *Solve.*

        (a)   ~109° about the left most C, sp^3; ~120° about the right-hand C, sp^2

        (b)   The doubly bonded O can be viewed as sp^2, the other as sp^3; the nitrogen is sp^3 with approximately 109° bond angles.

        (c)   9 σ bonds, 1 π bond

9.57   (a)   In a localized π bond, the electron density is concentrated strictly between the two atoms forming the bond. In a delocalized π bond, parallel p orbitals on more than two adjacent atoms overlap and the electron density is spread over all the atoms that contribute p orbitals to the network. There are still two regions of overlap, above and below the σ framework of the molecule.

        (b)   The existence of more than one resonance form is a good indication that a molecule will have delocalized π bonding.

        (c) $$\left[\,\ddot{\text{O}}{=}\overset{\cdot\cdot}{\text{N}}{-}\ddot{\text{O}}{:}\,\right]^{-} \longleftrightarrow \left[\,{:}\ddot{\text{O}}{-}\overset{\cdot\cdot}{\text{N}}{=}\ddot{\text{O}}\,\right]^{-}$$

        The existence of more than one resonance form for NO_2 indicates that the π bond is delocalized. From an orbital perspective, the electron-domain geometry around N is trigonal planar, so the hybridization at N is sp^2. This leaves a p orbital on N and one on each O atom perpendicular to the trigonal plane of the molecule, in the correct orientation for delocalized π overlap. Physically, the two N–O bond lengths are equal, indicating that the two N–O bonds are equivalent, rather than one longer single bond and one shorter double bond.

Molecular Orbitals

9.59   (a)   Both atomic and molecular orbitals have a characteristic energy and shape (region where there is a high probability of finding an electron). Each atomic or molecular orbital can hold a maximum of two electrons. Atomic orbitals are localized on single atoms and their energies are the result of interactions between the subatomic particles in a single atom. MOs can be delocalized over several or even all the atoms in a molecule and their energies are influenced by interactions between electrons on several atoms.

       (b)   There is a net stabilization (lowering in energy) that accompanies bond formation because the bonding electrons in H_2 are strongly attracted to both H nuclei.

       (c)   Two

9.61   (a)

       (b)   There is one electron in H_2^+.

       (c)   ⬛ σ_{1s}^*

             ⬛ σ_{1s}

       (d)   Bond order = 1/2 (1-0) = 1/2

       (e)   Fall apart. The stability of H_2^+ is due to the lower energy state of the σ bonding molecular orbital relative to the energy of a H 1s atomic orbital. If the single electron in H_2^+ is excited to the $\sigma^w{}_{1s}$ orbital, its energy is higher than the energy of an H 1s atomic orbital and H_2^+ will decompose into a hydrogen atom and a hydrogen ion.

$$H_2^+ \xrightarrow{h\nu} H + H^+.$$

9.63   *Analyze/Plan.* In a σ molecular orbital, the electron density is spherically symmetric about the internuclear axis and is concentrated along this axis. In a π MO, the electron density is concentrated above and below the internuclear axis and zero along it.
       Solve.

       (a)

(b)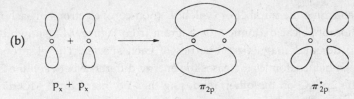

$P_x + P_x$ π_{2p} π^*_{2p}

(c) σ_{2p} is lower in energy than π_{2p} due to greater extent of orbital overlap in the σ
MO. $\sigma_{2p} < \pi_{2p} < \pi^*_{2p} < \sigma^*_{2p}$

9.65 (a) When comparing the same two bonded atoms, the greater the bond order, the
shorter the bond length and the greater the bond energy. That is, bond order and
bond energy are directly related, while bond order and bond length are inversely
related. When comparing different bonded nuclei, there are no simple
relationships (see Solution 8.92).

(b) Be_2, 4 e⁻ Be_2^+, 3 e⁻

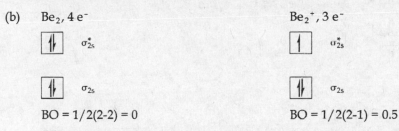

BO = 1/2(2-2) = 0   BO = 1/2(2-1) = 0.5

Be_2 has a bond order of zero and is not energetically favored over isolated Be
atoms; it is not expected to exist. Be_2^+ has a bond order of 0.5 and is slightly
lower in energy than isolated Be atoms. It will probably exist under special
experimental conditions, but be unstable.

9.67 (a), (b) Substances with no unpaired electrons are weakly repelled by a magnetic field.
This property is called *diamagnetism*.

(c) O_2^{2-}, Be_2^{2+} (see Figure 9.45)

9.69

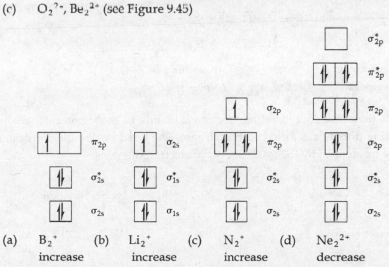

(a) B_2^+ (b) Li_2^+ (c) N_2^+ (d) Ne_2^{2+}
increase   increase   increase   decrease

Addition of an electron increases bond order if it occupies a bonding MO and
decreases stability if it occupies an antibonding MO.

9.71   *Analyze/Plan.* Determine the number of "valence" (non-core) electrons in each molecule or ion. Use the homonuclear diatomic MO diagram from Figure 9.42 (shown below) to calculate bond order and magnetic properties of each species. The electronegativity difference between heteroatomics increases the energy difference between the 2s AO on one atom and the 2p AO on the other, rendering the "no interaction" MO diagram in Figure 9.42 appropriate.   *Solve.*

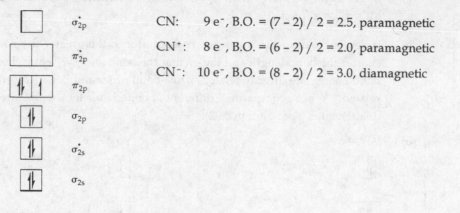

CN: $9 e^-$, B.O. $= (7 - 2) / 2 = 2.5$, paramagnetic

CN^+: $8 e^-$, B.O. $= (6 - 2) / 2 = 2.0$, paramagnetic

CN^-: $10 e^-$, B.O. $= (8 - 2) / 2 = 3.0$, diamagnetic

9.73   (a) $3s, 3p_x, 3p_y, 3p_z$ (b) π_{3p} (c)   Two

(d)   If the MO diagram for P_2 is similar to that of N_2, P_2 will have no unpaired electrons and be diamagnetic.

Additional Exercises

9.76

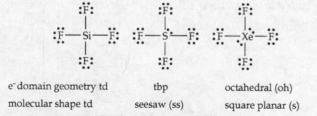

e$^-$domain geometry td   tbp   octahedral (oh)

molecular shape td   seesaw (ss)   square planar (s)

Although there are four bonding electron domains in each molecule, the number of nonbonding domains is different in each case. The bond angles and thus the molecular shape are influenced by the total number of electron domains.

9.79   (a) CO_2, 16 valence e^- (b) NCS^-, 16 valence e^-

$$\ddot{O} = C = \ddot{O}$$

2σ   2π

$$\left[\ddot{N} = C = \ddot{\underset{..}{S}} \right]^-$$

2σ   2π

two other resonance structures

(for any of the resonance structures)

(c) H_2CO, 12 valence e⁻   (d)   HCO(OH), 18 valence e⁻

3σ, 1π

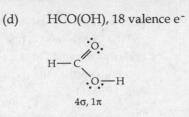

4σ, 1π

9.82   The compound on the right has a dipole moment. In the square planar trans structure on the left, all equivalent bond dipoles can be oriented opposite each other, for a net dipole moment of zero.

9.85

(a)   The molecule is not planar. The CH_2 planes at each end are twisted 90° from one another.

(b)   Allene has no dipole moment.

(c)   The bonding in allene would not be described as delocalized. The π electron clouds of the two adjacent C=C are mutually perpendicular. The mechanism for delocalization of π electrons is mutual overlap of parallel p atomic orbitals on adjacent atoms. If adjacent π electron clouds are mutually perpendicular, there is no overlap and no delocalization of π electrons.

9.91 σ_{2p}^* ☐

π_{2p}^* [↑ |]

σ_{2p} [↑]

π_{2p} [↑↓ | ↑↓]

σ_{2s}^* [↑↓]

σ_{2s} [↑↓]

N_2 in the ground state has a B.O. of 3; in the first excited state (at left) it has a B.O. of 2. Owing to the reduction in bond order, N_2 in the first excited state would be more reactive, have a smaller bond energy and a longer N–N separation.

9.95   (a)   CO, 10 e⁻, 5 e⁻ pair

:C≡O:

(b)   The bond order for CO, as predicted by the MO diagram in Figure 9.48, is 1/2[8 − 2] = 3.0. A bond order of 3.0 agrees with the triple bond in the Lewis structure.

(c)   Applying the MO diagram in Figure 9.48 to the CO molecule, the highest energy electrons would occupy the π_{2p} MOs. That is, π_{2p} would be the HOMO, highest occupied molecular orbital. If the true HOMO of CO is a σ-type MO, the order of the π_{2p} and σ_{2p} orbitals must be reversed. Figure 9.44 shows how the interaction of the 2s orbitals on one atom and the 2p orbitals on the other atom can affect the relative energies of the resulting MOs. This 2s–2p interaction in CO is significant enough so that the σ_{2p} MO is higher in energy than the π_{2p} MOs, and the σ_{2p} is the HOMO.

(d)   We expect the atomic orbitals of the more electronegative element to have lower energy than those of the less electronegative element. When atoms of the two elements combine, the lower energy atomic orbitals make a greater contribution to the bonding MOs and the higher energy atomic orbitals make a larger contribution to the antibonding orbitals. Thus, the π_{2p} bonding MOs will have a greater contribution from the more electronegative O atom.

Integrative Exercises

9.96   (a)   Assume 100 g of compound

$$2.1\,\text{g H} \times \frac{1\,\text{mol H}}{1.008\,\text{g H}} = 2.1\,\text{mol H}; 2.1/2.1 = 1$$

$$29.8\,\text{g N} \times \frac{1\,\text{mol N}}{14.01\,\text{g N}} = 2.13\,\text{mol N}; 2.13/2.1 \approx 1$$

$$68.1\,\text{g O} \times \frac{1\,\text{mol O}}{16.00\,\text{g O}} = 4.26\,\text{mol O}; 4.26/2.1 \approx 2$$

The empirical formula is HNO_2; formula weight = 47. Since the approximate molar mass is 50, the molecular formula is HNO_2.

(b)   Assume N is central, since it is unusual for O to be central, and part (d) indicates as much. HNO_2: 18 valence e^-

$$\ddot{O}=\ddot{N}-\ddot{O}-H \longleftrightarrow :\ddot{O}-\ddot{N}=\ddot{O}-H$$
$$\qquad\qquad\qquad\qquad\quad -1 \quad 0 \quad +1$$

The second resonance form is a minor contributor due to unfavorable formal charges.

(c)   The electron domain geometry around N is trigonal planar with an O–N–O angle of approximately 120°. If the resonance structure on the right makes a significant contribution to the molecular structure, all four atoms would lie in a plane. If only the left structure contributes, the H could rotate in and out of the molecular plane. The relative contributions of the two resonance structures could be determined by measuring the O–N–O and N–O–H bond angles.

(d)   3 VSEPR e^- domains around N, sp^2 hybridization

(e)   3 σ, 1 π for both structures (or for H bound to N).

9.99   (a)   Three electron domains around each central C atom, sp^2 hybridization

(b)   A 180° rotation around the C=C double bond is required to convert the trans isomer into the cis isomer. A 90° rotation around the bond eliminates all overlap of the p orbitals that form the π bond and it is broken.

(c)   **average bond enthalpy**

C=C   614 kJ/mol

C–C   348 kJ/mol

The difference in these values, 266 kJ/mol, is the average bond enthalpy of a C–C π bond. This is the amount of energy required to break 1 mol of C–C π bonds. The energy per molecule is

$$266 \text{ kJ/mol} \times \frac{1000 \text{ J}}{1 \text{ kJ}} \times \frac{1 \text{ mol}}{6.022 \times 10^{23} \text{ molecules}} = 4.417 \times 10^{-19}$$

$$= 4.42 \times 10^{-19} \text{ J/molecule}$$

(d) $\lambda = hc/E = \dfrac{6.626 \times 10^{-34} \text{ J} \bullet \text{s} \times 2.998 \times 10^{8} \text{ m/s}}{4.417 \times 10^{-19} \text{ J}} = 4.50 \times 10^{-7} \text{ m} = 450 \text{ nm}$

(e)   Yes, 450 nm light is in the visible portion of the spectrum. A cis-trans isomerization in the retinal portion of the large molecule rhodopsin is the first step in a sequence of molecular transformations in the eye that leads to vision. The sequence of events enables the eye to detect visible photons, in other words, to see.

9.101

(g) ⟶ 6C(g) + 6H(g)

ΔH = 6D(C–H) + 3D(C–C) + 3D(C=C) – 0

    = 6(413 kJ) + 3(348 kJ) + 3(614 kJ)

    = 5364 kJ

(The products are isolated atoms; there is no bond making.)

According to Hess' law:

$\Delta H^{\circ} = 6\Delta H_f^{\circ} \text{ C(g)} + 6\Delta H_f^{\circ} \text{ H(g)} - \Delta H_f^{\circ} \text{ C}_6\text{H}_6\text{(g)}$

    = 6(718.4 kJ) + 6(217.94 kJ) – (82.9 kJ)

    = 5535 kJ

The difference in the two results, 171 kJ/mol C_6H_6 is due to the resonance stabilization in benzene. That is, because the π electrons are delocalized, the molecule has a lower overall energy than that predicted for the presence of 3 localized C–C and C=C bonds. Thus, the amount of energy actually required to decompose 1 mole of C_6H_6(g), represented by the Hess' law calculation, is greater than the sum of the localized bond enthalpies (not taking resonance into account) from the first calculation above.

9.103   (a) $3d_{z^2}$

(b)   Ignoring the donut of the d_{z^2} orbital

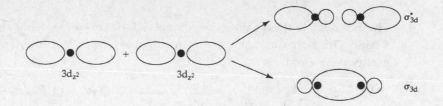

(c)   A node is generated in σ_{3d}^* because antibonding MOs are formed when AO lobes with opposite phases interact. Electron density is excluded from the internuclear region and a node is formed in the MO.

(d)   Sc: $[Ar]4s^2 3d^1$ Omitting the core electrons, there are six e^- in the energy level diagram.

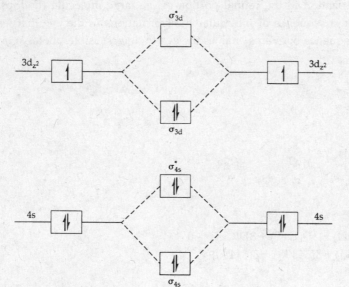

(e)   The bond order in Sc_2 is $1/2\,(4 - 2) = 1.0$.

10 Gases

Visualizing Concepts

10.1   (a) $V_1/T_1 = V_2/T_2$ (Charles' Law)   (b) $P_1V_1 = P_2V_2$ (Boyle's Law)

                   $V_1/300 \text{ K} = V_2/500 \text{ K}$ $1 \text{ atm} \times V_1 = 2 \text{ atm} \times V_2$

                   $V_2 = 5/3 \, V_1$ $V_2 = 1/2 \, V_1$

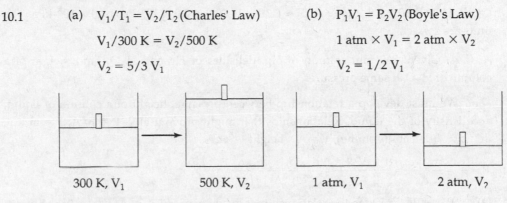

       300 K, V_1 500 K, V_2 1 atm, V_1 2 atm, V_2

10.4   Over time, the gases will mix perfectly. Each bulb will contain 4 blue and 3 red atoms. The "blue" gas has the greater partial pressure after mixing, because it has the greater number of particles (at the same T and V as the "red" gas.)

10.7   (a)   At constant temperature, the "average" (really root mean square) speed of a collection of gas particles is inversely related to molar mass; the lighter the particle, the faster it moves. Therefore, curve B represents He and curve A represents O_2. Curve B has the higher avg. molecular speed and He is the lighter gas. Curve A has the lower avg. molecular speed and O_2 is the heavier gas

           (b)   For the same gas "avg" kinetic energy ($1/2 \, mv^2$), and therefore "avg" speed (v) is directly related to Kelvin temperature. Curve A is the lower temperature and curve B is the higher temperature.

Gas Characteristics; Pressure

10.9   In the gas phase molecules are far apart, while in the liquid they are touching.

           (a)   A gas is much less dense than a liquid because most of the volume of a gas is empty space.

           (b)   A gas is much more compressible because of the distance between molecules.

           (c)   Gaseous molecules are so far apart that there is no barrier to mixing, regardless of the identity of the molecule. All mixtures of gases are homogeneous. Liquid molecules are touching. In order to mix, they must displace one another. Similar molecules displace each other and form homogeneous mixtures. Very dissimilar molecules form heterogeneous mixtures.

10.11 *Analyze.* Given: mass, area. Find: pressure. *Plan.* P=F/A = m × a/A; use this relationship, paying attention to units. *Solve.*

$$1\,Pa = \frac{1\,N}{m^2} = \frac{1\,kg \bullet m}{s^2} \times \frac{1}{m^2} = \frac{1\,kg}{m \bullet s^2} \quad \text{Change mass to kg and area to } m^2.$$

$$P = \frac{m \times a}{A} = \frac{130\,lb}{0.50\,in^2} \times \frac{9.81\,m}{1\,s^2} \times \frac{0.454\,kg}{1\,lb} \times \frac{39.4^2\,in^2}{1\,m^2} = 1.8 \times 10^6 \frac{kg}{m \bullet s^2}$$

$$= 1.8 \times 10^6\,Pa = 1.8 \times 10^3\,kPa$$

Check. $[1.30 \times 10 \times 0.5 \times (40)^2/0.5] \approx (130 \times 16,000) \approx 2.0 \times 10^6\,Pa \approx 2.0 \times 10^3\,kPa$. The units are correct.

10.13 *Analyze.* Given: 760 mm column of Hg, densities of Hg and H_2O. Find: height of a column of H_2O at same pressure.

Plan. We must develop a relationship between pressure, height of a column of liquid, and density of the liquid. Relationships that might prove useful: P = F/A; F = m × a; m = d × V(density)(volume); V = A × height   *Solve.*

$$P = \frac{F}{A} = \frac{m \times a}{A} = \frac{d \times V \times a}{A} = \frac{d \times A \times h \times a}{A} = d \times h \times a$$

(a) $P_{Hg} = P_{H_2O}$; Using the relationship derived above: $(d \times h \times a)_{H_2O} = (d \times h \times a)_{Hg}$

Since a, the acceleration due to gravity, is equal in both liquids,
$(d \times h)_{H_2O} = (d \times h)_{Hg}$

$1.00\,g/mL \; \times h_{H_2O} = 13.6\,g/mL \; \times 760\,mm$

$$h_{H_2O} = \frac{13.6\ g/mL \times 760\,mm}{1.00\ g/mL} = 1.034 \times 10^4 = 1.03 \times 10^4\,mm = 10.3\,m$$

(b)   Pressure due to H_2O:

1 atm = 1.034×10^4 mm H_2O (from part (a))

$$36\,ft\,H_2O \times \frac{12\,in}{1\,ft} \times \frac{2.54\,cm}{1\,in} \times \frac{10\,mm}{1\,cm} \times \frac{1\,atm}{1.034 \times 10^4\,mm} = 1.061 = 1.1\,atm$$

$P_{total} = P_{atm} + P_{H_2O} = 0.95\,atm + 1.061\,atm = 2.011 = 2.0\,atm$

10.15 (a)   The tube can have any cross-sectional area. (The height of the Hg column in a barometer is independent of the cross-sectional area. See the expression for pressure derived in Solution 10.13.)

(b)   At equilibrium, the force of gravity per unit area acting on the mercury column at the level of the outside mercury is not equal to the force of gravity per unit area acting on the atmosphere. (F = ma; the acceleration due to gravity is equal for the two substances, but the mass of Hg for a given cross-sectional area is different than the mass of air for this same area.)

(c)   The column of mercury is held up by the pressure of the atmosphere applied to the exterior pool of mercury.

10.17 *Analyze/Plan.* Follow the logic in Sample Exercise 10.1.   *Solve.*

(a)　　$265 \text{ torr} \times \dfrac{1 \text{ atm}}{760 \text{ torr}} = 0.349 \text{ atm}$

(b)　　$265 \text{ torr} \times \dfrac{1 \text{ mm Hg}}{1 \text{ torr}} = 265 \text{ mm Hg}$

(c)　　$265 \text{ torr} \times \dfrac{1.01325 \times 10^5 \text{ Pa}}{760 \text{ torr}} = 3.53 \times 10^4 \text{ Pa}$

(d)　　$265 \text{ torr} \times \dfrac{1.01325 \times 10^5 \text{ Pa}}{760 \text{ torr}} \times \dfrac{1 \text{ bar}}{1 \times 10^5 \text{ Pa}} = 0.353 \text{ bar}$

10.19　*Analyze/Plan.* Follow the logic in Sample Exercise 10.1.　*Solve.*

(a)　　$30.45 \text{ in Hg} \times \dfrac{25.4 \text{ mm}}{1 \text{ in}} \times \dfrac{1 \text{ torr}}{1 \text{ mm Hg}} = 773.4 \text{ torr}$

　　　　[The result has 4 sig figs because 25.4 mm/in is considered to be an exact number. (Section 1.5)]

(b)　　The pressure in Chicago is greater than **standard atmospheric pressure**, 760 torr, so it makes sense to classify this weather system as a "high pressure system."

10.21　*Analyze/Plan.* Follow the logic in Sample Exercise 10.2.　*Solve.*

(i)　　The Hg level is lower in the open end than the closed end, so the gas pressure is less than atmospheric pressure.

　　　　$P_{gas} = 0.985 \text{ atm} - \left(52 \text{ cm} \times \dfrac{1 \text{ atm}}{76 \text{ cm}} \right) = 0.30 \text{ atm}$

(ii)　　The Hg level is higher in the open end, so the gas pressure is greater than atmospheric pressure.

　　　　$P_{gas} = 0.985 \text{ atm} + \left(67 \text{ mm Hg} \times \dfrac{1 \text{ atm}}{760 \text{ mm Hg}} \right) = 1.073 \text{ atm}$

(iii)　This is a closed-end manometer, so $P_{gas} = h$.

　　　　$P_{gas} = 10.3 \text{ cm} \times \dfrac{1 \text{ atm}}{76 \text{ cm}} = 0.136 \text{ atm}$

The Gas Laws

10.23　*Analyze/Plan.* Given certain changes in gas conditions, predict the effect on other conditions. Consider the gas law relationships in Section 10.3.　*Solve.*

(a)　　P and V are inversely proportional at constant T. If the volume decreases by a factor of 4, the pressure increases by a factor of 4.

(b)　　P and T are directly proportional at constant V. If T decreases by a factor of 2, P also decreases by a factor of 2.

(c)　　P and n are directly proportional at constant V and T. If n decreases by a factor of 2, P also decreases by a factor of 2.

10.25   (a)   Avogadro's hypothesis states that equal volumes of gases at the same temperature and pressure contain equal numbers of molecules. Since molecules react in the ratios of small whole numbers, it follows that the volumes of reacting gases (at the same temperature and pressure) are in the ratios of small whole numbers.

(b)   Since the two gases are at the same temperature and pressure, the ratio of the numbers of atoms is the same as the ratio of volumes. There are 1.5 times as many Xe atoms as Ne atoms.

The Ideal-Gas Equation

(In *Solutions to Exercises*, the symbol for molar mass is MM.)

10.27   (a)   PV = nRT; P in atmospheres, V in liters, n in moles, T in kelvins

(b)   An ideal gas exhibits pressure, volume, and temperature relationships which are described by the equation PV = nRT. (An ideal gas obeys the ideal-gas equation.)

10.29   *Analyze/Plan.* PV = nRT. At constant volume and temperature, P is directly proportional to n.

Solve. For samples with equal masses of gas, the gas with MM = 30 will have twice as many moles of particles and twice the pressure. Thus, flask A contains the gas with MM = 30 and flask B contains the gas with MM = 60.

10.31   *Analyze/Plan.* Follow the strategy for calculations involving many variables given in Section 10.4.   *Solve.*

$$T = \frac{PV}{nR} = 2.00\,atm \times \frac{1.00\,L}{0.500\,mol} \times \frac{K \cdot mol}{0.08206\,L \cdot atm} = 48.7\,K$$

K = 27°C + 273 = 300 K

$$n = \frac{PV}{RT} = 0.300\,atm \times \frac{0.250\,L}{300\,K} \times \frac{K \cdot mol}{0.08206\,L \cdot atm} = 3.05 \times 10^{-3}\,mol$$

$$650\,torr \times \frac{1\,atm}{760\,torr} = 0.85526 = 0.855\,atm$$

$$V = \frac{nRT}{P} = 0.333\,mol \times \frac{350\,K}{0.85526\,atm} \times \frac{0.08206\,L \cdot atm}{K \cdot mol} = 11.2\,L$$

585 mL = 0.585 L

$$P = \frac{nRT}{V} = 0.250\,mol \times \frac{295\,K}{0.585\,L} \times \frac{0.08206\,L \cdot atm}{K \cdot mol} = 10.3\,atm$$

P	V	N	T
2.00 atm	1.00 L	0.500 mol	**48.7 K**
0.300 atm	0.250 L	**3.05 × 10⁻³ mol**	27°C
650 torr	**11.2 L**	0.333 mol	350 K
10.3 atm	585 mL	0.250 mol	295 K

10.33   *Analyze/Plan.* Follow the strategy for calculations involving many variables.   *Solve.*

$n = g/MM; PV = nRT; PV = gRT/MM; g = MM \times PV/RT$

$P = 1.0$ atm, $T = 23°C = 296$ K, $V = 2.0 \times 10^5$ m^3. Change m^3 to L, then calculate grams (or kg).

$$2.0 \times 10^5 \, m^3 \times \frac{10^3 \, dm^3}{1 \, m^3} \times \frac{1 \, L}{1 \, dm^3} = 2.0 \times 10^8 \, L \, H_2$$

$$g = \frac{2.02 \, g \, H_2}{1 \, mol \, H_2} \times \frac{K \cdot mol}{0.08206 \, L \cdot atm} \times \frac{1.0 \, atm \times 2.0 \times 10^8 \, L}{296 \, K} = 1.7 \times 10^7 \, g = 1.7 \times 10^4 \, kg \, H_2$$

10.35   *Analyze/Plan.* Follow the strategy for calculations involving many variables.   *Solve.*

$$V = 2.50 \, L; T = 273 + 37°C = 310 \, K; P = 735 \, torr \times \frac{1 \, atm}{760 \, torr} = 0.96710 = 0.967 \, atm$$

$PV = nRT$, $n = PV/RT$, number of molecules (#) $= n \times 6.022 \times 10^{23}$

$$\# = \frac{0.9671 \, atm \times 2.50 \, L}{310 \, K} \times \frac{K \cdot mol}{0.08206 \, L \cdot atm} \times \frac{6.022 \times 10^{23} \, molecules}{mol}$$

$$= 5.72 \times 10^{22} \, molecules$$

10.37   *Analyze/Plan.* Follow the strategy for calculations involving many variables.   *Solve.*

(a) $P = \dfrac{nRT}{V}; n = 0.29 \, kg \, O_2 \times \dfrac{1000 \, g}{1 \, kg} \times \dfrac{1 \, mol \, O_2}{32.00 \, g \, O_2} = 9.0625 = 9.1 \, mol; V = 2.3 \, L;$

       $T = 273 + 9°C = 282 \, K$

$$P = \frac{9.0625 \, mol}{2.3 \, L} \times \frac{0.08206 \, L \cdot atm}{K \cdot mol} \times 282 \, K = 91 \, atm$$

(b) $V = \dfrac{nRT}{P}; = \dfrac{9.0625 \, mol}{0.95 \, atm} \times \dfrac{0.08206 \, L \cdot atm}{K \cdot mol} \times 299 \, K = 2.3 \times 10^2 \, L$

10.39   *Analyze/Plan.* Follow the strategy for calculations involving many variables.   *Solve.*

$V = 8.70 \, L, T = 24°C = 297 \, K, P = 895 \, torr \times \dfrac{1 \, atm}{760 \, torr} = 1.1776 = 1.18 \, atm = 5.0 \times 10^{-3}$

(a) $g = \dfrac{MM \times PV}{RT}; g = \dfrac{70.91 \, g \, Cl_2}{1 \, mol \, Cl_2} \times \dfrac{K \cdot mol}{0.08206 \, L \cdot atm} \times \dfrac{1.1776 \, atm}{297 \, K} \times 8.70 \, L$

                                     $= 29.8 \, g \, Cl_2$

(b) $V_2 = \dfrac{P_1 V_1 T_2}{T_1 P_2} = \dfrac{895 \, torr \times 8.70 \, L \times 273 \, K}{297 \, K \times 760 \, torr} = 9.42 \, L$

(c) $T_2 = \dfrac{P_2 V_2 T_1}{P_1 V_1} = \dfrac{876 \, torr \times 15.00 \, L \times 297 \, K}{895 \, torr \times 8.70 \, L} = 501 \, K$

(d) $P_2 = \dfrac{P_1 V_1 T_2}{V_2 T_1} = \dfrac{895 \, torr \times 8.70 \, L \times 331 \, K}{6.00 \, L \times 297 \, K} = 1.45 \times 10^3 \, torr = 1.90 \, atm$

10.41   *Analyze.* Given: mass of cockroach, rate of O_2 consumption, temperature, percent O_2 in air, volume of air. Find: mol O_2 consumed per hour; mol O_2 in 1 quart of air; mol O_2 consumed in 48 hr.

(a)   *Plan/Solve.* V of O_2 consumed = rate of consumption × mass × time. n = PV/RT.

$$5.2\,g \times 1\,hr \times \frac{0.8\,mL\,O_2}{1\,g \cdot hr} = 4.16 = 4\,mL\,O_2 \text{ consumed}$$

$$n = \frac{PV}{RT} = 1\,atm \times \frac{K \cdot mol}{0.08206\,L \cdot atm} \times \frac{0.00416\,L}{297\,K} = 1.71 \times 10^{-4} = 2 \times 10^{-4}\,mol\ O_2$$

(b)   *Plan/Solve.* qt air → L air → L O_2 available. mol O_2 available = PV/RT.
mol O_2/hr (from part (a)) → total mol O_2 consumed. Compare O_2 available and O_2 consumed.

$$1\,qt\,air \times \frac{0.946\,L}{1\,qt} \times 0.21\,O_2 \text{ in air} = 0.199\,L\,O_2 \text{ available}$$

$$n = 1\,atm \times \frac{K \cdot mol}{0.08206\,L \cdot atm} \times \frac{0.199\,L}{297\,K} = 8.16 \times 10^{-3} = 8 \times 10^{-3}\,mol\ O_2 \text{ available}$$

$$\text{roach uses} \frac{1.71 \times 10^{-4}\,mol}{1\,hr} \times 48\,hr = 8.19 \times 10^{-3} = 8 \times 10^{-3}\,mol\ O_2 \text{ consumed}$$

Not only does the roach use 20% of the available O_2, it needs all the O_2 in the jar.

Further Applications of the Ideal-Gas Equation

10.43   (c) $Cl_2(g)$ is the most dense at 1.00 at and 298 K. Gas density is directly proportional to molar mass and pressure, and inversely proportional to temperature (Equation [10.10]). For gas samples at the same conditions, molar mass determines density. Of the three gases listed, Cl_2 has the largest molar mass.

10.45   (c) Because the helium atoms are of lower mass than the average air molecule, the helium gas is less dense than air. The balloon thus weighs less than the air displaced by its volume.

10.47   *Analyze/Plan.* Conditions (P, V, T) and amounts of gases are given. Rearrange the relationship $PV \times MM = gRT$ to obtain the desired of quantity, paying attention (as always!) to units.   *Solve.*

(a) $d = \dfrac{MM \times P}{RT}$; MM = 46.0 g/mol; P = 0.970 atm, T = 35°C = 308 K

$$d = \frac{46.0\,g\,NO_2}{1\,mol} \times \frac{K \cdot mol}{0.08206\,L \cdot atm} \times \frac{0.970\,atm}{308\,K} = 1.77\,g/L$$

(b) $MM = \dfrac{gRT}{PV} = \dfrac{2.50\,g}{0.875\,L} \times \dfrac{0.08206\,L \cdot atm}{K \cdot mol} \times \dfrac{308\,K}{685\,torr} \times \dfrac{760\,torr}{1\,atm} = 80.1\,g/mol$

10.49   *Analyze/Plan.* Given: mass, conditions (P, V, T) of unknown gas. Find: molar mass. $MM = gRT/PV$. *Solve.*

$$MM = \frac{gRT}{PV} = \frac{1.012\,g}{0.354\,L} \times \frac{0.08206\,L \cdot atm}{K \cdot atm} \times \frac{372\,K}{742\,torr} \times \frac{760\,torr}{1\,atm} = 89.4\,g/mol$$

10.51   *Analyze/Plan.* Follow the logic in Sample Exercise 10.9.   *Solve.*

$$\text{mol O}_2 = \frac{PV}{RT} = 3.5 \times 10^{-6} \text{ torr} \times \frac{1\,\text{atm}}{760\,\text{torr}} \times \frac{\text{K} \cdot \text{atm}}{0.08206\,\text{L} \cdot \text{atm}} \times \frac{0.382\,\text{L}}{300\,\text{K}} = 7.146 \times 10^{-11}$$

$$= 7.1 \times 10^{-11} \text{ mol O}_2$$

$$7.146 \times 10^{-11} \text{ mol O}_2 \times \frac{2\,\text{mol Mg}}{1\,\text{mol O}_2} \times \frac{24.3\,\text{g Mg}}{1\,\text{mol Mg}} = 3.5 \times 10^{-9} \text{ g Mg}$$

10.53   *Analyze/Plan.* g glucose $\rightarrow$ mol glucose $\rightarrow$ mol CO_2 $\rightarrow$ V CO_2 *Solve.*

$$24.5\,\text{g} \times \frac{1\,\text{mol glucose}}{180.1\,\text{g}} \times \frac{6\,\text{mol CO}_2}{1\,\text{mol glucose}} = 0.8162 = 0.816 \text{ mol CO}_2$$

$$V = \frac{nRT}{P} = 0.8162 \text{ mol} \times \frac{0.08206\,\text{L} \cdot \text{atm}}{\text{K} \cdot \text{mol}} \times \frac{310\,\text{K}}{0.970\,\text{atm}} = 21.4 \text{ L CO}_2$$

10.55   *Analyze/Plan.* The gas sample is a mixture of $H_2(g)$ and $H_2O(g)$. Find the partial pressure of $H_2(g)$ and then the moles of $H_2(g)$ and $Zn(s)$. *Solve.*

$$P_t = 738 \text{ torr} = P_{H_2} + P_{H_2O}$$

From Appendix B, the vapor pressure of H_2O at 24°C = 22.38 torr

$$P_{H_2} = (738 \text{ torr} - 22.38 \text{ torr}) \times \frac{1\,\text{atm}}{760\,\text{torr}} = 0.9416 = 0.942 \text{ atm}$$

$$n_{H_2} = \frac{P_{H_2}V}{RT} = 0.9416 \text{ atm} \times \frac{\text{K} \cdot \text{mol}}{0.08206\ \text{L} \cdot \text{atm}} \times \frac{0.159\,\text{L}}{297\,\text{K}} = 0.006143 = 0.00614 \text{ mol H}_2$$

$$0.006143 \text{ mol H}_2 \times \frac{1\,\text{mol Zn}}{1\,\text{mol H}_2} \times \frac{65.39\,\text{g Zn}}{1\,\text{mol Zn}} = 0.402 \text{ g Zn}$$

Partial Pressures

10.57   (a)   When the stopcock is opened, the volume occupied by $N_2(g)$ increases from 2.0 L to 5.0 L. At constant T, $P_1V_1 = P_2V_2$. 1.0 atm × 2.0 L = P_2 × 5.0 L; P_2 = 0.40 atm

      (b)   When the gases mix, the volume of $O_2(g)$ increases from 3.0 L to 5.0 L. At constant T, $P_1V_1 = P_2V_2$. 2.0 atm × 3.0 L = P_2 × 5.0 L; P_2 = 1.2 atm

      (c) $P_t = P_{N_2} + P_{O_2} = 0.40 \text{ atm} + 1.2 \text{ atm} = 1.6 \text{ atm}$

10.59   *Analyze.* Given: amount, V, T of three gases. Find: P of each gas, total P.

      Plan. P = nRT/V; $P_t = P_1 + P_2 + P_3 + \cdots$ *Solve.*

      (a) $P_{He} = \dfrac{nRT}{V} = 0.538 \text{ mol} \times \dfrac{0.08206\,\text{L} \cdot \text{atm}}{\text{K} \cdot \text{atm}} \times \dfrac{298\,\text{K}}{7.00\,\text{L}} = 1.88 \text{ atm}$

           $P_{Ne} = \dfrac{nRT}{V} = 0.315 \text{ mol} \times \dfrac{0.08206\,\text{L} \cdot \text{atm}}{\text{K} \cdot \text{atm}} \times \dfrac{298\,\text{K}}{7.00\,\text{L}} = 1.10 \text{ atm}$

           $P_{Ar} = \dfrac{nRT}{V} = 0.103 \text{ mol} \times \dfrac{0.08206\,\text{L} \cdot \text{atm}}{\text{K} \cdot \text{atm}} \times \dfrac{298\,\text{K}}{7.00\,\text{L}} = 0.360 \text{ atm}$

      (b) P_t = 1.88 atm + 1.10 atm + 0.360 atm = 3.34 atm

10.61 *Analyze.* Given: mass CO_2 at V, T; pressure of air at same V, T. Find: partial pressure of CO_2 at these conditions, total pressure of gases at V, T.

  Plan. $g\,CO_2 \to mol\,CO_2 \to P_{CO_2}$ (via $P = nRT/V$); $P_t = P_{CO_2} + P_{air}$ *Solve.*

$$5.50\ g\ CO_2 \times \frac{1\,mol\,CO_2}{44.01\ g\ CO_2} = 0.12497 = 0.125\ mol\,CO_2; T = 273 + 24°C = 297\ K$$

$$P_{CO_2} = 0.12497\ mol \times \frac{297\ K}{10.0\ L} \times \frac{0.08206\ L \cdot atm}{K \cdot mol} = 0.30458 = 0.305\ atm$$

$$P_{air} = 705\ torr \times \frac{1\,atm}{760\ torr} = 0.92763 = 0.928\ atm$$

$$P_t = P_{CO_2} + P_{air} = 0.30458 + 0.92763 = 1.23221 = 1.232\ atm$$

  (Result has 3 decimal places and 4 sig figs.)

10.63 *Analyze/Plan.* The partial pressure of each component is equal to the mole fraction of that gas times the total pressure of the mixture. Find the mole fraction of each component and then its partial pressure. *Solve.*

  $n_t = 0.75\ mol\,N_2 + 0.30\ mol\,O_2 + 0.15\ mol\,CO_2 = 1.20\ mol$

$$\chi_{N_2} = \frac{0.75}{1.20} = 0.625 = 0.63; P_{N_2} = 0.625 \times 1.56\ atm = 0.98\ atm$$

$$\chi_{O_2} = \frac{0.30}{1.20} = 0.250 = 0.25; P_{O_2} = 0.250 \times 1.56\ atm = 0.39\ atm$$

$$\chi_{CO_2} = \frac{0.15}{1.20} = 0.125 = 0.13; P_{CO_2} = 0.125 \times 1.56\ atm = 0.20\ atm$$

10.65 *Analyze/Plan.* Mole fraction = pressure fraction. Find the desired mole fraction of O_2 and change to mole percent. *Solve.*

$$\chi_{O_2} = \frac{P_{O_2}}{P_t} = \frac{0.21\,atm}{8.38\,atm} = 0.025; mole\ \% = 0.025 \times 100 = 2.5\%$$

10.67 *Analyze/Plan.* $N_2(g)$ and $O_2(g)$ undergo changes of conditions and are mixed. Calculate the new pressure of each gas and add them to obtain the total pressure of the mixture.

  $P_2 = P_1 V_1 T_2 / V_2 T_1; P_T = P_{N_2} + P_{O_2}.$ *Solve.*

$$P_{N_2} = \frac{P_1 V_1 T_2}{V_2 T_1} = \frac{4.75\ atm \times 1.00\ L \times 293\ K}{10.0\ L \times 299\ K} = 0.46547 = 0.465\ atm$$

$$P_{O_2} = \frac{P_1 V_1 T_2}{V_2 T_1} = \frac{5.25\ atm \times 5.00\ L \times 293\ K}{10.0\ L \times 299\ K} = 2.5723 = 2.57\ atm$$

$$P_t = 0.46547\ atm + 2.5723\ atm = 3.0378 = 3.04\ atm$$

Kinetic-Molecular Theory; Graham's Law

10.69　(a)　Increase in temperature at constant volume, decrease in volume, increase in pressure

　　　(b)　Decrease in temperature

　　　(c)　Increase in volume, decrease in pressure

　　　(d)　Increase in temperature

10.71　The fact that gases are readily compressible supports the assumption that most of the volume of a gas sample is empty space.

10.73　*Analyze/Plan.* We have samples of two different gases at different pressures and temperatures. Compare the two samples by considering the postulates of the kinetic-molecular theory that pertain to the quantities in (a)–(d). *Solve.*

　　　(a)　$n \propto P/T$ (V/R is the same for A and B.) Since P is greater and T is smaller for vessel A, it has more molecules.

　　　(b)　Vessel A has more molecules but the molar mass of CO is smaller than the molar mass of SO_2, so we need to calculate the masses. Since volume is not specified, calculate g/L.

$$\frac{g_A}{V} = \frac{MM \times P}{RT} = \frac{28.01\,g\,CO}{1\,mol\,CO} \times \frac{K \bullet mol}{0.08206\,L \bullet atm} \times \frac{1\,atm}{273\,K} = 1.25\,g\,CO/L$$

$$\frac{g_B}{V} = \frac{MM \times P}{RT} = \frac{64.07\,g\,SO_2}{1\,mol\,SO_2} \times \frac{K \bullet mol}{0.08206\,L \bullet atm} \times \frac{0.5\,atm}{293\,K} = 1.33\,g\,SO_2/L$$

Vessel B has more mass.

　　　(c)　Vessel B is at a higher temperature so the average kinetic energy of its molecules is higher.

　　　(d)　The two factors that affect rms speed are temperature and molar mass. The molecules in vessel A have smaller molar mass but are at the lower temperature, so we must calculate the rms speeds.

Mathematically, according to Equation [10.24],

$$\frac{u_A}{u_B} = \sqrt{\frac{T_A/MM_A}{T_B/MM_B}} = \sqrt{\frac{273/28.01}{293/64.07}} = 1.46$$

The ratio is greater than 1; vessel A has the greater rms speed.

10.75　(a)　*Plan.* The larger the molar mass, the slower the average speed (at constant temperature).

　　　　Solve. In order of increasing speed (and decreasing molar mass):
　　　　$HBr < NF_3 < SO_2 < CO < Ne$

　　　(b)　*Plan.* Follow the logic of Sample Exercise 10.14. *Solve.*

$$u = \sqrt{\frac{3RT}{M}} = \left(\frac{3 \times 8.314\,kg \bullet m^2/s^2 \bullet K \bullet mol \times 298\,K}{71.0 \times 10^{-3}\,kg/mol} \right)^{1/2} = 324\,m/s$$

10.77 *Plan.* The heavier the molecule, the slower the rate of effusion. Thus, the order for increasing rate of effusion is in the order of decreasing mass. *Solve.*

rate $^2H^{37}Cl$ < rate $^1H^{37}Cl$ < rate $^2H^{35}Cl$ < rate $^1H^{35}Cl$

10.79 *Analyze.* Given: relative effusion rates of two gases at same temperature. Find: molecular formula of one of the gases. *Plan.* Use Graham's law to calculate the formula weight of arsenic (III) sulfide, and thus the molecular formula. *Solve.*

$$\frac{\text{rate (sulfide)}}{\text{rate (Ar)}} = \left[\frac{39.9}{\text{MM (sulfide)}}\right]^{1/2} = 0.28$$

MM (sulfide) = $(39.9 / 0.28)^2$ = 510 g/mol (two significant figures)

The empirical formula of arsenic(III) sulfide is As_2S_3, which has a formula mass of 246.1. Twice this is 490 g/mol, close to the value estimated from the effusion experiment. Thus, the formula of the vapor phase molecule is As_4S_6.

Nonideal-Gas Behavior

10.81 (a) Nonideal gas behavior is observed at very high pressures and/or low temperatures.

(b) The real volumes of gas molecules and attractive intermolecular forces between molecules cause gases to behave nonideally.

(c) The ratio PV/RT is equal to the number of moles of particles in an ideal-gas sample; this number should be a constant for all pressure, volume, and temperature conditions. If the value of this ratio changes with increasing pressure, the gas sample is not behaving ideally. That is, the gas is not behaving according to the ideal-gas equation.

Negative deviations indicate fewer "effective" particles in the sample, a result of attractive forces among particles. Positive deviations indicate more "effective" particles, a result of the real volume occupied by the particles.

10.83 *Plan.* The constants a and b are part of the correction terms in the van der Waals equation. The smaller the values of a and b, the smaller the corrections and the more ideal the gas. *Solve.*

Ar (a = 1.34, b = 0.0322) will behave more like an ideal gas than CO_2 (a = 3.59, b = 0.0427) at high pressures.

10.85 *Analyze.* Conditions and amount of $CCl_4(g)$ are given. *Plan.* Use ideal-gas equation and van der Waals equation to calculate pressure of gas at these conditions. *Solve.*

(a) $P = 1.00 \text{ mol} \times \dfrac{0.08206 \text{ L} \cdot \text{atm}}{\text{K} \cdot \text{mol}} \times \dfrac{313 \text{ K}}{28.0 \text{ L}} = 0.917 \text{ atm}$

(b) $P = \dfrac{nRT}{V - nb} - \dfrac{an^2}{V^2} = \dfrac{1.00 \times 0.08206 \times 313}{28.0 - (1.00 \times 0.1383)} - \dfrac{20.4(1.00)^2}{(28.0)^2} = 0.896 \text{ atm}$

Check. The van der Waals result indicates that the real pressure will be less than the ideal pressure. That is, intermolecular forces reduce the effective number of particles and the real pressure. This is reasonable for 1 mole of gas at relatively low temperature and pressure.

10 Gases

Additional Exercises

10.87 A mercury barometer with water trapped in its tip would not read the correct pressure. The standard relationship between height of an Hg column and atmospheric pressure (760 mm Hg = 1 atm) assumes that there is a vacuum in the closed end of the barometer and that gravity is the only downward force on the Hg column. Water at the top of the Hg column would establish a vapor pressure, which would exert additional downward pressure and partially counterbalance the pressure of the atmosphere. The Hg column would read lower than the actual atmospheric pressure.

10.90 $PV = nRT$, $n = PV/RT$. Since RT is constant, n is proportional to PV.

Total available n = $(15.0 \text{ L} \times 1.00 \times 10^2 \text{ atm}) - (15.0 \text{ L} \times 1.00 \text{ atm}) = 1485$

$$= 1.49 \times 10^3 \text{ L} \cdot \text{atm}$$

Each balloon holds 2.00 L × 1.00 atm = 2.00 L•atm

$$\frac{1485 \text{ L} \cdot \text{atm available}}{2.00 \text{ L} \cdot \text{atm/balloon}} = 742.5 = 742 \text{ balloons}$$

(Only 742 balloons can be filled completely, with a bit of He left over.)

10.93 If the air in the room is at STP, the partial pressure of O_2 is 0.2095 × 1 atm = 0.2095 atm. Since the gases in air are perfectly mixed, the volume of O_2 is the volume of the room.

$$V = 10.0 \text{ ft} \times 8.0 \text{ ft} \times 8.0 \text{ ft} \times \frac{(12)^3 \text{ in}^3}{\text{ft}^3} \times \frac{(2.54)^3 \text{ cm}^3}{\text{in}^3} \times \frac{1 \text{ L}}{1000 \text{ cm}^3} = 1.812 \times 10^4$$

$$= 1.8 \times 10^4 \text{ L}$$

$$g = \frac{MM \times PV}{RT} = \frac{32.00 \text{ g O}_2}{\text{mol O}_2} \times \frac{K \cdot \text{mol}}{0.08026 \text{ L} \cdot \text{atm}} \times \frac{0.2095 \text{ atm} \times 1.812 \times 10^4 \text{ L}}{273 \text{ K}} = 5.4 \times 10^3 \text{ g O}_2$$

10.96 (a) $n = \dfrac{PV}{RT} = 0.980 \text{ atm} \times \dfrac{K \cdot \text{mol}}{0.08206 \text{ L} \cdot \text{atm}} \times \dfrac{0.524 \text{ L}}{347 \text{ K}} = 0.018034 = 0.0180 \text{ mol air}$

$$\text{mol O}_2 = 0.018034 \text{ mol air} \times \frac{0.2095 \text{ mol O}_2}{1 \text{ mol air}} = 0.003778 = 0.00378 \text{ mol O}_2$$

(b) $C_8H_{18}(l) + 25/2 \; O_2(g) \rightarrow 8CO_2(g) + 9H_2O(g)$

(The H_2O produced in an automobile engine is in the gaseous state.)

$$0.003778 \text{ mol O}_2 \times \frac{1 \text{ mol C}_8H_{18}}{12.5 \text{ mol O}_2} \times \frac{114.2 \text{ g C}_8H_{18}}{1 \text{ mol C}_8H_{18}} = 0.0345 \text{ g C}_8H_{18}$$

10.99 $MM_{avg} = \dfrac{dRT}{P} = \dfrac{1.104 \text{ g}}{1 \text{ L}} \times \dfrac{0.08206 \text{ L} \cdot \text{atm}}{K \cdot \text{mol}} \times \dfrac{300 \text{ K}}{435 \text{ torr}} \times \dfrac{760 \text{ torr}}{1 \text{ atm}} = 47.48 = 47.5 \text{ g/mol}$

χ = mole fraction O_2; $1 - \chi$ = mole fraction Kr

47.48 g = χ(32.00) + (1 – χ)(83.80)

36.3 = 51.8 χ; χ = 0.701; 70.1% O_2

10.102 (a) The effect of intermolecular attraction becomes more significant as a gas is compressed to a smaller volume at constant temperature. This compression causes the pressure, and thus the number of intermolecular collisions, to increase. Intermolecular attraction causes some of these collisions to be inelastic, which amplifies the deviation from ideal behavior.

(b) The effect of intermolecular attraction becomes less significant as the temperature of a gas is increased at constant volume. When the temperature of a gas is increased at constant volume, the pressure of the gas, the number of intermolecular collisions, and the average kinetic energy of the gas particles increases. This higher average kinetic energy means that a larger fraction of the molecules has sufficient kinetic energy to overcome intermolecular attractions, even though there are more total collisions. This increases the fraction of elastic collisions, and the gas more closely obeys the ideal-gas equation.

Integrative Exercises

10.105   (a) $MM = \dfrac{gRT}{VP} = \dfrac{1.56\,g}{1.00\,L} \times \dfrac{0.08206\,L \bullet atm}{K \bullet mol} \times \dfrac{323\,K}{0.984\,atm} = 42.0\,g/mol$

Assume 100 g cyclopropane

$100\,g \times 0.857\,C = 85.7\,g\,C \times \dfrac{1\,mol\,C}{12.01\,g} = \dfrac{7.136\,mol\,C}{7.136} = 1\,mol\,C$

$100\,g \times 0.143\,H = 14.3\,g\,H \times \dfrac{1\,mol\,H}{1.008\,g} = \dfrac{14.19\,mol\,H}{7.136} = 2\,mol\,H$

The empirical formula of cyclopropane is CH_2 and the empirical formula weight is 12 + 2 = 14 g. The ratio of molar mass to empirical formula weight, 42.0 g/14 g, is 3; therefore, there are three empirical formula units in one cyclopropane molecule. The molecular formula is $3 \times (CH_2) = C_3H_6$.

(b) Ar is a monoatomic gas. Cyclopropane molecules are larger and more structurally complex, even though the molar masses of Ar and C_3H_6 are similar. If both gases are at the same relatively low temperature, they have approximately the same average kinetic energy, and the same ability to overcome intermolecular attractions. We expect intermolecular attractions to be more significant for the more complex C_3H_6 molecules, and that C_3H_6 will deviate more from ideal behavior at the conditions listed. This conclusion is supported by the $\underline{a}$ values in Table 10.3. The $\underline{a}$ values for CH_4 and CO_2, more complex molecules than Ar atoms, are larger than the value for Ar. If the pressure is high enough for the volume correction in the van der Waals equation to dominate behavior, the larger C_3H_6 molecules definitely deviate more than Ar atoms from ideal behavior.

10.108 $n = \dfrac{PV}{RT} = 1.00\,atm \times \dfrac{K \bullet mol}{0.08206\,L \bullet atm} \times \dfrac{2.7 \times 10^{12}\,L}{273\,K} = 1.205 \times 10^{11} = 1.2 \times 10^{11}\,mol\,CH_4$

$CH_4(g) + 2O_2(g) \rightarrow CO_2(g) + 2H_2O(l)$ $\Delta H° = -890.4\,kJ$

(At STP, H_2O is in the liquid state.)

$\Delta H_{rxn}^{\circ} = \Delta H_f^{\circ}CO_2(g) + 2\Delta H_f^{\circ}H_2O(l) - \Delta H_f^{\circ}CH_4(g) - \Delta H_f^{\circ}O_2(g)$

$\Delta H_{rxn}^{\circ} = -393.5\,kJ + 2(-285.83\,kJ) - (-74.8\,kJ) - 0 = -890.4\,kJ$

$$\frac{-890.4 \text{ kJ}}{1 \text{ mol CH}_4} \times 1.205 \times 10^{11} \text{ mol CH}_4 = -1.073 \times 10^{14} = -1.1 \times 10^{14} \text{ kJ}$$

The negative sign indicates heat evolved by the combustion reaction.

10.112   After reaction, the flask contains $IF_5(g)$ and whichever reactant is in excess. Determine the limiting reactant, which regulates the moles of IF_5 produced and moles of excess reactant.

$$I_2(s) + 5F_2(g) \rightarrow 2 \, IF_5(g)$$

$$10.0 \text{ g } I_2 \times \frac{1 \text{ mol } I_2}{253.8 \text{ g } I_2} \times \frac{5 \text{ mol } F_2}{1 \text{ mol } I_2} = 0.1970 = 0.197 \text{ mol } F_2$$

$$10.0 \text{ g } F_2 \times \frac{1 \text{ mol } F_2}{38.00 \text{ g } F_2} = 0.2632 = 0.263 \text{ mol } F_2 \text{ available}$$

I_2 is the limiting reactant; F_2 is in excess.

0.263 mol F_2 available – 0.197 mol F_2 reacted = 0.066 mol F_2 remain.

$$10.0 \text{ g } I_2 \times \frac{1 \text{ mol } I_2}{253.8 \text{ g } I_2} \times \frac{2 \text{ mol } IF_5}{1 \text{ mol } I_2} = 0.0788 \text{ mol } IF_5 \text{ produced}$$

(a) $P_{IF_5} = \dfrac{nRT}{V} = 0.0788 \text{ mol} \times \dfrac{0.08206 \text{ L} \cdot \text{atm}}{\text{K} \cdot \text{mol}} \times \dfrac{398 \text{ K}}{5.00 \text{ L}} = 0.515 \text{ atm}$

(b) $\chi_{IF_5} = \dfrac{\text{mol } IF_5}{\text{mol } IF_5 + \text{mol } F_2} = \dfrac{0.0788}{0.0788 + 0.066} = 0.544$

11 Intermolecular Forces, Liquids and Solids

Visualizing Concepts

11.1   The diagram best describes a **liquid**. In the diagram, the particles are close together, mostly touching but there is no regular arrangement or order. This rules out a gaseous sample, where the particles are far apart, and a crystalline solid, which has a regular repeating structure in all three directions.

11.4   (a)   385 mm Hg. Find 30°C on the horizontal axis, and follow a vertical line from this point to its intersection with the red vapor pressure curve. Follow a horizontal line from the intersection to the vertical axis and read the vapor pressure.

      (b)   22°C. Reverse the procedure outlined in part (a). Find 300 torr on the vertical axis, follow it to the curve and down to the value on the horizontal axis.

      (c)   47°C. The normal boiling point of a liquid is the temperature at which its vapor pressure is 1 atm, or 760 mm Hg. On this diagram, the vapor pressure curve ends at this point, approximately 47°C.

11.7   (a)   Nb: $6 \times 1/2 = 3$; O: $12 \times 1/4 = 3$

      (b)   NbO

      (c)   This is primarily an ionic solid, because Nb is a metal and O is a nonmetal. There may be some covalent character to the $Nb \cdots O$ bonds.

Kinetic-Molecular Theory

11.9   (a)   solid < liquid < gas

      (b)   gas < liquid < solid

11.11   (a)   Gases are more compressible than liquids because there is much empty space between gas particles.

      (b)   The solid and liquid forms of a substance are called *condensed phases* because in both states there is very little space between particles; the volume of the substance is condensed.

      (c)   Liquids have greater ability to flow because their average kinetic energy is on the same order of magnitude as the average attractive energy. This is easy to see for the solid and liquid forms of the same substance, where intermolecular attractive forces are the same, but the average kinetic energy of the liquid is greater because its temperature is higher.

Intermolecular Forces

11.13 (a) London-dispersion forces

(b) dipole-dipole and London-dispersion forces

(c) dipole-dipole or in certain cases hydrogen bonding

11.15 (a) Br_2 is a nonpolar covalent molecule, so only London-dispersion forces must be overcome to convert the liquid to a gas.

(b) CH_3OH is a polar covalent molecule that experiences London-dispersion, dipole-dipole, and hydrogen-bonding (O–H bonds) forces. All of these forces must be overcome to convert the liquid to a gas.

(c) H_2S is a polar covalent molecule that experiences London-dispersion and dipole-dipole forces, so these must be overcome to change the liquid into a gas. (H–S bonds do not lead to hydrogen-bonding interactions.)

11.17 (a) *Polarizability* is the ease with which the charge distribution (electron cloud) in a molecule can be distorted to produce a transient dipole.

(b) Te is most polarizable because its valence electrons are farthest from the nucleus and least tightly held.

(c) Polarizability increases as molecular size (and thus molecular weight) increases. In order of increasing polarizability: $CH_4 < SiH_4 < SiCl_4 < GeCl_4 < GeBr_4$

(d) The magnitude of London-dispersion forces and thus the boiling points of molecules increase as polarizability increases. The order of increasing boiling points is the order of increasing polarizability:
$CH_4 < SiH_4 < SiCl_4 < GeCl_4 < GeBr_4$

11.19 *Analyze/Plan.* For molecules with similar structures, the strength of dispersion forces increases with molecular size (molecular weight and number of electrons in the molecule).

Solve: (a) H_2S (b) CO_2 (c) SiH_4

11.21 Both hydrocarbons experience dispersion forces. Rod-like butane molecules can contact each other over the length of the molecule, while spherical 2-methylpropane molecules can only touch tangentially. The larger contact surface of butane produces greater polarizability and a higher boiling point.

11.23 (a) Molecules with N–H, O–H, and F–H bonds form hydrogen bonds with like molecules.

(b) **CH_3NH_2** and **CH_3OH** have N–H and O–H bonds, respectively. (CH_3F has C–F and C–H bonds, but no H–F bonds.)

11.25 (a) HF has the higher boiling point because hydrogen bonding is stronger than dipole-dipole forces.

(b) $CHBr_3$ has the higher boiling point because it has the higher molar mass, which leads to greater polarizability and stronger dispersion forces.

(c)   ICl has the higher boiling point because it is a polar molecule. For molecules with similar structures and molar masses, dipole-dipole forces are stronger than dispersion forces.

11.27   (a)   Hydrogen bonding occurs in both water and ice. In water, the molecules are as close together as possible, but in constant motion relative to each other. Hydrogen bonds are constantly broken and new ones formed. In ice, molecules are fixed relative to each other. They adopt an ordered structure that maximizes the number of hydrogen bonding interactions. For one molecule, the two lone pairs interact with H atoms of adjacent molecules and both H atoms interact with lone pairs of two other adjacent molecules. Each molecule participates in four hydrogen bonding interactions oriented tetrahedrally with respect to each other. Maintaining these four interactions per molecule creates a network structure (Figure 11.10) with more open space between molecules than in the liquid state and a corresponding lower density for the solid than the liquid.

(b)   The temperature of a substance is a measure of the average kinetic energy of the sample. In order to increase the kinetic energy of water molecules, strong attractive hydrogen bonds must be broken. A large amount of heat must be added to water to break hydrogen bonds and increase its temperature.

Viscosity and Surface Tension

11.29   (a)   Surface tension and viscosity are the result of intermolecular attractive forces or cohesive forces among molecules in a liquid sample. As temperature increases, the number of molecules with sufficient kinetic energy to overcome these attractive forces increases, and viscosity and surface tension decrease.

(b)   Surface tension and viscosity are both directly related to the strength of intermolecular attractive forces. The same attractive forces that cause surface molecules to be difficult to separate cause molecules elsewhere in the sample to resist movement relative to one another. Liquids with high surface tension have intermolecular attractive forces sufficient to produce a high viscosity as well.

11.31   (a) $CHBr_3$ has a higher molar mass, is more polarizable, and has stronger dispersion forces, so the surface tension is greater [see Solution 11.25(b)].

(b)   As temperature increases, the viscosity of the oil decreases because the average kinetic energy of the molecules increases [Solution 11.29(a)].

(c)   Adhesive forces between polar water and nonpolar car wax are weak, so the large surface tension of water draws the liquid into the shape with the smallest surface area, a sphere.

Changes of State

11.33   (a)   freezing, exothermic

(b)   evaporation or vaporization, endothermic

(c)   deposition, exothermic

11.35　The heat energy required to increase the kinetic energy of molecules enough to melt the solid does not produce a large separation of molecules. The specific order is disrupted, but the molecules remain close together. On the other hand, when a liquid is vaporized, the intermolecular forces which maintain close molecular contacts must be overcome. Because molecules are being separated, the energy requirement is higher than for melting.

11.37　*Analyze.* The heat required to vaporize 50 g of H_2O equals the heat lost by the cooled water.

　　　Plan. Using the enthalpy of vaporization, calculate the heat required to vaporize 50 g of H_2O in this temperature range. Using the specific heat capacity of water, calculate the mass of water than can be cooled 13°C if this much heat is lost.

　　　Solve. Evaporation of 50 g of water requires:

$$50\,g\,H_2O \times \frac{2.4\,kJ}{1\,g\,H_2O} = 1.2 \times 10^2\,kJ \text{ or } 1.2 \times 10^5\,J$$

　　　Cooling a certain amount of water by 13°C:

$$1.2 \times 10^5\,J \times \frac{1\,g \cdot K}{4.184\,J} \times \frac{1}{13°C} = 2206 = 2.2 \times 10^3\,g\,H_2O$$

　　　Check. The units are correct. A surprisingly large mass of water (2200 g ≈ 2.2 L) can be cooled by this method.

11.39　*Analyze/Plan.* Follow the logic in Sample Exercise 11.4. *Solve.*

　　　Heat the solid from –120°C to –114°C (153 K to 159 K), using the specific heat of the solid.

$$75.0\,g\,C_2H_5OH \times \frac{0.97\,J}{g \cdot K} \times 6\,K \times \frac{1\,kJ}{1000\,J} = 0.4365 = 0.4\,kJ$$

　　　At –114°C (159 K), melt the solid, using its enthalpy of fusion.

$$75.0\,g\,C_2H_5OH \times \frac{1\,mol\,C_2H_5OH}{46.07\,g\,C_2H_5OH} \times \frac{5.02\,kJ}{1\,mol} = 8.172 = 8.17\,kJ$$

　　　Heat the liquid from –114°C to 78°C (159 K to 351 K), using the specific heat of the liquid.

$$75.0\,g\,C_2H_5OH \times \frac{2.3\,J}{g \cdot K} \times 192\,K \times \frac{1\,kJ}{1000\,J} = 33.12 = 33\,kJ$$

　　　At 78°C (351 K), vaporize the liquid, using its enthalpy of vaporization.

$$75.0\,g\,C_2H_5OH \times \frac{1\,mol\,C_2H_5OH}{46.07\,g\,C_2H_5OH} \times \frac{38.56\,kJ}{1\,mol} = 62.77 = 62.8\,kJ$$

　　　The total energy required is 0.4365 kJ + 8.172 kJ + 33.12 kJ + 62.77 kJ = 104.50 = 105 kJ. (The result has zero decimal places, from 33 kJ required to heat the liquid.)

　　　Check. The relative energies of the various steps are reasonable; vaporization is the largest.

11.41   (a)   The critical pressure is the pressure required to cause liquefaction at the critical temperature.

         (b)   The critical temperature is the highest temperature at which a gas can be liquefied, regardless of pressure. As the force of attraction between molecules increases, the critical temperature of the compound increases.

         (c)   The temperature of $N_2(l)$ is 77 K. All of the gases in Table 11.5 have critical temperatures higher than 77 K, so all of them can be liquefied at this temperature, given sufficient pressure.

Vapor Pressure and Boiling Point

11.43   (a)   No effect.

         (b)   No effect.

         (c)   Vapor pressure decreases with increasing intermolecular attractive forces because fewer molecules have sufficient kinetic energy to overcome attractive forces and escape to the vapor phase.

         (d)   Vapor pressure increases with increasing temperature because average kinetic energies of molecules increases.

11.45   (a)   *Analyze/Plan.* Given the molecular formulae of several substances, determine the kind of intermolecular forces present, and rank the strength of these forces. The weaker the forces, the more volatile the substance.   *Solve.*

$$CBr_4 < CHBr_3 < CH_2Br_2 < CH_2Cl_2 < CH_3Cl < CH_4$$

The weaker the intermolecular forces, the higher the vapor pressure, the more volatile the compound. The order of increasing volatility is the order of decreasing strength of intermolecular forces. By analogy to the boiling points of HCl and HBr (Section 11.2), the trend will be dominated by dispersion forces, even though four of the molecules ($CHBr_3$, CH_2Br_2, CH_2Cl_2 and CH_3Cl) are polar. Thus, the order of increasing volatility is the order of decreasing molar mass and decreasing strength of dispersion forces.

         (b) $CH_4 < CH_3Cl < CH_2Cl_2 < CH_2Br_2 < CHBr_3 < CBr_4$

Boiling point increases as the strength of intermolecular forces increases, so the order of boiling points is the order of increasing strength of forces. This is the order of decreasing volatility and the reverse of the order in part (a).

11.47   (a)   The water in the two pans is at the same temperature, the boiling point of water at the atmospheric pressure of the room. During a phase change, the temperature of a system is constant. All energy gained from the surroundings is used to accomplish the transition, in this case to vaporize the liquid water. The pan of water that is boiling vigorously is gaining more energy and the liquid is being vaporized more quickly than in the other pan, but the temperature of the phase change is the same.

    (b)   Vapor pressure does not depend on either volume or surface area of the liquid. As long as the containers are at the same temperature, the vapor pressures of water in the two containers are the same.

11.49   The boiling point is the temperature at which the vapor pressure of a liquid equals atmospheric pressure.

    (a)   The boiling point of diethyl ether at 400 torr is ~17°C, or, at 17°C, the vapor pressure of diethyl ether is 400 torr.

    (b)   The vapor pressure of ethyl alcohol at 70°C is approximately 510 torr. Thus, at 70°C ethyl alcohol would boil at an external pressure of 510 torr.

    (c)   The vapor pressure of water at 25°C is 23.76 torr, and at 26°C is 25.21 torr. At a pressure of 25 torr, the boiling point of water is much closer to 26°C then 25°C. To two significant figures, the boiling point is 26°C.

    (d)   A pressure of 1.2 atm corresponds to $1.2 \, \text{atm} \times \dfrac{760 \, \text{torr}}{1 \, \text{atm}} = 912 = 9.1 \times 10^2$ torr.

According to Appendix B, the vapor pressure of water reaches 912 torr somewhere between 104–106°C. By linear interpolation, the boiling point should be near

$$104°C + \left[\frac{(912 - 875)\,\text{torr}}{(938 - 875)\,\text{torr}} \times 2°C \right] = 105.2°C = 105°C$$

Phase Diagrams

11.51   (a)   The *critical point* is the temperature and pressure beyond which the gas and liquid phases are indistinguishable.

    (b)   The gas/liquid line ends at the critical point because at conditions beyond the critical temperature and pressure, there is no distinction between gas and liquid. In experimental terms, a gas cannot be liquefied at temperatures higher than the critical temperature, regardless of pressure.

11.53   (a)   The water vapor would condense to form a solid at a pressure of around 4 torr. At higher pressure, perhaps 5 atm or so, the solid would melt to form liquid water. This occurs because the melting point of ice, which is 0°C at 1 atm, decreases with increasing pressure.

    (b)   In thinking about this exercise, keep in mind that the **total** pressure is being maintained at a constant 0.50 atm. That pressure is composed of water vapor pressure and some other pressure, which could come from an inert gas. At 100°C and 0.50 atm, water is in the vapor phase. As it cools, the water vapor will condense to the liquid at the temperature where the vapor pressure of liquid water is 0.50 atm. From Appendix B, we see that condensation occurs at approximately 82°C. Further cooling of the liquid water results in freezing to the solid at approximately 0°C. The freezing point of water increases with decreasing pressure, so at 0.50 atm, the freezing temperature is very slightly above 0°C.

11.55   (a)

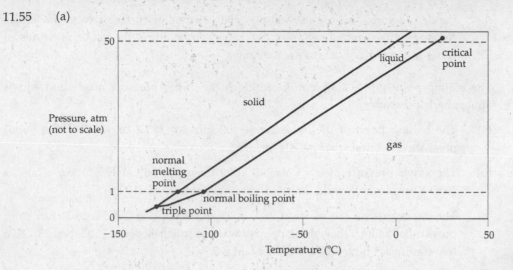

(b)   The solid-liquid line on the phase diagram is normal and the melting point of Xe(s) increases with increasing pressure. This means that Xe(s) is denser than Xe(l).

(c)   Cooling Xe(g) at 100 torr will cause deposition of the solid. A pressure of 100 torr is below the pressure of the triple point, so the gas will change directly to the solid upon cooling.

Structures of Solids

11.57   In a crystalline solid, the component particles (atoms, ions, or molecules) are arranged in an ordered repeating pattern. In an amorphous solid, there is no orderly structure. Quartz glass (Figure 11.30(b)) is an example of an amorphous solid. Paraffin wax is another example of an amorphous solid. Also, most plastics show no long range order and are amorphous overall, although they can show regions of order (see Section 12.2).

11.59   Ca: Ca atoms occupy the 8 corners of the cube.

                   8 corners × 1/8 sphere/corner = 1 Ca atom

       O: O atoms occupy the centers of the 6 faces of the cube.

                   6 faces × 1/2 atom/face = 3 O atoms

       Ti: There is 1 Ti atom at the body center of the cube.

       Formula: $CaTiO_3$

11.61   *Analyze.* Given the cubic unit cell edge length and arrangement of Ir atoms, calculate the atomic radius and the density of the metal. *Plan.* There is space between the atoms along the unit cell edge, but they touch along the face diagonal. Use the geometry of the right equilateral triangle to calculate the atomic radius. From the definition of density and paying attention to units, calculate the density of Ir(s).   *Solve.*

(a)   The length of the face diagonal of a face-centered cubic unit cell is four times the radius of the atom and $\sqrt{2}$ times the unit cell dimension or edge length, usually designated *a* for cubic unit cells.

134

$$4\,r = \sqrt{2}\;a; r = \sqrt{2}\,a/4 = \frac{\sqrt{2} \times 3.833\,\text{Å}}{4} = 1.3552 = 1.355\,\text{Å}$$

(b)   The density of iridium is the mass of the unit cell contents divided by the unit cell volume. There are 4 Ir atoms in a face-centered cubic unit cell.

$$\frac{4\,\text{Ir atoms}}{(3.833 \times 10^{-8}\,\text{cm})^3} \times \frac{192.22\,\text{g Ir}}{6.022 \times 10^{23}\,\text{Ir atoms}} = 22.67\,\text{g/cm}^3$$

Check. The units of density are correct. Note that Ir is quite dense.

11.63   *Analyze.* Given the atomic arrangement, length of the cubic unit cell edge and density of the solid, calculate the atomic weight of the element. *Plan.* If we calculate the mass of a single unit cell, and determine the number of atoms in one unit cell, we can calculate the mass of a single atom and of a mole of atoms.   *Solve.*

The volume of the unit cell is $(2.86 \times 10^{-8}\,\text{cm})^3$. The mass of the unit cell is:

$$\frac{7.92\,\text{g}}{\text{cm}^3} \times \frac{(2.86 \times 10^{-8})^3\,\text{cm}^3}{\text{unit cell}} = 1.853 \times 10^{-22}\,\text{g/unit cell}$$

There are two atoms of the element present in the body-centered cubic unit cell. Thus the atomic weight is:

$$\frac{1.853 \times 10^{-22}\,\text{g}}{\text{unit cell}} \times \frac{1\,\text{unit cell}}{2\,\text{atoms}} \times \frac{6.022 \times 10^{23}\,\text{atoms}}{1\,\text{mol}} = 55.8\,\text{g/mol}$$

Check. The result is a reasonable atomic weight and the units are correct. The element could be iron.

11.65   (a)   Each sphere is in contact with 12 nearest neighbors; its coordination number is thus 12.

(b)   Each sphere has a coordination number of six.

(c)   Each sphere has a coordination number of eight.

11.67   *Analyze.* Given the atomic arrangement and density of the solid, calculate the unit cell edge length. *Plan.* Calculate the mass of a single unit cell and then use density to find the volume of a single unit cell. The edge length is the cube-root of the volume of a cubic cell. *Solve.* There are four PbSe units in the unit cell. The unit cell edge is designated a.

$$8.27\,\text{g/cm}^3 = \frac{4\,\text{PbSe units}}{a^3} \times \frac{286.2\,\text{g}}{6.022 \times 10^{23}\,\text{PbSe units}} \times \left(\frac{1\,\text{Å}}{1 \times 10^{-8}\,\text{cm}}\right)^3$$

$a^3 = 229.87\,\text{Å}^3, a = 6.13\,\text{Å}$

11.69   (a)   The U ions in UO_2 are represented by the smaller spheres in Figure 11.42(c). The chemical formula requires twice as many O^{2-} ions as U^{4+} ions. There are eight complete large spheres and four total $(8 \times 1/8 + 6 \times 1/2)$ small spheres, so the small ones must represent U^{4+}. (It is probably true that O^{2-} has a physically larger radius than U^{4+}, but the elements' large separation on the periodic chart makes the relative radii difficult to estimate from trends.)

(b)　According to Figure 11.42(c), there are four UO_2 units in the "fluorite" unit cell.

$$\frac{4\,UO_2\ units}{(5.468\ \text{Å})^3} \times \frac{270.03\ g}{6.022 \times 10^{23}\ UO_2\ units} \times \left(\frac{1\ \text{Å}}{1 \times 10^{-8}\ cm}\right)^3 = 10.97\ g/cm^3$$

Bonding in Solids

11.71　(a)　hydrogen bonding, dipole-dipole forces, London dispersion forces

(b)　covalent chemical bonds (mainly)

(c)　ionic bonds (mainly)　　(d)　metallic bonds

11.73　In molecular solids, relatively weak intermolecular forces (hydrogen bonding, dipole-dipole, dispersion) bind the molecules in the lattice, so relatively little energy is required to disrupt these forces. In covalent-network solids, covalent bonds join atoms into an extended network. Melting or deforming a covalent-network solid means breaking these covalent bonds, which requires a large amount of energy.

11.75　Because of its relatively high melting point and properties as a conducting solution, the solid must be ionic.

11.77　(a)　Xe — greater atomic weight, stronger dispersion forces

(b)　SiO_2 — covalent-network lattice versus weak dispersion forces

(c)　KBr — strong ionic versus weak dispersion forces

(d)　C_6Cl_6 — both are influenced by dispersion forces, C_6Cl_6 has the higher molar mass.

Additional Exercises

11.79　(a)　decrease　　(b)　increase　　(c)　increase　　(d)　increase

(e)　increase　　(f)　increase　　(g)　increase

11.82　(a)　In dibromomethane, CH_2Br_2, the dispersion force contribution will be larger than for CH_2Cl_2, because bromine is more polarizable than the lighter element chlorine. At the same time, the dipole-dipole contribution for CH_2Cl_2 is greater than for CH_2Br_2 because CH_2Cl_2 has a larger dipole moment.

(b)　Just the opposite comparisons apply to CH_2F_2, which is less polarizable and has a higher dipole moment than CH_2Cl_2.

11.85　The two O–H groups in ethylene glycol are involved in many hydrogen bonding interactions, leading to its high boiling point and viscosity, relative to pentane, which experiences only dispersion forces.

11.88　(a)　Sweat, or salt water, on the surface of the body vaporizes to establish its typical vapor pressure at atmospheric pressure. Since the atmosphere is a totally open system, typical vapor pressure is never reached, and the sweat evaporates continuously. Evaporation is an endothermic process. The heat required to vaporized sweat is absorbed from your body, helping to keep it cool.

(b)   The vacuum pump reduces the pressure of the atmosphere (air + water vapor) above the water. Eventually, atmospheric pressure equals the vapor pressure of water and the water boils. Boiling is an endothermic process, and the temperature drops if the system is not able to absorb heat from the surroundings fast enough. As the temperature of the water decreases, the water freezes. (On a molecular level, the evaporation of water removes the molecules with the highest kinetic energies from the liquid. This decrease in average kinetic energy is what we experience as a temperature decrease.)

11.92   In a metallic solid such as gold, the atoms are held in their very orderly arrangement by metallic bonding, the result of valence electrons delocalized throughout the three-dimensional lattice. A large amount of kinetic energy is required to disrupt this delocalized bonding network and allow the atoms to translate relative to each, so the melting point of Au(s) is quite high. Xe atoms are held in a cubic close-packed arrangement by London-dispersion forces much weaker than metallic bonding. Very little kinetic energy is required for Xe atoms to overcome these forces and melt, so the melting point of Xe is quite low.

11.95   The most effective diffraction of light by a grating occurs when the wavelength of light and the separation of the slits in the grating are similar. When X-rays are diffracted by a crystal, layers of atoms serve as the "slits." The most effective diffraction occurs when the distances between layers of atoms are similar to the wavelength of the X-rays. Typical interlayer distances in crystals range from 2 Å to 20 Å. Visible light, 400–700 nm or 4,000 to 7,000 Å, is too long to be diffracted effectively by crystals. Molybdenum X-rays of 0.71 Å are on the same order of magnitude as interlayer distances in crystals and are diffracted.

Integrative Exercises

11.98   *Analyze.* Given: mass % of Al, Mg, O; density, unit cell edge length. Find: number of each type of atom. *Plan.* We are not given the type of cubic unit cell, primitive, body centered, face-centered. So we must calculate the number of formula units in the unit cell, using density, cell volume, and formula weight. Begin by determining the empirical formula and formula weight from mass % data.

Solve. Assume 100 g spinel.

$$37.9\,\text{g Al} \times \frac{1\,\text{mol Al}}{26.98\,\text{g Al}} = 1.405\,\text{mol Al}; 1.405/0.7036 \approx 2$$

$$17.1\,\text{g Mg} \times \frac{1\,\text{mol Mg}}{24.305\,\text{g Mg}} = 0.7036\,\text{mol Mg}; 0.7036/0.7036 = 1$$

$$45.0\,\text{g O} \times \frac{1\,\text{mol O}}{16.00\,\text{g Al}} = 2.813\,\text{mol O}; 2.813/0.7036 \approx 4$$

The empirical formula is Al_2MgO_4; formula weight = 142.3 g/mol

Calculate the number of formula units per unit cell.

809 pm = 809×10^{-12} m = 8.09×10^{-8} cm; V = $(8.09 \times 10^{-8})^3$ cm^3

$$\frac{3.57\,g}{cm^3} \times (8.09 \times 10^{-8})^3\,cm^3 \times \frac{1\,mol}{142.3\,g} \times \frac{6.022 \times 10^{23}\,units}{mol} = 7.999 = 8$$

There are 8 formula units per unit cell, for a total of 16 Al atoms, 8 Mg atoms, and 32 O atoms.

[The relationship between density (d), unit cell volume (V), number of formula units (Z), formula weight (FW), and Avogadro's number (N) is a useful one. It can be rearranged to calculate any single variable, knowing values for the others. For densities in g/cm^3 and unit cell volumes in cm^3 the relationship is $Z = (N \times d \times V)/FW$.]

11.101

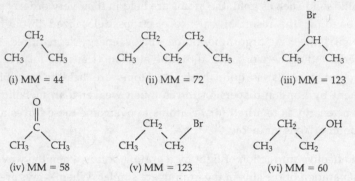

(i) MM = 44   (ii) MM = 72   (iii) MM = 123

(iv) MM = 58   (v) MM = 123   (vi) MM = 60

It is useful to draw the structural formulas because intermolecular forces are determined by the size and shape (structure) of molecules.

(a)   *Molar mass*: compounds (i) and (ii) have similar rod-like structures; (ii) has a longer rod. The longer chain leads to greater molar mass, stronger London-dispersion forces and higher heat of vaporization.

(b)   *Molecular shape*: compounds (iii) and (v) have the same chemical formula and molar mass but different molecular shapes (they are structural isomers). The more rod-like shape of (v) leads to more contact between molecules, stronger dispersion forces and higher heat of vaporization.

(c)   *Molecular polarity*: rod-like hydrocarbons (i) and (ii) are essentially nonpolar, owing to free rotation about C–C σ bonds, while (iv) is quite polar, owing to the C=O group. (iv) has a smaller molar mass than (ii) but a larger heat of vaporization, which must be due to the presence of dipole-dipole forces in (iv). [Note that (iii) and (iv), with similar shape and molecular polarity, have very similar heats of vaporization.]

(d)   *Hydrogen-bonding interactions*: molecules (v) and (vi) have similar structures, but (vi) has hydrogen bonding and (v) does not. Even though molar mass and thus dispersion forces are larger for (v), (vi) has the higher heat of vaporization. This must be due to hydrogen bonding interactions.

11.105 $P = \dfrac{nRT}{V} = \dfrac{g\,RT}{M\,V}$; $T = 273.15 + 26.0^{\circ}C = 299.15 = 299.2\,K$; $V = 5.00\,L$

$g\,C_6H_6(g) = 7.2146 - 5.1493 = 2.0653\,g\,C_6H_6(g)$

$$P(vapor) = \frac{2.0653\,g}{78.11\,g/mol} \times \frac{299.15\,K}{5.00\,L} \times \frac{0.08206\,L \bullet atm}{K \bullet mol} \times \frac{760\,torr}{1\,atm} = 98.660 = 98.7\,torr$$

138

12 Modern Materials

Visualizing Concepts

12.1   *Analyze/Plan.* Given energy diagrams of band structures for two materials, predict which material is a metal. Consider the conductivity of a metal and how it relates to band structure. Review the band structures in Figure 12.2.

Solve. The band structure of material A is that of a metal. Metals are excellent conductors of electricity and current; they have essentially no barrier to electron flow and a correspondingly small or zero energy band gap. This describes the band structure of material A.

12.4   *Analyze/Plan.* Given diagrams with rod-like (columnar) molecules in different orientations, decide which diagram represents molecules in a liquid crystalline phase. Recall the molecular orientations that define solids, liquid crystals and isotropic liquids.

Solve. Liquid crystals exhibit molecular order in at least one dimension. Solids display three-dimensional order, and liquids are randomly ordered with close molecular contacts. Diagram (a) has columnar molecules with their long directions oriented parallel to the vertical direction of the box, a one-dimensional order characteristic of a liquid crystalline phase. Diagram (b) has the close but randomly oriented molecules characteristic of an isotropic liquid.

Classes of Materials

12.7   *Analyze.* Given: formula of pure substance. Find: metal, semiconductor, insulator.

Plan. Use the periodic table and specific examples as a guideline. Metals are on the left of the periodic chart, the s-block, d-block, and p-block elements up to metalloids. Elemental semiconductors are Si, Ge, and C(graphite). Compound semiconductors are often combinations of Group 3A and 5A elements. Other nonmetals and most ionic and molecular compounds are insulators.   *Solve.*

(a)   GaN — semiconductor; Gp 3A + Gp 5A

(b)   B — insulator; nonmetal, not an elemental semiconductor

(c)   ZnO — semiconductor, like CdS, average 4 e$^-$ per atom

(d)   Pb — metal; p-block metal left of metalloid line

12.9   *Analyze.* Given: GaAs. Find: dopant to make n-type semiconductor.

Plan. An n-type semiconductor has extra negative charges. If the dopant replaces a few Ga atoms, it should have more valence electrons than Ga, Group 3A.

Solve. The obvious choice is a Group 4A element, either Ge or Si. Ge would be closer to Ga in bonding atomic radius (Figure 7.5).

12.11   (a)   False. Semiconductors conduct some electricity, while insulators do not. Semiconductors must have a smaller barrier to electron mobility, a smaller band gap.

(b)   True. Dopants create either more electrons or more "holes," both of which increase the conductivity of the semiconductor.

(c)   True. Metals conduct electricity (and heat) because delocalized electrons in the lattice provide a mechanism for charge mobility.

(d)   True. Metal oxides are ionic substances with essentially localized electrons. There is a large barrier to charge mobility, a band gap so large that metal oxides are insulators.

2.13   A superconducting material offers no resistance to the flow of electrical current; *superconductivity* is the frictionless flow of electrons. Superconductive materials could transmit electricity with no heat loss and therefore much greater efficiency than current carriers. Because of the Meisner effect, they are also potential materials for magnetically levitated trains.

12.15   Below 39 K, MgB_2 conducts electricity with zero resistivity, the definition of a superconductor. Above 39 K, the material is not superconducting. The sharp drop in resistivity of MgB_2 near 39 K is the superconducting transition temperature, T_c.

12.17   Superconductivity acts on electrons that are conduction electrons. Thus you need to have electrons available to be conductors. Insulators have no electrons available for conduction.

Materials for Structure

12.19   *n*-decane does not have a sufficiently high chain length or molecular mass to be considered a polymer.

12.21   *Analyze.* Given two types of reactant molecules, we are asked to write a condensation reaction with an ester product. *Plan.* A condensation reaction occurs when two smaller molecules combine to form a larger molecule and a small molecule, often water. Consider the structures of the two reactants and how they could combine to join the larger fragments and split water.   *Solve.*

A carboxylic acid contains the $-\overset{O}{\underset{}{C}}-OH$ functional group; an alcohol contains the

–OH functional group. These can be arranged to form the $-\overset{O}{\underset{}{C}}-O-C$ ester functional group and H_2O. Condensation reaction to form an ester:

$CH_3-\overset{O}{\underset{}{C}}-[O-H + H]-O-CH_2-CH_3 \longrightarrow CH_3-\overset{O}{\underset{}{C}}-O-CH_2CH_3 + H_2O$
acetic acid   ethanol   ethyl acetate

If a dicarboxylic acid (two –COOH groups, usually at opposite ends of the molecule) and a dialcohol (two –OH groups, usually at opposite ends of the molecule) are combined, there is the potential for propagation of the polymer chain at both ends of both monomers. Polyethylene terephthalate (Table 12.4) is an example of a polyester formed from the monomers ethylene glycol and terephthalic acid.

12.23 *Analyze/Plan.* Decide whether the given polymer is an addition or condensation polymer. Select the smallest repeat unit and deconstruct it into the monomer(s) with the specific functional group(s) that would form the stated polymer.   *Solve.*

(a)

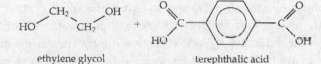

vinyl chloride (chloroethylene or chloroethene)

(b)

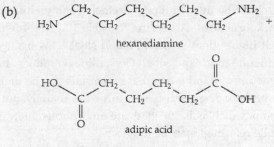

hexanediamine

adipic acid

(Formulas given in Equation [12.3].)

(c)

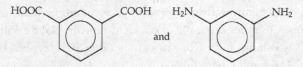

ethylene glycol   terephthalic acid

12.25 *Plan/Solve.* When nylon polymers are made, H_2O is produced as the C–N bonds are formed. Reversing this process (adding H_2O across the C–N bond), we see that the monomers used to produce Nomex™ are:

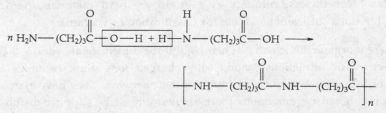

and

12.27 *Analyze/Plan.* Given the formula of a monomer, write the equation for condensation polymerization. The monomers are aligned so that the caroboxyl end of one monomer joins the amine end of another molecule.   *Solve.*

$$n\ H_2N-(CH_2)_3C\overset{O}{\underset{}{\|}}\boxed{-O-H + H-}\overset{H}{\underset{}{N}}-(CH_2)_3C\overset{O}{\underset{}{\|}}-OH \longrightarrow$$

$$\left[-NH-(CH_2)_3C\overset{O}{\underset{}{\|}}-NH-(CH_2)_3C\overset{O}{\underset{}{\|}}-\right]_n$$

12.29 Most of a polymer backbone is composed of σ bonds. The geometry around individual atoms is tetrahedral with bond angles of 109°, so the polymer is not flat, and there is relatively free rotation around the σ bonds. The flexibility of the molecular chains causes flexibility of the bulk material. Flexibility is enhanced by molecular features that inhibit order, such as branching, and diminished by features that encourage order, such as cross-linking or delocalized π electron density.

Cross-linking is the formation of chemical bonds between polymer chains. It reduces flexibility of the molecular chains and increases the hardness of the material. Cross-linked polymers are less chemically reactive because of the links.

12.31   The function of the material (polymer) determines whether high molecular mass and high degree of crystallinity are desirable properties. If the material will be formed into containers or pipes, the rigidity and structural strength associated with high molecular mass are required. If the polymer will be used as a flexible wrapping or as a garment material, high molecular mass and rigidity are undesirable properties.

12.33   Ceramics are not readily recyclable because of their extremely high melting points and rigid ionic or covalent-network structures. According to Table 12.2, the melting points of ceramic materials are much higher than those of Al and steel. This makes recycling ceramics technologically difficult and expensive. Crystalline ceramics have rigid, precise three-dimensional structures. If and when these materials can be melted, either covalent or ionic bonds are broken. The precise structures are usually not reformed upon cooling. Recyclable ceramics such as bottle glass are amorphous; there is no exact repeating structure that must be reformed after melting.

12.35   Very small, uniformly sized and shaped particles are required for the production of a strong ceramic object by sintering. During sintering, the small ceramic particles are heated to a high temperature below the melting point of the solid. This high temperature initiates condensation reactions between molecules at the surfaces of the spheres; the spheres are then connected by chemical bonds between atoms in different spheres. The more uniform the particle size and the greater the total surface area of the solid, the more chemical bonds are formed, and the stronger the ceramic object.

12.37   The ceramics are: MgO, soda lime glass, ZrB_2, Al_2O_3, and TaC. The criteria are a combination of chemical formula (with corresponding bonding characteristics) and Knoop values. Ceramics are ionic or covalent-network solids with fairly large hardness values. $CaCO_3$ is ionic, but carbonates are not one of the typical types of ceramics listed in Section 12.1. Ag and Cr are metals.

Hardness alone is not a sufficient criteria for ceramics. The range of hardness values for the ceramics in this group is large; Cr, a metal, lies in the middle of the range, as could other metals. Nonetheless, ceramics as a group are hard materials; hardness is a necessary, but not a sufficient condition for classification as a ceramic.

12.39   Ceramics are inorganic ionic solids. Their rigid three-dimensional order is the result of strong electrostatic attractions among fully charged ions. Bulk ceramics are hard, because it is difficult to budge an ion from this ionic network. They have extremely high melting points because tremendous thermal energy must be supplied before the ions have sufficient kinetic energy to break away from the network and move relative to each other.

Materials for Medicine

12.41   Is the neoprene biocompatible: is the surface smooth enough and is the chemical composition appropriate so that there are no inflammatory reactions in the body? Does neoprene meet the physical requirements of a flexible lead: will it remain resistant to degradation by body fluids over a long time period; will it maintain elasticity over the same time period? Can neoprene be prepared in sufficiently pure form (free of trace amounts of monomer, catalyst, etc.) so that it can be classified as medical grade?

12.43   Current vascular-graft materials cannot be lined with cells similar to those in the native artery. The body detects that the graft is "foreign" and platelets attach to the inside surfaces, causing blood clots. The inside surfaces of the future vascular implants need to accommodate a lining of cells that do not attract or attach to platelets.

12.45   In order for skin cells in a culture medium to develop into synthetic skin, a mechanical matrix must be present that holds the cells in contact with one another and allows them to differentiate. The matrix must be mechanically strong, biocompatible, and biodegradable. It probably has polar functional groups that are capable of hydrogen bonding with biomolecules in the tissue cells.

Materials for Electronics

12.47   Silicon is the semiconductor material of choice for integrated circuits because it is abundant, cheap, nontoxic, can be highly purified, and grows nearly perfect enormous crystals. That is, it has appropriate chemical and physical properties and is cost effective.

12.49   The bonding atomic radius of a Si atom is 1.11 Å (Figure 7.5), so the diameter is 2.22 Å.

$$60 \text{ nm} \times \frac{1 \times 10^{-9} \text{ m}}{1 \text{ nm}} \times \frac{1 \text{ Å}}{1 \times 10^{-10} \text{ m}} \times \frac{1 \text{ Si atom}}{2.22 \text{ Å}} = 2.7 \times 10^{2} \text{ Si atoms.}$$

12.51   *Analyze.* Given: 1.1 eV. Find: wavelength in meters that corresponds to the energy 1.1 eV. *Plan.* Use dimensional analysis to find wavelength.

Solve. 1 eV = 1.602×10^{-19} J (inside-back cover of text); $\lambda = hc/E$

$$\lambda = 6.626 \times 10^{-34} \text{ J} \cdot \text{s} \times \frac{3.00 \times 10^{8} \text{ m}}{\text{s}} \times \frac{1}{1.1 \text{ eV}} \times \frac{1 \text{ eV}}{1.602 \times 10^{-19} \text{ J}} = 1.128 \times 10^{-6}$$

$$= 1.1 \times 10^{-6} \text{ m}$$

Si can absorb energies > 1.1 eV, or wavelengths ≤ 1.1×10^{-6} m. The range of wavelengths in the solar spectrum at sea level is 3×10^{-6} to 2×10^{-7} m, or 30×10^{-7} to 2×10^{-7} m, a span of 28×10^{-7} m. Si can absorb 11×10^{-7} to 2×10^{-7} m, a span of 9×10^{-7} m.

This represents $\dfrac{9 \times 10^{-7}}{28 \times 10^{-7}} \times 100 = 32\%$ of the wavelengths in the solar spectrum.

According to the diagram, these wavelengths represent much more than 32% of the total flux.

Materials for Optics

12.53   Both an ordinary liquid and a nematic liquid crystal phase are fluids; they are converted directly to the solid phase upon cooling. The nematic phase is cloudy and more viscous than an ordinary liquid. Upon heating, the nematic phase is converted to an ordinary liquid.

12.55   In the solid state, there is three-dimensional order; the relative orientation of the molecules is fixed and repeating in all three dimensions. Essentially no translational or rotational motion is allowed. When a substance changes to the nematic liquid-crystalline phase, the molecules remain aligned in one dimension (the long dimension of the molecule). Translational motion is allowed, but rotational motion is restricted. Transformation to the isotropic-liquid phase destroys the one-dimensional order. Free translational and rotational motion result in random molecular orientations that change continuously.

12.57   The presence of polar groups or nonbonded electron pairs leads to relatively strong dipole-dipole interactions between molecules. These are a significant part of the orienting forces necessary for liquid crystal formation.

12.59   In the nematic phase, molecules are aligned in one dimension, the long dimension of the molecule. In a smectic phase (A or C), molecules are aligned in two dimensions. Not only are the long directions of the molecules aligned, but the ends are also aligned. The molecules are organized into layers; the height of the layer is related to the length of the molecule.

12.61   A nematic phase is composed of sheets of molecules aligned along their lengths, but with no additional order within the sheet or between sheets. A cholesteric phase also contains this kind of sheet, but with some ordering between sheets. In a cholesteric phase, there is a characteristic angle between molecules in one sheet and those in an adjacent sheet. That is, one sheet of molecules is twisted at some characteristic angle relative to the next, producing a "screw" axis perpendicular to the sheets.

12.63   *Plan/Solve.* Follow the logic in Solution 12.51.

$$\lambda = hc/E = 6.626 \times 10^{-34}\ \text{J} \bullet \text{s} \times \frac{3.00 \times 10^8\ \text{m}}{\text{s}} \times \frac{1}{2.2\ \text{eV}} \times \frac{1\ \text{eV}}{1.602 \times 10^{-19}\ \text{J}} = 5.640 \times 10^{-7}$$

$$= 5.6 \times 10^{-7}\ \text{m} = 560\ \text{nm}$$

Materials for Nanotechnology

12.65   Continuous energy bands of molecular orbitals require a large number of atoms contributing a large number of atomic orbitals to the molecular orbital scheme. If a solid has dimensions 1–10 nm, nanoscale dimensions, there may not be enough contributing atomic orbitals to produce continuous energy bands of molecular orbitals.

12.67   (a)   False. As particle size decreases, the band gap increases. The smaller the particle, the fewer AOs that contribute to the MO scheme, the more localized the bonding and the larger the band gap.

(b)   False. The wavelength of emitted light corresponds to the energy of the band gap. As particle size decreases, band gap increases and wavelength decreases ($E = hc/\lambda$).

12.69   *Analyze.* Given: Au, 4 atoms per unit cell, 4.08 Å cell edge, volume of sphere = $4/3\,\pi\,r^3$. Find: Au atoms in 20 nm diameter sphere.

Plan. Relate the number of Au atoms in the volume of 1 cubic unit cell to the number of Au atoms in a 20 nm diameter sphere. Change units to Å (you could just as well have chosen nm as the common unit), calculate the volumes of the unit cell and sphere, and use a ratio to calculate atoms in the sphere.

Solve. vol of unit cell = $(4.08\ \text{Å})^3 = 67.9173 = 67.9\ \text{Å}^3$

$$20\ \text{nm diameter} = 10\ \text{nm radius}; 10\ \text{nm} \times \frac{1 \times 10^{-9}\ \text{m}}{1\,\text{nm}} \times \frac{1\,\text{Å}}{1 \times 10^{-10}\ \text{m}} = 100\ \text{Å radius}$$

(Note that 1 nm = 10 Å.)

vol. of sphere = $4/3 \times 3.14159 \times (100\ \text{Å})^3 = 4.18879 \times 10^8 = 4.19 \times 10^6\ \text{Å}^3$

$$\frac{4\ \text{Au atoms}}{67.9173\ \text{Å}^3} = \frac{x\ \text{Au atoms}}{4.18879 \times 10^6\ \text{Å}}; x = 2.46699 \times 10^6 = 2.47 \times 10^5\ \text{Au atoms}$$

Additional Exercises

12.72   A dipole moment (permanent, partial charge separation) roughly parallel to the long dimension of the molecule would cause the molecules to reorient when an electric field is applied perpendicular to the usual direction of molecular orientation.

12.77   At the temperature where a substance changes from the solid to the liquid-crystalline phase, kinetic energy sufficient to overcome most of the long range order in the solid has been supplied. A few van der Waals forces have sufficient attractive energy to impose the one-dimensional order characteristic of the liquid-crystalline state. Very little additional kinetic energy (and thus a relatively small increase in temperature) is required to overcome these aligning forces and produce an isotropic liquid.

12.81 $TiCl_4(g) + 2SiH_4(g) \rightarrow TiSi_2(s) + 4HCl(g) + 2H_2(g)$

As a ceramic, TiSi2 will have a three dimensional network structure similar to that of Si. [Si(s) has a diamond-like covalent-network structure, Figure 11.41.] At the surface of the thin film there will be Ti atoms and Si atoms with incomplete valences that can and will chemically bond with Si atoms on the surface of the substrate. This kind of bonding would not be possible with a Cu thin film. Strong adherence to the surface is an essential component of thin film performance.

Integrative Exercises

12.83   (a)

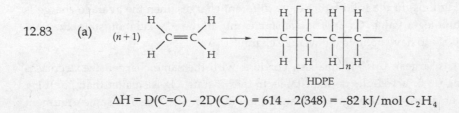

$$\Delta H = D(C{=}C) - 2D(C{-}C) = 614 - 2(348) = -82\ \text{kJ/mol}\ C_2H_4$$

(b)

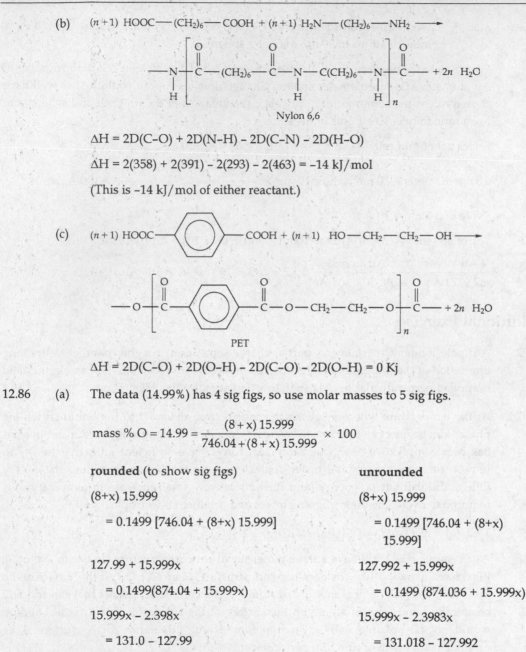

Nylon 6,6

$\Delta H = 2D(C-O) + 2D(N-H) - 2D(C-N) - 2D(H-O)$

$\Delta H = 2(358) + 2(391) - 2(293) - 2(463) = -14$ kJ/mol

(This is –14 kJ/mol of either reactant.)

(c) $(n+1)$ HOOC——⬡——COOH + $(n+1)$ HO—CH_2—CH_2—OH ⟶

PET

$\Delta H = 2D(C-O) + 2D(O-H) - 2D(C-O) - 2D(O-H) = 0$ Kj

12.86   (a)   The data (14.99%) has 4 sig figs, so use molar masses to 5 sig figs.

$$\text{mass \% O} = 14.99 = \frac{(8+x)\,15.999}{746.04 + (8+x)\,15.999} \times 100$$

rounded (to show sig figs)   **unrounded**

$(8+x)\,15.999$ $(8+x)\,15.999$

    $= 0.1499\,[746.04 + (8+x)\,15.999]$ $= 0.1499\,[746.04 + (8+x)$ 15.999]

$127.99 + 15.999x$ $127.992 + 15.999x$

    $= 0.1499(874.04 + 15.999x)$ $= 0.1499\,(874.036 + 15.999x)$

$15.999x - 2.398x$ $15.999x - 2.3983x$

    $= 131.0 - 127.99$ $= 131.018 - 127.992$

$13.601x = 3.0;\ x = 0.22$ $13.6007x = 3.026;\ x = 0.2225$

(b)   Hg and Cu both have more than one stable oxidation state. If different Cu ions (or Hg ions) in the solid lattice have different charges, then the average charge is a noninteger value. Ca and Ba are stable only in the +2 oxidation state; they are unlikely to have noninteger average charge.

(c) Ba^{2+} is largest; Cu^{2+} is smallest. For ions with the same charge, size decreases going up or across the periodic table. In the +2 state, Hg is smaller than Ba. If Hg has an average charge greater than 2+, it will be smaller yet. The same argument is true for Cu and Ca.

13 Properties of Solutions

Visualizing Concepts

13.1 The energy of the ion-solvent interaction is greater for Li^+ than Na^+. The smaller size of the Li^+ ion means that ion-dipole interactions with polar water molecules are stronger.

13.6 Vitamin B_6 is likely to be largely water soluble. The three –OH groups and the $—\ddot{N}—$ can enter into many hydrogen bonding interactions with water. The relatively small molecular size indicates that dispersion forces will not play a large role in intermolecular interactions and the hydrogen bonding will dominate. Vitamin E is likely to be largely fat soluble. The long, rod-like hydrocarbon chain will lead to strong dispersion forces among vitamin E and mostly nonpolar fats. Although vitamin E has one –OH and one $—O—$ group, the long hydrocarbon chain prevents water from surrounding and separating the vitamin E molecules, reducing its water-solubility.

13.9 A detergent for solubilizing large hydrophobic proteins (or any other large nonpolar solute, such as greasy dirt) needs a hydrophobic part to interact with the solute, and a hydrophilic part to interact with water. In n-octyl glycoside, the eight-carbon n-octyl chain has strong dispersion interactions with the hydrophobic (nonpolar) protein. The –OH groups on the glycoside (sugar) ring form strong hydrogen bonds with water. This causes the glycoside to dissolve, dragging the hydrophobic protein along with it.

The Solution Process

13.11 If the enthalpy released due to solute-solvent attractive forces (ΔH_3) is at least as large as the enthalpy required to separate the solute particles (ΔH_1), the overall enthalpy of solution (ΔH_{soln}) will be either slightly endothermic (owing to $+\Delta H_2$) or exothermic. Even if ΔH_{soln} is slightly endothermic, the increase in disorder due to mixing will cause a significant amount of solute to dissolve. If the magnitude of ΔH_3 is small relative to the magnitude of ΔH_1, ΔH_{soln} will be large and endothermic (energetically unfavorable) and not much solute will dissolve.

13.13 *Analyze/Plan.* Decide whether the solute and solvent in question are ionic, polar covalent, or nonpolar covalent. Draw Lewis structures as needed. Then state the appropriate type of solute-solvent interaction. *Solve.*

(a) CCl_4, nonpolar; benzene, nonpolar; dispersion forces

(b) methanol, polar with hydrogen bonding; water, polar with hydrogen bonding; hydrogen bonding

(c)   KBr, ionic; water, polar; ion-dipole forces

(d)   HCl, polar; CH_3CN, polar; dipole-dipole forces

13.15 (a)   Lattice energy is the amount of energy required to completely separate a mole of solid ionic compound into its gaseous ions (Section 8.2). For ionic solutes, this corresponds to ΔH_1 (solute-solute interactions) in Equation [13.1].

(b)   In Equation [13.1], ΔH_3 is always exothermic. Formation of attractive interactions, no matter how weak, always lowers the energy of the system, relative to the energy of the isolated particles.

13.17 (a) ΔH_{soln} is determined by the relative magnitudes of the "old" solute-solute (ΔH_1) and solvent-solvent (ΔH_2) interactions and the new solute-solvent interactions (ΔH_3); $\Delta H_{soln} = \Delta H_1 + \Delta H_2 + \Delta H_3$. Since the solute and solvent in this case experience very similar London dispersion forces, the energy required to separate them individually and the energy released when they are mixed are approximately equal.

$\Delta H_1 + \Delta H_2 \approx -\Delta H_3$. Thus, ΔH_{soln} is nearly zero.

(b)   Mixing hexane and heptane produces a homogeneous solution from two pure substances, and the randomness of the system increases. Since no strong intermolecular forces prevent the molecules from mixing, they do so spontaneously due to the increase in disorder.

Saturated Solutions; Factors Affecting Solubility

13.19 (a)   Supersaturated

(b)   Add a seed crystal. Supersaturated solutions exist because not enough solute molecules are properly aligned for crystallization to occur. A seed crystal provides a nucleus of already aligned molecules, so that ordering of the dissolved particles is more facile.

13.21 *Analyze/Plan.* On Figure 13.17, find the solubility curve for the appropriate solute. Find the intersection of 40°C and 40 g solute on the graph. If this point is below the solubility curve, more solute can dissolve and the solution is unsaturated. If the intersection is on or above the curve, the solution is saturated.   *Solve.*

(a)   unsaturated   (b)   saturated   (c)   saturated   (d)   unsaturated

13.23 The liquids water and glycerol form homogenous mixtures (solutions), regardless of the relative amounts of the two components. Glycerol has an –OH group on each C atom in the molecule. This structure facilitates strong hydrogen bonding similar to that in water. Like dissolves like and the two liquids are miscible in all proportions.

13.25 (a)   Dispersion interactions among nonpolar $CH_3(CH_2)_{16}$ –chains dominate the properties of stearic acid. It is more soluble in nonpolar CCl_4 than polar (hydrogen bonding) water, despite the presence of the –COOH group.

(b)

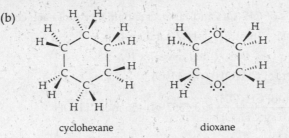

cyclohexane   dioxane

Dioxane can act as a hydrogen bond acceptor, so it will be more soluble than cyclohexane in water.

13.27   *Analyze/Plan.* Hexane is a nonpolar hydrocarbon that experiences dispersion forces with other C_6H_{14} molecules. Solutes that primarily experience dispersion forces will be more soluble in hexane.   *Solve.*

     (a) CCl_4 is more soluble because dispersion forces among nonpolar CCl_4 molecules are similar to dispersion forces in hexane. Ionic bonds in $CaCl_2$ are unlikely to be broken by weak solute-solvent interactions. For $CaCl_2$, ΔH_1 is too large, relative to ΔH_3.

     (b)   Benzene, C_6H_6, is also a nonpolar hydrocarbon and will be more soluble in hexane. Glycerol experiences hydrogen bonding with itself; these solute-solute interactions are less likely to be overcome by weak solute-solvent interactions.

     (c)   Octanoic acid, $CH_3(CH_2)_6COOH$, will be more soluble than acetic acid CH_3COOH. Both solutes experience hydrogen bonding by –COOH groups, but octanoic acid has a long, rod-like hydrocarbon chain with dispersion forces similar to those in hexane, facilitating solubility in hexane.

13.29   (a)   Carbonated beverages are stored with a partial pressure of $CO_2(g)$ greater than 1 atm above the liquid. A sealed container is required to maintain this CO_2 pressure.

     (b)   Since the solubility of gases increases with decreasing temperature, some $CO_2(g)$ will remain dissolved in the beverage if it is kept cool.

13.31   *Analyze/Plan.* Follow the logic in Sample Exercise 13.3.   *Solve.*

$S_{He} = 3.7 \times 10^{-4} \, M/atm \times 1.5 \, atm = 5.6 \times 10^{-4} \, M$

$S_{N_2} = 6.0 \times 10^{-4} \, M/atm \times 1.5 \, atm = 9.0 \times 10^{-4} \, M$

Concentrations of Solutions

13.33   *Analyze/Plan.* Follow the logic in Sample Exercise 13.4.   *Solve.*

     (a) $\text{mass \%} = \dfrac{\text{mass solute}}{\text{total mass solution}} \times 100 = \dfrac{10.6 \, g \, Na_2SO_4}{10.6 \, g \, Na_2SO_4 + 483 \, g \, H_2O} \times 100 = 2.15\%$

     (b) $\text{ppm} = \dfrac{\text{mass solute}}{\text{total mass solution}} \times 10^6; \dfrac{2.86 \, g \, Ag}{1 \, ton \, ore} \times \dfrac{1 \, ton}{2000 \, lb} \times \dfrac{1 \, lb}{453.6 \, g} \times 10^6 = 3.15 \, ppm$

13.35   *Analyze/Plan.* Given masses of CH_3OH and H_2O, calculate moles of each component.

(a)   Mole fraction CH_3OH = (mol CH_3OH)/(total mol)

(b)   mass % CH_3OH = [(g CH_3OH)/(total mass)] × 100

(c)   molality CH_3OH = (mol CH_3OH)/(kg H_2O).   *Solve.*

(a) $14.6 \text{ g } CH_3OH \times \dfrac{1 \text{ mol } CH_3OH}{32.04 \text{ g } CH_3OH} = 0.4557 = 0.456 \text{ mol } CH_3OH$

$184 \text{ g } H_2O \times \dfrac{1 \text{ mol } H_2O}{18.02 \text{ g } H_2O} = 10.211 = 10.2 \text{ mol } H_2O$

$\chi_{CH_3OH} = \dfrac{0.4557}{0.4557 + 10.211} = 0.04272 = 0.0427$

(b) $\text{mass \% } CH_3OH = \dfrac{14.6 \text{ g } CH_3OH}{14.6 \text{ g } CH_3OH + 184 \text{ g } H_2O} \times 100 = 7.35\% \ CH_3OH$

(c) $m = \dfrac{0.4557 \text{ mol } CH_3OH}{0.184 \text{ kg } H_2O} = 2.477 = 2.48 \text{ m } CH_3OH$

13.37   *Analyze/Plan.* Given mass solute and volume solution, calculate mol solute, then molarity = mol solute/L solution. Or, for dilution, $M_c \times L_c = M_d \times L_d$ *Solve.*

(a) $M = \dfrac{\text{mol solute}}{\text{L soln}}; \dfrac{0.540 \text{ g } Mg(NO_3)_2}{0.2500 \text{ L soln}} \times \dfrac{1 \text{ mol } Mg(NO_3)_2}{148.3 \text{ g } Mg(NO_3)_2} = 1.46 \times 10^{-2} \text{ M } Mg(NO_3)_2$

(b) $\dfrac{22.4 \text{ g } LiClO_4 \bullet 3H_2O}{0.125 \text{ L soln}} \times \dfrac{1 \text{ mol } LiClO_4 \bullet 3H_2O}{160.4 \text{ g } LiClO_4 \bullet 3H_2O} = 1.12 \text{ } M \text{ } LiClO_4 \bullet 3H_2O$

(c) $M_c \times L_c = M_d \times L_d; 3.50 \text{ } M \text{ } HNO_3 \times 0.0250 \text{ L} = ?M \text{ } HNO_3 \times 0.250 \text{ L}$
        250 mL of 0.350 M HNO_3

13.39   *Analyze/Plan.* Follow the logic in Sample Exercise 13.5.   *Solve.*

(a) $m = \dfrac{\text{mol solute}}{\text{kg solvent}}; \dfrac{8.66 \text{ g } C_6H_6}{23.6 \text{ g } CCl_4} \times \dfrac{1 \text{ mol } C_6H_6}{78.11 \text{ g } C_6H_6} \times \dfrac{1000 \text{ g } CCl_4}{1 \text{ kg } CCl_4} = 4.70 \text{ } m \text{ } C_6H_6$

(b)   The density of H_2O = 0.997 g/mL = 0.997 kg/L.

$\dfrac{4.80 \text{ g } NaCl}{0.350 \text{ L } H_2O} \times \dfrac{1 \text{ mol } NaCl}{58.44 \text{ g } NaCl} \times \dfrac{1 \text{ L } H_2O}{0.997 \text{ kg } H_2O} = 0.235 \text{ } m \text{ } NaCl$

13.41   *Analyze/Plan.* Assume 1 L of solution. Density gives the total mass of 1 L of solution. The g H_2SO_4/L are also given in the problem. Mass % = (mass solute/total mass solution) × 100. Calculate mass solvent from mass solution and mass solute. Calculate moles solute and solvent and use the appropriate definitions to calculate mole fraction, molality, and molarity.   *Solve.*

(a) $\dfrac{571.6 \text{ g } H_2SO_4}{1 \text{ L soln}} \times \dfrac{1 \text{ L soln}}{1329 \text{ g soln}} = 0.430098 \text{ g } H_2SO_4/\text{g soln}$

mass percent is thus 0.4301 × 100 = 43.01% H_2SO_4

(b)   In a liter of solution there are 1329 – 571.6 = 757.4 = 757 g H_2O.

$$\frac{571.6 \text{ g } H_2SO_4}{98.09 \text{ g/mol}} = 5.827 \text{ mol } H_2SO_4 \; ; \frac{757.4 \text{ g } H_2O}{18.02 \text{ g/mol}} = 42.03 = 42.0 \text{ mol } H_2O$$

$$\chi_{H_2SO_4} = \frac{5.827}{42.03 + 5.827} = 0.122$$

(The result has 3 sig figs because 42.0 mol H_2O limits the denominator to 3 sig figs.)

(c) $\text{molality} = \dfrac{5.827 \text{ mol } H_2SO_4}{0.7574 \text{ kg } H_2O} = 7.693 = 7.69 \; m \; H_2SO_4$

(d) $\text{molarity} = \dfrac{5.827 \text{ mol } H_2SO_4}{1 \text{ L soln}} = 5.827 \; M \; H_2SO_4$

13.43   *Analyze/Plan.* Given: 98.7 mL of $CH_3CN(l)$, 0.786 g/mL; 22.5 mL CH_3OH, 0.791 g/mL. Use the density and volume of each component to calculate mass and then moles of each component. Use the definitions to calculate mole fraction, molality, and molarity. *Solve.*

(a) $\text{mol } CH_3CN = \dfrac{0.786 \text{ g}}{1 \text{ mL}} \times 98.7 \text{ mL} \times \dfrac{1 \text{ mol } CH_3CN}{41.05 \text{ g } CH_3CN} = 1.8898 = 1.89 \text{ mol}$

$\text{mol } CH_3OH = \dfrac{0.791 \text{ g}}{1 \text{ mL}} \times 22.5 \text{ mL} \times \dfrac{1 \text{ mol } CH_3OH}{32.04 \text{ g } CH_3OH} = 0.5555 = 0.556 \text{ mol}$

$\chi_{CH_3OH} = \dfrac{0.5555 \text{ mol } CH_3OH}{1.8898 \text{ mol } CH_3CN + 0.5555 \text{ mol } CH_3OH} = 0.227$

(b)   Assuming CH_3OH is the solute and CH_3CN is the solvent,

$98.7 \text{ mL } CH_3CN \times \dfrac{0.786 \text{ g}}{1 \text{ mL}} \times \dfrac{1 \text{ kg}}{1000 \text{ g}} = 0.07758 = 0.0776 \text{ kg } CH_3CN$

$m_{CH_3OH} = \dfrac{0.5555 \text{ mol } CH_3OH}{0.07758 \text{ kg } CH_3CN} = 7.1602 = 7.16 \; m \; CH_3OH$

(c)   The total volume of the solution is 121.2 mL, assuming volumes are additive.

$M = \dfrac{0.5555 \text{ mol } CH_3OH}{0.1212 \text{ L solution}} = 4.58 \; M \; CH_3OH$

13.45   *Analyze/Plan.* Given concentration and volume of solution use definitions of the appropriate concentration units to calculate amount of solute; change amount to moles if needed.   *Solve.*

(a) $\text{mol} = M \times L; \dfrac{0.250 \text{ mol } SrBr_2}{1 \text{ L soln}} \times 0.600 \text{ L} = 0.150 \text{ mol } SrBr_2$

(b)   Assume that for dilute aqueous solutions, the mass of the solvent is the mass of solution. Use proportions to get mol KCl.

$\dfrac{0.180 \text{ mol KCl}}{1 \text{ kg } H_2O} = \dfrac{x \text{ mol KCl}}{0.0864 \text{ kg } H_2O}; x = 1.56 \times 10^{-2} \text{ mol KCl}$

(c)   Use proportions to get mass of glucose, then change to mol glucose.

$$\frac{6.45 \text{ g } C_6H_{12}O_6}{100 \text{ g soln}} = \frac{x \text{ g } C_6H_{12}O_6}{124.0 \text{ g soln}}; x = 8.00 \text{ g } C_6H_{12}O_6$$

$$8.00 \text{ g } C_6H_{12}O_6 \times \frac{1 \text{ mol } C_6H_{12}O_6}{180.2 \text{ g } C_6H_{12}O_6} = 4.44 \times 10^{-2} \text{ mol } C_6H_{12}O_6$$

13.47   *Analyze/Plan.* When preparing solution, we must know amount of solute and solvent. Use the appropriate concentration definition to calculate amount of solute. If this amount is in moles, use molar mass to get grams; use mass in grams directly. Amount of solvent can be expressed as total volume or mass of solution. Combine mass solute and solvent to produce the required amount (mass or volume) of solution.   *Solve.*

(a) $\text{mol} = M \times L;$ $\dfrac{1.50 \times 10^{-2} \text{ mol KBr}}{1 \text{ L soln}} \times 0.75 \text{ L} \times \dfrac{119.0 \text{ g KBr}}{1 \text{ mol KBr}} = 1.3 \text{ g KBr}$

Weigh out 1.5 g KBr, dissolve in water, dilute with stirring to 0.75 L (750 mL).

(b)   Mass of solution is required, but density is not specified. Use molality to calculate mass fraction, and then the masses of solute and solvent needed for 125 g of solution.

$$\frac{0.180 \text{ mol KBr}}{1000 \text{ g } H_2O} \times \frac{119.0 \text{ g KBr}}{1 \text{ mol KBr}} = 21.42 = 21.4 \text{ g KBr/kg } H_2O$$

Thus, mass fraction $= \dfrac{21.42 \text{ g KBr}}{1000 + 21.42} = 0.02097 = 0.0210$

In 125 g of the 0.180 *m* solution, there are

$$(125 \text{ g soln}) \times \frac{0.02097 \text{ g KBr}}{1 \text{ g soln}} = 2.621 = 2.62 \text{ g KBr}$$

Weigh out 2.62 g KBr, dissolve it in 125 – 2.62 = 122.38 = 122 g H_2O to make exactly 125 g of 0.180 *m* solution.

(c)   Using solution density, calculate the total mass of 1.85 L of solution, and from the mass % of KBr, the mass of KBr required.

$$1.85 \text{ L soln} \times \frac{1000 \text{ mL}}{1 \text{ L}} \times \frac{1.10 \text{ g soln}}{1 \text{ mL}} = 2035 = 2.04 \times 10^3 \text{ g soln}$$

$0.120 \ (2035 \text{ g soln}) = 244.2 = 244 \text{ g KBr}$

Dissolve 244 g KBr in water, dilute with stirring to 1.85 L.

(d)   Calculate moles KBr needed to precipitate 16.0 g AgBr. $AgNO_3$ is present in excess.

$$16.0 \text{ g AgBr} \times \frac{1 \text{ mol AgBr}}{187.8 \text{ g AgBr}} \times \frac{1 \text{ mol KBr}}{1 \text{ mol AgBr}} = 0.08520 = 0.0852 \text{ mol KBr}$$

$$0.0852 \text{ mol KBr} \times \frac{1 \text{ L soln}}{0.150 \text{ mol KBr}} = 0.568 \text{ L soln}$$

Weigh out 0.0852 mol KBr (10.1 g KBr), dissolve it in a small amount of water, and dilute to 0.568 L.

13.49　　*Analyze/Plan.* Assume a solution volume of 1.00 L. Calculate the mass of 1.00 L of solution and the mass of HNO_3 in 1.00 L of solution. Mass % = (mass solute/mass solution) × 100.　*Solve.*

$$1.00\,L \times \frac{1000\,mL}{1\,L} \times \frac{1.42\,g\,soln}{mL\,soln} = 1.42 \times 10^3\,g\,soln$$

$$16\,M = \frac{16\,mol\,HNO_3}{1\,L\,soln} \times \frac{63.02\,g\,HNO_3}{1\,mol\,HNO_3} = 1008 = 1.0 \times 10^3\,g\,HNO_3$$

$$mass\,\% = \frac{1008\,g\,HNO_3}{1.42 \times 10^3\,g\,soln} \times 100 = 71\%\,HNO_3$$

13.51　　*Analyze.* Given: 80.0% Cu, 20.0% Zn by mass; density = 8750 kg/m^3. Find: (a) *m* of Zn (b) *M* of Zn

　　(a)　*Plan.* In the brass alloy, Zn is the solute (lesser component) and Cu is the solvent (greater component). *m* = mol Zn/ kg Cu. Assume 1 $m^3 \to$ 8750 kg of brass. 80.0% is Cu, 20.0% is Zn. Change g Zn $\to$ mol Zn and solve for *m*.　*Solve.*

$$8750\,kg\,brass \times \frac{80\,g\,Cu}{100\,g\,brass} = 7.00 \times 10^3\,kg\,Cu$$

8750 kg brass – 7000 kg Cu = 1750 kg Zn

$$1750\,kg\,Zn \times \frac{1000\,g}{kg} \times \frac{1\,mol\,Zn}{65.39\,g\,Zn} = 26{,}762.5 = 2.68 \times 10^4\,mol\,Zn$$

$$m = \frac{2.676 \times 10^4\,mol\,Zn}{7000\,kg\,Cu} = 3.82\,m\,Zn$$

　　(b)　*Plan.* *M* = mol Zn/L brass. Use mol Zn from part (a). Change 1 $m^3 \to$ L brass and calculate *M*.　*Solve.*

$$1\,m^3 \times \frac{(10)^3\,dm^3}{m^3} \times \frac{1\,L}{1\,dm^3} = 1000\,L$$

$$M = \frac{2.676 \times 10^4\,mol\,Zn}{1000\,L\,brass} = 26.76 = 26.8\,M\,Zn$$

13.53　　*Analyze.* Given: 4.6% CO_2 by volume (in air), 1 atm total pressure. Find: partial pressure and molarity of CO_2 in air.

　　Plan. 4.6% CO_2 by volume means 4.6 mL of CO_2 could be isolated from 100 mL of air, at the same temperature and pressure. According to Avogadro's Law, equal volumes of gases at the same temperature and pressure contain equal numbers of moles. By inference, the volume ratio of CO_2 to air, 4.6/100 or 0.046, is also the mole ratio.　*Solve.*

$$P_{CO_2} = \chi_{CO_2} \times P_t = 0.046\,(1\,atm) = 0.046\,atm$$

M = mol CO_2/L air = n/V.　PV = nRT,　M = n/V = P/RT

$$M_{CO_2} = \frac{P_{CO_2}}{RT} = \frac{0.046\,atm}{310\,K} \times \frac{K \bullet mol}{0.08206\,L \bullet atm} = 1.8 \times 10^{-3}\,M$$

Colligative Properties

13.55 freezing point depression, $\Delta T_f = K_f(m)$; boiling point elevation, $\Delta T_b = K_b(m)$;

osmotic pressure, $\pi = M\,RT$; vapor pressure lowering, $P_A = \chi_A P_A{}^\circ$

13.57 The vapor pressure over the sucrose solution is higher than the vapor pressure over the glucose solution. Since sucrose has a greater molar mass, 10 g of sucrose contains fewer particles than 10 g of glucose. The solution that contains fewer particles, the sucrose solution, will have the higher vapor pressure.

13.59 (a) *Analyze/Plan.* H_2O vapor pressure will be determined by the mole fraction of H_2O in the solution. The vapor pressure of pure H_2O at 338 K (65°C) = 187.5 torr. *Solve.*

$$\frac{22.5\,g\,C_{12}H_{22}O_{11}}{342.3\,g/mol} = 0.06573 = 0.0657\,mol; \quad \frac{200.0\,g\,H_2O}{18.02\,g/mol} = 11.09878 = 11.10\,mol$$

$$P_{H_2O} = \chi_{H_2O}\,P^\circ_{H_2O} = \frac{11.09878\,mol\,H_2O}{11.09878 + 0.06573} \times 187.5\,torr = 186.4\,torr$$

(b) *Analyze/Plan.* For this problem, it will be convenient to express Raoult's law in terms of the lowering of the vapor pressure of the solvent, ΔP_A.

$\Delta P_A = P_A{}^\circ - \chi_A P_A{}^\circ = P_A{}^\circ (1 - \chi_A)$. $1 - \chi_A = \chi_B$, the mole fraction of the *solute* particles

$\Delta P_A = \chi_B P_A{}^\circ$; the vapor pressure of the solvent (A) is lowered according to the mole fraction of solute (B) particles present.   *Solve.*

$$P_{H_2O}\ at\ 40^\circ C = 55.3\,torr; \quad \frac{340\,g\,H_2O}{18.02\,g/mol} = 18.868 = 18.9\,mol\,H_2O$$

$$\chi_{C_3H_8O_2} = \frac{2.88\,torr}{55.3\,torr} = \frac{y\,mol\,C_3H_8O_2}{y\,mol\,C_3H_8O_2 + 18.868\,mol\,H_2O} = 0.05208 = 0.0521$$

$$0.05208 = \frac{y}{y + 18.868}; 0.05208\,y + 0.98263 = y; 0.94792\,y = 0.98263,$$

$$y = 1.0366 = 1.04\,mol\,C_3H_8O_2$$

This result has 3 sig figs because (18.9 × 0.0521 = 0.983) has 3 sig figs.

$$1.0366\,mol\,C_3H_8O_2 \times \frac{76.09\,g\,C_3H_8O_2}{mol\,C_3H_8O_2} = 78.88 = 78.9\,g\,C_3H_8O_2$$

13.61 *Analyze/Plan.* At 63.5°C, $P^\circ_{H_2O} = 175\,torr$, $P^\circ_{Eth} = 400\,torr$. Let G = the mass of H_2O and/or C_2H_5OH.   *Solve.*

(a) $$\chi_{Eth} = \dfrac{\dfrac{G}{46.07\,g\,C_2H_5OH}}{\dfrac{G}{46.07\,g\,C_2H_5OH} + \dfrac{G}{18.02\,g\,H_2O}}$$

Multiplying top and bottom of the right side of the equation by $1/G$ gives:

$$\chi_{Eth} = \frac{1/46.07}{1/46.07 + 1/18.02} = \frac{0.02171}{0.02171 + 0.05549} = 0.2812$$

(b) $P_t = P_{Eth} + P_{H_2O}$; $P_{Eth} = \chi_{Eth} \times P^{\circ}_{Eth}$; $P_{H_2O} = \chi_{H_2O} P^{\circ}_{H_2O}$

$\chi_{Eth} = 0.2812$, $P_{Eth} = 0.2812\,(400\ torr) = 112.48 = 112\ torr$

$\chi_{H_2O} = 1 - 0.2812 = 0.7188$; $P_{H_2O} = 0.7188(175\ torr) = 125.8 = 126\ torr$

$P_t = 112.5\ torr + 125.8\ torr = 238.3 = 238\ torr$

(c) χ_{Eth} in vapor $= \dfrac{P_{Eth}}{P_{total}} = \dfrac{112.5\ torr}{238.3\ torr} = 0.4721 = 0.472$

13.63   (a)   Because NaCl is a soluble ionic compound and a strong electrolyte, there are 2 mol dissolved particles for every 1 mol of NaCl solute. $C_6H_{12}O_6$ is a molecular solute, so there is 1 mol of dissolved particles per mol solute. Boiling point elevation is directly related to total moles of dissolved particles; 0.10 m NaCl has more dissolved particles so its boiling point is higher than 0.10 m $C_6H_{12}O_6$.

(b)   *Analyze/Plan.* $\Delta T = K_b\, m$; K_b for H_2O is 0.51 °C/m (Table 13.4)   *Solve.*

$$0.10\ m\ NaCl: \Delta T = \frac{0.51°C}{m} \times 0.20\ m = 0.102\,°C;\ T_b = 100.0 + 0.102 = 100.1°C$$

$$0.10\ m\ C_6H_{12}O_6 : \Delta T = \frac{0.51°C}{m} \times 0.10\ m = 0.051\,°C;\ T_b = 100.0 + 0.051 = 100.1°C$$

Check. Because K_b for H_2O is so small, there is little real difference in the boiling points of the two solutions.

(c)   In solutions of strong electrolytes like NaCl, electrostatic attractions between ions lead to ion pairing. Ion pairing reduces the effective number of particles in solution, decreasing the **change** in boiling point. The actual boiling point is then lower than the calculated boiling point for a 0.1 M solution.

13.65   For dilute aqueous solutions such as these, M is essentially equal to m. For the purposes of comparison, assume we can use M.

The more solute particles, the higher the boiling point. Since LiBr and $Zn(NO_3)_2$ are electrolytes, the particle concentrations in these solutions are 0.10 M and 0.15 M, respectively (although ion-ion attractive forces may decrease the real concentrations somewhat). Thus, the order of increasing boiling points is:

0.050 M LiBr < 0.120 M glucose < 0.050 M $Zn(NO_3)_2$

13.67   *Analyze/Plan.* $\Delta T = K\,(m)$; first, calculate the **molality** of each solution. *Solve.*

(a)   0.22 m

(b) $2.45\ mol\ CHCl_3 \times \dfrac{119.4\ g\ CHCl_3}{mol\ CHCl_3} = 292.53\ g = 0.293\ kg$;

$$\frac{0.240\ mol\ C_{10}H_8}{0.29253\ kg\ CHCl_3} = 0.8204 = 0.820\ m$$

(c) $2.04 \text{ g KBr} \times \dfrac{1 \text{ mol KBr}}{119.0 \text{ g KBr}} \times \dfrac{2 \text{ mol particles}}{1 \text{ mol KBr}} = 0.03429 = 0.0343 \text{ mol particles}$

$4.82 \text{ g C}_6\text{H}_{12}\text{O}_6 \times \dfrac{1 \text{ mol C}_6\text{H}_{12}\text{O}_6}{180.2 \text{ g C}_6\text{H}_{12}\text{O}_6} = 0.02675 = 0.0268 \text{ mol particles}$

$m = \dfrac{(0.03429 + 0.02675) \text{ mol particles}}{0.188 \text{ kg H}_2\text{O}} = 0.32465 = 0.325 \, m$

Solve. Then, f.p. = $T_f - K_f(m)$; b.p. = $T_b + K_b(m)$; T in °C

	m	T_f	$-K_f(m)$	f.p.	T_b	$+K_b(m)$	b.p.
(a)	0.22	−114.6	−1.99(0.22) = −0.44	−115.0	78.4	1.22(0.22) = 0.27	78.7
(b)	0.820	−63.5	−4.68(0.820) = −3.84	−67.3	61.2	3.63(0.820) = 2.98	64.2
(c)	0.325	0.0	−1.86(0.325) = −0.604	−0.6	100.0	0.51(0.325) = 0.17	100.2

13.69   *Analyze/Plan.* $\pi = M\,RT$; T = 25°C + 273 = 298 K; M = mol $C_9H_8O_4$/L soln   *Solve.*

$M = \dfrac{44.2 \text{ mg C}_9\text{H}_8\text{O}_4}{0.358 \text{ L}} \times \dfrac{1 \text{ g}}{1000 \text{ mg}} \times \dfrac{1 \text{ mol C}_9\text{H}_8\text{O}_4}{180.2 \text{ g C}_9\text{H}_8\text{O}_4} = 6.851 \times 10^{-4} = 6.85 \times 10^{-4} \, M$

$\pi = \dfrac{6.851 \times 10^{-4} \text{ mol}}{\text{L}} \times \dfrac{0.08206 \text{ L} \cdot \text{atm}}{\text{K} \cdot \text{mol}} \times 298 \text{ K} = 0.01675 = 0.0168 \text{ atm}$

13.71   *Analyze/Plan.* Follow the logic in Sample Exercise 13.12 to calculate the molar mass of adrenaline based on the boiling point data. Use the structure to obtain the molecular formula and molar mass. Compare the two values.   *Solve.*

$\Delta T_b = K_b \, m$; $m = \dfrac{\Delta T_b}{K_b} = \dfrac{+0.49}{5.02} = 0.0976 = 0.098 \, m$ adrenaline

$m = \dfrac{\text{mol adrenaline}}{\text{kg CCl}_4} = \dfrac{\text{g adrenaline}}{\text{MM adrenaline} \times \text{kg CCl}_4}$

$\text{MM adrenaline} = \dfrac{\text{g adrenaline}}{m \times \text{kg CCl}_4} = \dfrac{0.64 \text{ g adrenaline}}{0.0976 \, m \times 0.0360 \text{ kg CCl}_4} = 1.8 \times 10^2 \text{ g/mol adrenaline}$

The molecular formula is $C_9H_{13}NO_3$, MM = 183 g/mol. The values agree to 2 sig figs, the precision of the experimental value.

13.73   *Analyze/Plan.* Follow the logic in Sample Exercise 13.13.   *Solve.*

$\pi = MRT$; $M = \dfrac{\pi}{RT}$;   T = 25°C + 273 = 298 K

$M = 0.953 \text{ torr} \times \dfrac{1 \text{ atm}}{760 \text{ torr}} \times \dfrac{\text{K} \cdot \text{mol}}{0.08206 \text{ L} \cdot \text{atm}} \times \dfrac{1}{298 \text{ K}} = 5.128 \times 10^{-5} = 5.13 \times 10^{-5} \, M$

$\text{mol} = M \times \text{L} = 5.128 \times 10^{-5} \times 0.210 \text{ L} = 1.077 \times 10^{-5} = 1.08 \times 10^{-5} \text{ mol lysozyme}$

$\text{MM} = \dfrac{\text{g}}{\text{mol}} = \dfrac{0.150 \text{ g}}{1.077 \times 10^{-5} \text{ mol}} = 1.39 \times 10^4 \text{ g/mol lysozyme}$

13.75   (a)   *Analyze/Plan.* $i = \pi$ (measured) / π (calculated for a nonelectrolyte);

π (calculated) $= M\,RT$.   *Solve.*

$$\pi\,(\text{calculated}) = \frac{0.010\,\text{mol}}{\text{L}} \times \frac{0.08206\,\text{L} \bullet \text{atm}}{\text{mol} \bullet \text{K}} \times 298\,\text{K} = 0.2445 = 0.24\,\text{atm}$$

$i = 0.674\,\text{atm}/0.2445\,\text{atm} = 2.756 = 2.8$

(b)   The van't Hoff factor is the effective number of particles per mole of solute. The closer the measured i value is to a theoretical integer value, the more ideal the solution. Ion-pairing and other interparticle attractive forces reduce the effective number of particles in solution and reduce the measured value of i. The more concentrated the solution, the greater the ion-pairing and the smaller the measured value of i.

Colloids

13.77   (a)   In the gaseous state, the particles are far apart and intermolecular attractive forces are small. When two gases combine, all terms in Equation [13.1] are essentially zero and the mixture is always homogeneous.

(b)   The outline of a light beam passing through a colloid is visible, whereas light passing through a true solution is invisible unless collected on a screen. This is the Tyndall effect. To determine whether Faraday's (or anyone's) apparently homogeneous dispersion is a true solution or a colloid, shine a beam of light on it and see if the light is scattered.

13.79   (a)   hydrophobic   (b)   hydrophilic   (c)   hydrophobic

(d)   hydrophobic (but stabilized by adsorbed charges)

13.81   Proteins form hydrophilic colloids because they carry charges on their surface (Figure 13.28). When electrolytes are added to a suspension of proteins, the dissolved ions form ion pairs with the protein surface charges, effectively neutralizing them. The protein's capacity for ion-dipole interactions with water is diminished and the colloid separates into a protein layer and a water layer.

Additional Exercises

13.83   The outer periphery of the BHT molecule is mostly hydrocarbon-like groups, such as $-CH_3$. The one $-OH$ group is rather buried inside, and probably does little to enhance solubility in water. Thus, BHT is more likely to be soluble in the nonpolar hydrocarbon hexane, C_6H_{14}, than in polar water.

13.85   Assume that the density of the solution is 1.00 g/mL.

(a) $4\,\text{ppm}\,O_2 = \dfrac{4\,\text{mg}\,O_2}{1\,\text{kg soln}} = \dfrac{4 \times 10^{-3}\,\text{g}\,O_2}{1\,\text{L soln}} \times \dfrac{1\,\text{mol}\,O_2}{32.0\,\text{g}\,O_2} = 1.25 \times 10^{-4} = 1 \times 10^{-4}\,M$

(b) $S_{O_2} = kP_{O_2}$; $P_{O_2} = S_{O_2}/k = \dfrac{1.25 \times 10^{-4}\,\text{mol}}{\text{L}} \times \dfrac{\text{L} \bullet \text{atm}}{1.71 \times 10^{-3}\,\text{mol}} = 0.0731 = 0.07\,\text{atm}$

$0.0731\,\text{atm} \times \dfrac{760\,\text{mm Hg}}{1\,\text{atm}} = 55.6 = 60\,\text{mm Hg}$

13.88   *Plan.* The definition of ppm is (mass solute/mass solution) $\times 10^6$. Use the ratio of (mass K^+/mass KCl) to calculate (mass K^+/mass solution) $\times 10^6$.   *Solve.*

$$260 \text{ ppm KCl} = \frac{260 \text{ g KCl}}{1\times10^6 \text{ g solution}} \times \frac{39.10 \text{ g K}^+}{74.55 \text{ g KCl}} = \times \frac{136 \text{ g K}^+}{1 \times 10^6 \text{ g solution}} = 136 \text{ ppm K}^+$$

Note that even though 1 mol KCl contains 1 mol K , the ppm concentration of KCl and K^+ are not equal. This is because ppm is a mass-based, not mole-based, concentration unit.

13.91   (a) $m = \dfrac{\text{mol Na(s)}}{\text{kg Hg(l)}}$; $1.0 \text{ cm}^3 \text{Na(s)} \times \dfrac{0.97 \text{ g}}{1 \text{ cm}^3} \times \dfrac{1 \text{ mol}}{23.0 \text{ g Na}} = 0.0422 = 0.042 \text{ mol Na}$

$$20.0 \text{ cm}^3 \text{Hg(l)} \times \frac{13.6 \text{ g}}{1 \text{ cm}^3} \times \frac{1 \text{ kg}}{1000 \text{ g}} = 0.272 \text{ kg Hg(l)};$$

$$m = \frac{0.0422 \text{ mol Na}}{0.272 \text{ kg Hg(l)}} = 0.155 = 0.16 \text{ m Na}$$

(b) $M = \dfrac{\text{mol Na(s)}}{\text{L soln}} = \dfrac{0.0422 \text{ mol Na}}{0.021 \text{ L soln}} = 2.01 = 2.0 \, M \text{ Na}$

(c)   Clearly, molality and molarity are not the same for this amalgam. Only in the instance that one kg solvent and the mass of one liter solution are nearly equal do the two concentration units have similar values. In this example, one kg Hg has a volume much less than one liter.

13.95   The compound with the larger *i* value is the stronger electrolyte.

$i = \dfrac{\Delta T_f(\text{measured})}{\Delta T_f(\text{calculated})}$ The idealized value is 3 for both salts.

$Hg(NO_3)_2$: $m = \dfrac{10.0 \text{ g Hg(NO}_3)_2}{1.00 \text{ kg H}_2\text{O}} \times \dfrac{1 \text{ mol Hg(NO}_3)_2}{324.6 \text{ g Hg(NO}_3)_2} = 0.0308 \, m$

ΔT_f (nonelectrolyte) = –1.86(0.0308) = –0.0573°C

$i = \dfrac{-0.162° \text{ C}}{-0.0573° \text{ C}} = 2.83$

$HgCl_2$: $m = \dfrac{10.0 \text{ g HgCl}_2}{1.00 \text{ kg H}_2\text{O}} \times \dfrac{1 \text{ mol HgCl}_2}{271.5 \text{ g HgCl}_2} = 0.0368 \, m$

ΔT_f (nonelectrolyte) = –1.86(0.0368) = –0.0685°C

$i = \dfrac{-0.0685}{-0.0685} = 1.00$

With an *i* value of 2.83, $Hg(NO_3)_2$ is almost completely dissociated into ions; with an *i* value of 1.00, the $HgCl_2$ behaves essentially like a nonelectrolyte. Clearly, $Hg(NO_3)_2$ is the stronger electrolyte.

13.98 $M = \dfrac{\pi}{RT} = \dfrac{57.1 \text{ torr}}{298 \text{ K}} \times \dfrac{1 \text{ atm}}{760 \text{ torr}} \times \dfrac{\text{K} \cdot \text{mol}}{0.08206 \text{ L} \cdot \text{atm}} = 3.072 \times 10^{-3} = 3.07 \times 10^{-3} \, M$

$$\frac{0.036 \, g \, solute}{100 \, g \, H_2O} \times \frac{1000 \, g \, H_2O}{1 \, kg \, H_2O} = 0.36 \, g \, solute/kg \, H_2O$$

Assuming molarity and molality are the same in this dilute solution, we can then say 0.36 g solute = 3.072×10^{-3} mol; MM = 117 g/mol. Because the salt is completely ionized, the formula weight of the lithium salt is **twice** this calculated value, or **234 g/mol**. The organic portion, $C_nH_{2n+1}O_2^-$, has a formula weight of 234–7 = 227 g. Subtracting 32 for the oxygens, and 1 to make the formula C_nH_{2n}, we have C_nH_{2n}, MM = 194 g/mol. Since each CH_2 unit has a mass of 14, n ≈ 194/14 ≈ 14. The formula for our salt is $LiC_{14}H_{29}O_2$.

Integrative Exercises

13.99   Since these are very dilute solutions, assume that the density of the solution ≈ the density of H_2O ≈ 1.0 g/mL at 25°C. Then, 100 g solution = 100 g H_2O = 0.100 kg H_2O.

(a) CF_4 : $\dfrac{0.0015 \, g \, CF_4}{0.100 \, kg \, H_2O} \times \dfrac{1 \, mol \, CF_4}{88.00 \, g \, CF_4} = 1.7 \times 10^{-4} \, m$

  $CClF_3$: $\dfrac{0.009 \, g \, CClF_3}{0.100 \, kg \, H_2O} \times \dfrac{1 \, mol \, CClF_3}{104.46 \, g \, CClF_3} = 8.6 \times 10^{-4} \, m = 9 \times 10^{-4} \, m$

  CCl_2F_2 : $\dfrac{0.028 \, g \, CCl_2F_2}{0.100 \, kg \, H_2O} \times \dfrac{1 \, mol \, CCl_2F_2}{120.9 \, g \, CCl_2F_2} = 2.3 \times 10^{-3} \, m$

  $CHClF_2$: $\dfrac{0.30 \, g \, CHClF_2}{0.100 \, kg \, H_2O} \times \dfrac{1 \, mol \, CHClF_2}{86.47 \, g \, CHClF_2} = 3.5 \times 10^{-2} \, m$

(b) $m = \dfrac{mol \, solute}{kg \, solvent}$; $M = \dfrac{mol \, solute}{L \, solution}$

Molality and molarity are numerically similar when kilograms solvent and liters solution are nearly equal. This is true when solutions are dilute, so that the density of the solution is essentially the density of the solvent, and when the density of the solvent is nearly 1 g/mL. That is, for dilute aqueous solutions such as the ones in this problem, $M \approx m$.

(c)   Water is a polar solvent; the solubility of solutes increases as their polarity increases. All the fluorocarbons listed have tetrahedral molecular structures. CF_4, a symmetrical tetrahedron, is nonpolar and has the lowest solubility. As more different atoms are bound to the central carbon, the electron density distribution in the molecule becomes less symmetrical and the molecular polarity increases. The most polar fluorocarbon, $CHClF_2$, has the greatest solubility in H_2O. It may act as a weak hydrogen bond acceptor for water.

(d) $S_g = k \, P_g$. Assume $M = m$ for $CHClF_2$. $P_g = 1$ atm

$$k = \frac{S_g}{P_g} = \frac{M}{P_g}; k = \frac{3.5 \times 10^{-2} \, M}{1.0 \, atm} = 3.5 \times 10^{-2} \, mol/L \bullet atm$$

This value is greater than the Henry's law constant for $N_2(g)$, because $N_2(g)$ is nonpolar and of lower molecular mass than $CHCIF_2$. In fact, the Henry's law constant for nonpolar CF_4, 1.7×10^{-4} mol/L •atm is similar to the value for N_2, 6.8×10^{-4} mol L•atm.

13.103   (a)

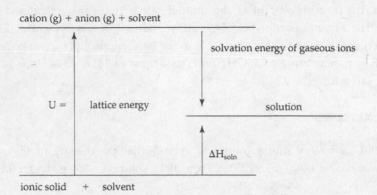

(b)   If the lattice energy (U) of the ionic solid (ion-ion forces) is too large relative to the solvation energy of the gaseous ions (ion-dipole forces), ΔH_{soln} will be too large and positive (endothermic) for solution to occur. This is the case for solutes like NaBr. Lattice energy is inversely related to the distance between ions, so salts with large cations like $(CH_3)_4N^+$ have smaller lattice energies than salts with simple cations like Na^+. The smaller lattice energy of $(CH_4)_3NBr$ causes it to be more soluble in nonaqueous polar solvents. Also, the $-CH_3$ groups in the large cation are capable of dispersion interactions with the $-CH_3$ (or other nonpolar groups) of the solvent molecules. This produces a more negative solvation energy for the salts with large cations.

Overall, for salts with larger cations, U is smaller (less positive), the solvation energy of the gaseous ions is more negative, and ΔH_{soln} is less endothermic. These salts are more soluble in polar nonaqueous solvents.

13.106   The process is spontaneous with no significant change in enthalpy, so we suspect that there is an increase in entropy. In general, dilute solutions of the same solute have greater entropy than concentrated ones, because the solute particles are more free to move about the solution. There are a greater number of equivalent environments available to the solute. In the limiting case that the more dilute solution is pure solvent, there is a definite increase in entropy as the concentrated solution is diluted. In Figure 13.23, there may be some entropy decrease as the dilute solution loses solvent, but this is more than offset by the entropy increase that accompanies dilution. There is a net increase in entropy of the system, going from the left to center panels in Figure 13.23.

14 Chemical Kinetics

Visualizing Concepts

14.1　(a)　X is a product, because its concentration increases with time.

　　　(b)　The average rate of reaction between any two points on the graph is the slope of the line connecting the two points. The average rate is greater between points 1 and 2 than between points 2 and 3 because they are different stages in the overall process. Points 1 and 2 are earlier in the reaction when more reactants are available, so the rate of formation of products is greater. As reactants are used up, the rate of X production decreases, and the average rate between points 2 and 3 is smaller.

14.3　*Analyze/Plan.* Using the relationship rate = $k[A]^x$, determine the value of x that produces a rate law to match the described situation. *Solve.*

　　　(a)　x = 0. The rate of reaction does not depend on [A], so the reaction is zero-order in A.

　　　(b)　x = 2. When [A] increases by a factor of 3, rate increases by a factor of $(3)^2 = 9$.

　　　(c)　x = 3. When [A] increases by a factor of 2, rate increases by a factor of $(2)^3 = 8$.

14.5　*Analyze.* Given concentrations of reactants and products at two times, as represented in the diagram, find $t_{1/2}$ for this first-order reaction.

　　　Plan. For a first order reaction, $t_{1/2} = 0.693/k$; $t_{1/2}$ depends only on k. Use equation [14.12] to solve for k. *Solve.*

　　　(a)　Since reactants and products are in the same container, use number of particles as a measure of concentration. The red dots are reactant A, and the blue are product B. $[A]_0 = 8$, $[A]_{30} = 2$, t = 30 min.

$$\ln \frac{[A]_t}{[A]_0} = -kt. \quad \ln(2/8) = -k(30 \text{ min}); \quad \frac{-1.3863}{-30 \text{min}} = k;$$

$$k = 0.046210 = 0.0462 \text{ min}^{-1}$$

$$t_{1/2} = 0.693/k = 0.693/0.046210 = 15 \text{ min}$$

　　　　　By examination, $[A]_0 = 8$, $[A]_{30} = 2$. After 1 half-life, [A] = 4; after a second half-life, [A] = 2. Thirty minutes represents exactly 2 half-lives, so $t_{1/2} = 15$ min. [This is more straightforward than the calculation, but a less general method.]

　　　(b)　After 4 half-lives, $[A]_t = [A]_0 \times 1/2 \times 1/2 \times 1/2 \times 1/2 = [A]_0/16$. In general, after n half-lives, $[A] = [A]_0/2^n$.

14.8   This is the profile of a two-step mechanism, A → B and B → C. There is one intermediate, B. Because there are two energy maxima, there are two transition states. The B → C step is faster, because its activation energy is smaller. The reaction is exothermic because the energy of the products is lower than the energy of the reactants.

Reaction Rates

14.11   (a)   *Reaction rate* is the change in the amount of products or reactants in a given amount of time; it is the speed of a chemical reaction.

   (b)   Rates depend on concentration of reactants, surface area of reactants, temperature and presence of catalyst.

   (c)   The stoichiometry of the reaction (mole ratios of reactants and products) must be known to relate rate of disappearance of reactants to rate of appearance of products.

14.13   *Analyze/Plan.* Given mol A at a series of times in minutes, calculate mol B produced, molarity of A at each time, change in M of A at each 10 min interval, and ΔM A/s. For this reaction, mol B produced equals mol A consumed. M of A or [A] = mol A/0.100 L. The average rate of disappearance of A for each 10 minute interval is

$$-\frac{\Delta[A]}{s} = -\frac{[A]_1 - [A]_0}{10 \text{ min}} \times \frac{1 \text{ min}}{60 \text{ s}}$$

Solve.

Time (min)	Mol A	(a) Mol B	[A]	Δ[A]	(b) Rate $-(\Delta$[A]/s)
0	0.065	0.000	0.65		
10	0.051	0.014	0.51	−0.14	2.3×10^{-4}
20	0.042	0.023	0.42	−0.09	2×10^{-4}
30	0.036	0.029	0.36	−0.06	1×10^{-4}
40	0.031	0.034	0.31	−0.05	0.8×10^{-4}

   (c) $\dfrac{\Delta M_B}{\Delta t} = \dfrac{(0.029 - 0.014) \text{ mol}/0.100 \text{ L}}{(30 - 10) \text{ min}} \times \dfrac{1 \text{ min}}{60 \text{ s}} = 1.25 \times 10^{-4} = 1.3 \times 10^{-4} \; M/s$

14.15   (a)   *Analyze/Plan.* Follow the logic in Sample Exercise 14.1.   *Solve.*

Time (sec)	Time Interval (sec)	Concentration (M)	ΔM	Rate (M/s)
0		0.0165		
2,000	2,000	0.0110	−0.0055	28×10^{-7}
5,000	3,000	0.00591	−0.0051	17×10^{-7}
8,000	3,000	0.00314	−0.00277	9.23×10^{-7}
12,000	4,000	0.00137	−0.00177	4.43×10^{-7}
15,000	3,000	0.00074	−0.00063	2.1×10^{-7}

(b) From the slopes of the lines in the figure at right, the rates are 12×10^{-7} M/s at 5000 s, 5.8 $\times 10^{-7}$ M/s at 8000 s.

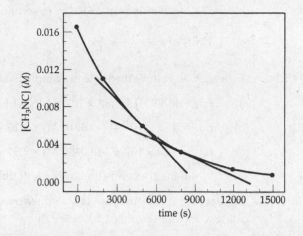

14.17 *Analyze/Plan.* Follow the logic in Sample Exercise 14.3.   *Solve.*

(a) $-\Delta[H_2O_2]/\Delta t = \Delta[H_2]/\Delta t = \Delta[O_2]/\Delta t$

(b) $-\Delta[N_2O]/2\Delta t = \Delta[N_2]/2\Delta t = \Delta[O_2]/\Delta t$

$-\Delta[N_2O]/\Delta t = \Delta[N_2]/\Delta t = 2\Delta[O_2]/\Delta t$

(c) $-\Delta[N_2]/\Delta t = \Delta[NH_3]/2\Delta t; -\Delta[H_2]/3\Delta t = \Delta[NH_3]/2\Delta t$

$-2\Delta[N_2]/\Delta t = \Delta[NH_3]/\Delta t; -\Delta[H_2]/\Delta t = 3\Delta[NH_3]/2\Delta t$

14.19 *Analyze/Plan.* Use Equation [14.4] to relate the rate of disappearance of reactants to the rate of appearance of products. Use this relationship to calculate desired quantities. *Solve.*

(a) $\Delta[H_2O]/2\Delta t = -\Delta[H_2]/2\Delta t = -\Delta[O_2]/\Delta t$

H_2 is burning, $-\Delta[H_2]/\Delta t = 0.85$ mol/s

O_2 is consumed, $-\Delta[O_2]/\Delta t = -\Delta[H_2]/2\Delta t = 0.85$ mol/s/2 = 0.43 mol/s

H_2O is produced, $+\Delta[H_2O]/\Delta t = -\Delta[H_2]/\Delta t = 0.85$ mol/s

(b) The change in total pressure is the sum of the changes of each partial pressure. NO and Cl_2 are disappearing and NOCl is appearing.

$-\Delta P_{NO}/\Delta t = -23$ torr/min

$-\Delta P_{Cl_2}/\Delta t = \Delta P_{NO}/2\Delta t = -12$ torr/min

$+\Delta P_{NOCl}/\Delta t = -\Delta P_{NO}/\Delta t = +23$ torr/min

$\Delta P_T/\Delta t = -23$ torr/min $- 12$ torr/min $+ 23$ torr/min $= -12$ torr/min

Rate Laws

14.21 *Analyze/Plan.* Follow the logic in Sample Exercises 14.4 and 14.5. *Solve.*

(a) If [A] is doubled, there will be no change in the rate or the rate constant. The overall rate is unchanged because [A] does not appear in the rate law; the rate constant changes only with a change in temperature.

(b)   The reaction is zero order in A, second order in B and second order overall.

(c)   Units of $k = \dfrac{M/s}{M^2} = M^{-1}\,s^{-1}$

14.23   *Analyze/Plan.* Follow the logic in Sample Exercise 14.4.   *Solve.*

(a)   rate = $k[N_2O_5] = 4.82 \times 10^{-3}\,s^{-1}\,[N_2O_5]$

(b)   rate = $4.82 \times 10^{-3}\,s^{-1}\,(0.0240\,M) = 1.16 \times 10^{-4}\,M/s$

(c)   rate = $4.82 \times 10^{-3}\,s^{-1}\,(0.0480\,M) = 2.31 \times 10^{-4}\,M/s$

When the concentration of N_2O_5 doubles, the rate of the reaction doubles.

14.25   *Analyze/Plan.* Write the rate law and rearrange to solve for k. Use the given data to calculate k, including units.   *Solve.*

(a, b)   rate = $k[CH_3Br][OH^-]$; $k = \dfrac{rate}{[CH_3Br][OH^-]}$

at 298 K, $k = \dfrac{0.0432\,M/s}{(5.0 \times 10^{-3}\,M)(0.050\,M)} = 1.7 \times 10^2\,M^{-1}s^{-1}$

(c)   Since the rate law is first order in $[OH^-]$, if $[OH^-]$ is tripled, the rate triples.

14.27   *Analyze/Plan.* Follow the logic in Sample Exercise 14.6.   *Solve.*

(a)   From the data given, when $[OCl^-]$ doubles, rate doubles. When $[I^-]$ doubles, rate doubles. The reaction is first order in both $[OCl^-]$ and $[I^-]$. rate = $[OCl^-][I^-]$

(b)   Using the first set of data:

$k = \dfrac{rate}{[OCl^-][I^-]} = \dfrac{1.36 \times 10^{-4}\,M/s}{(1.5 \times 10^{-3}\,M)(1.5 \times 10^{-3}\,M)} = 60.444 = 60\,M^{-1}\,s^{-1}$

(c) rate $= \dfrac{60.444}{M \bullet s}(2.0 \times 10^{-3}\,M)(5.0 \times 10^{-4}\,M) = 6.0444 \times 10^{-5} = 6.0 \times 10^{-5}\,M/s$

14.29   *Analyze/Plan.* Follow the logic in Sample Exercise 14.6 to deduce the rate law. Rearrange the rate law to solve for k and deduce units. Calculate a k value for each set of concentrations and then average the three values.   *Solve.*

(a)   Doubling $[NH_3]$ while holding $[BF_3]$ constant doubles the rate (experiments 1 and 2). Doubling $[BF_3]$ while holding $[NH_3]$ constant doubles the rate (experiments 4 and 5).

Thus, the reaction is first order in both BF_3 and NH_3; rate = $k[BF_3][NH_3]$.

(b)   The reaction is second order overall.

(c)   From experiment 1: $k = \dfrac{0.2130\,M/s}{(0.250\,M)(0.250\,M)} = 3.41\,M^{-1}\,s^{-1}$

(Any of the five sets of initial concentrations and rates could be used to calculate the rate constant k. The average of these 5 values is $k_{avg} = 3.408 = 3.41\,M^{-1}s^{-1}$)

(d)   rate = $3.41\,M^{-1}s^{-1}(0.100\,M)(0.500\,M) = 0.1704 = 0.170\,M/s$

14.31   *Analyze/Plan.* Follow the logic in Sample Exercise 4.6 to deduce the rate law. Rearrange the rate law to solve for k and deduce units. Calculate a k value for each set of concentrations and then average the three values.   *Solve.*

(a)   Increasing [NO] by a factor of 2.5 while holding [Br$_2$] constant (experiments 1 and 2) increases the rate by a factor 6.25 or (2.5)^2. Increasing [Br$_2$] by a factor of 2.5 while holding [NO] constant increases the rate by a factor of 2.5. The rate law for the appearance of NOBr is: rate = Δ[NOBr]/Δt = k[NO]^2[Br$_2$].

(b)   From experiment 1: $k_1 = \dfrac{24\ M/s}{(0.10\ M)^2\ (0.20\ M)} = 1.20 \times 10^4 = 1.2 \times 10^4\ M^{-2}\ s^{-1}$

$k_2 = 150/(0.25)^2(0.20) = 1.20 \times 10^4 = 1.2 \times 10^4\ M^{-2}\ s^{-1}$

$k_3 = 60/(0.10)^2(0.50) = 1.20 \times 10^4 = 1.2 \times 10^4\ M^{-2}\ s^{-1}$

$k_4 = 735/(0.35)^2(0.50) = 1.2 \times 10^4 = 1.2 \times 10^4\ M^{-2}\ s^{-1}$

$k_{avg} = (1.2 \times 10^4 + 1.2 \times 10^4 + 1.2 \times 10^4 + 1.2 \times 10^4)/4 = 1.2 \times 10^4\ M^{-2}\ s^{-1}$

(c)   Use the reaction stoichiometry and Equation 14.4 to relate the designated rates. Δ[NOBr]/2Δt = $-\Delta$[Br$_2$]/Δt; the rate of disappearance of Br$_2$ is half the rate of appearance of NOBr.

(d)   Note that the data are given in terms of appearance of NOBr.

$$\dfrac{-\Delta[Br_2]}{\Delta t} = \dfrac{k[NO]^2\ [Br_2]}{2} = \dfrac{1.2 \times 10^4}{2\ M^2\ s} \times (0.075\ M)^2 \times (0.25\ M) = 8.4\ M/s$$

Change of Concentration with Time

14.33   (a)   [A]$_0$ is the molar concentration of reactant A at time zero, the initial concentration of A. [A]$_t$ is the molar concentration of reactant A at time t. $t_{1/2}$ is the time required to reduce [A]$_0$ by a factor of 2, the time when [A]$_t$ = [A]$_0$/2. k is the rate constant for a particular reaction. k is independent of reactant concentration but varies with reaction temperature.

(b)   A graph of ln[A] vs time yields a straight line for a first-order reaction.

14.35   *Analyze/Plan.* The half-life of a first-order reaction depends only on the rate constant, $t_{1/2}$ = 0.693/k. Use this relationship to calculate k for a given $t_{1/2}$, and, at a different temperature, $t_{1/2}$ given k.   *Solve.*

(a) $t_{1/2}$ = 2.3 $\times$ 10^5 s; $t_{1/2}$ = 0.693/k, k = 0.693/$t_{1/2}$

$k = 0.693/2.3 \times 10^5\ s = 3.0 \times 10^{-6}\ s^{-1}$

14.37   *Analyze/Plan.* Follow the logic in Sample Exercise 14.7. In this reaction, pressure is a measure of concentration. In (a) we are given k, [A]$_0$, t and asked to find [A]$_t$, using Equation [14.13], the integrated form of the first-order rate law. In (b), [A$_t$] = 0.1[A$_0$], find t.   *Solve.*

(a) $\ln P_t = -kt + \ln P_0$; P$_0$ = 375 torr; t = 65 s

$\ln P_{65} = -4.5 \times 10^{-2}\ s^{-1}(65) + \ln(375) = -2.925 + 5.927 = 3.002$

$P_{65} = 20.12 = 20$ torr

(b) $P_t = 0.10\ P_0$; $\ln(P_t/P_0) = -kt$

     $\ln(0.10\ P_0/P_0) = -kt$, $\ln(0.10) = -kt$; $-\ln(0.10)/k = t$

     $t = -(-2.303)/4.5 \times 10^{-2}\ s^{-1} = 51.2 = 51\ s$

Check. From part (a), the pressure at 65 s is 20 torr, $P_t \sim 0.05\ P_0$. In part (b) we calculate the time where $P_t = 0.10\ P_0$ to be 51 s. This time should be smaller than 65 s, and it is. Data and results in the two parts are consistent.

14.39   *Analyze/Plan.* Given reaction order, various values for t and P_t, find the rate constant for the reaction at this temperature. For a first-order reaction, a graph of lnP vs t is linear with as slope of –k.   *Solve.*

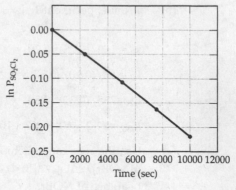

t(s)	$P_{SO_2Cl_2}$	$\ln P_{SO_2Cl_2}$
0	1.000	0
2500	0.947	–0.0545
5000	0.895	–0.111
7500	0.848	–0.165
10000	0.803	–0.219

Graph $\ln P_{SO_2Cl_2}$ vs. time. (Pressure is a satisfactory unit for a gas, since the concentration in moles/liter is proportional to P.) The graph is linear with slope $-2.19 \times 10^{-5}\ s^{-1}$

14.41   *Analyze/Plan.* Given: mol A, t. Change mol to M at various times. Make both first- and second-order plots to see which is linear.   *Solve.*

(a)

time(min)	mol A	[A] (M)	ln[A]	1/mol A
0	0.065	0.65	–0.43	1.5
10	0.051	0.51	–0.67	2.0
20	0.042	0.42	–0.87	2.4
30	0.036	0.36	–1.02	2.8
40	0.031	0.31	–1.17	3.2

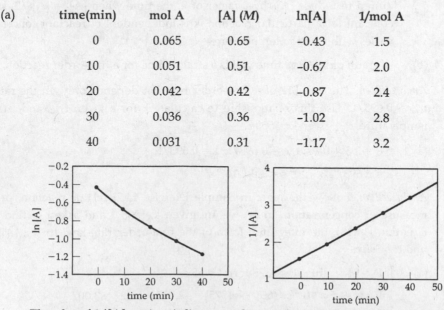

The plot of 1/[A] vs time is linear, so the reaction is second-order in [A].

166

(b)   For a second-order reaction, a plot of $1/[A]$ vs. t is linear with slope k.

$k = $ slope $= (3.2 - 2.0) \ M^{-1} / 30$ min $= 0.040 \ M^{-1}$ min^{-1}

(The best fit to the line yields slope $= 0.042 \ M^{-1}$ min^{-1}.)

(c) $t_{1/2} = 1/k[A]_0 = 1/(0.040 \ M^{-1}$ min$^{-1})(0.65 \ M) = 38.46 = 38$ min

(Using the "best-fit" slope, $t_{1/2} = 37$ min.)

14.43   *Analyze/Plan.* Follow the logic in Solution 14.41. Make both first and second order plots to see which is linear. *Solve.*

(a)

time(s)	[NO$_2$](M)	ln[NO$_2$]	1/[NO$_2$]
0.0	0.100	–2.303	10.0
5.0	0.017	–4.08	59
10.0	0.0090	–4.71	110
15.0	0.0062	–5.08	160
20.0	0.0047	–5.36	210

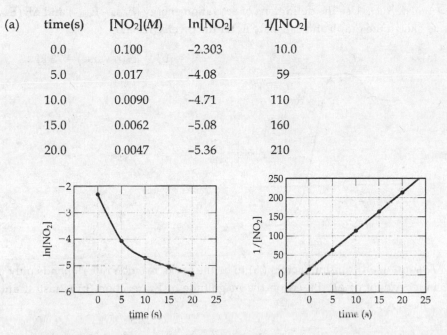

The plot of $1/[NO_2]$ vs time is linear, so the reaction is second order in NO$_2$.

(b)   The slope of the line is $(210 - 59) \ M^{-1} / 15.0$ s $= 10.07 = 10 \ M^{-1}$s$^{-1} = k$. (The slope of the best-fit line is $10.02 = 10 \ M^{-1}$s^{-1}.)

Temperature and Rate

14.45   (a)   The energy of the collision and the orientation of the molecules when they collide determine whether a reaction will occur.

(b)   According to the kinetic-molecular theory (Chapter 10), the higher the temperature, the greater the speed and kinetic energy of the molecules. Therefore, at a higher temperature, there are more total collisions and each collision is more energetic.

14.47 *Analyze/Plan.* Given the temperature and energy, use Equation [14.18] to calculate the fraction of Ar atoms that have at least this energy.   *Solve.*

$$f = e^{-E_a/RT} \quad E_a = 10.0 \text{ kJ/mol} = 1.00 \times 10^4 \text{ J/mol}; \ T = 400 \text{ K} (127°C)$$

$$-E_a/RT = -\frac{1.00 \times 10^4 \text{ J/mol}}{400 \text{ K}} \times \frac{\text{mol} \cdot \text{K}}{8.314 \text{ J}} = -3.0070 = -3.01$$

$$f = e^{-3.0070} = 4.9 \times 10^{-2}$$

At 400 K, approximately 1 out of 20 molecules has this kinetic energy.

14.49 *Analyze/Plan.* Use the definitions of activation energy ($E_{max} - E_{react}$) and ΔE ($E_{prod} - E_{react}$) to sketch the graph and calculate E_a for the reverse reaction.   *Solve.*

(a)   (b) E_a(reverse) = 73 kJ

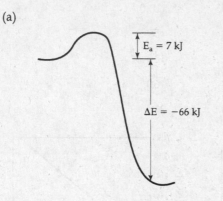

$$E_a = 7 \text{ kJ}$$

$$\Delta E = -66 \text{ kJ}$$

14.51 Assuming all collision factors (A) to be the same, reaction rate depends only on E_a; it is independent of ΔE. Based on the magnitude of E_a, reaction (b) is fastest and reaction (c) is slowest.

14.53 *Analyze/Plan.* Given k_1, at T_1, calculate k_2 at T_2. Change T to Kelvins, then use the Equation [14.21] to calculate k_2.   *Solve.*

$$T_1 = 20°C + 273 = 293 \text{ K}; \ T_2 = 60°C + 273 = 333 \text{ K}; \ k_1 = 2.75 \times 10^{-2}\text{s}^{-1}$$

(a) $$\ln\left(\frac{k_1}{k_2}\right) = \frac{E_a}{R}\left(\frac{1}{333} - \frac{1}{293}\right) = \frac{75.5 \times 10^3 \text{ J/mol}}{8.314 \text{ J/mol}}(-4.100 \times 10^{-4})$$

$$\ln(k_1/k_2) = -3.7229 = -3.7; \ k_1/k_2 = 0.0242 = 0.02; \ k_2 = \frac{0.0275 \text{ s}^{-1}}{0.0242} = 1.14 = 1 \text{ s}^{-1}$$

(b) $$\ln\left(\frac{k_1}{k_2}\right) = \frac{105 \times 10^3 \text{ J/mol}}{8.314 \text{ J/mol}}\left(\frac{1}{333} - \frac{1}{293}\right) = -5.1776 = -5.2$$

$$k_1/k_2 = 5.642 \times 10^{-3} = 6 \times 10^{-3}; \ k_2 = \frac{0.0275 \text{ s}^{-1}}{5.642 \times 10^{-3}} = 4.88 = 5 \text{ s}^{-1}$$

14.55   *Analyze/Plan.* Follow the logic in Sample Exercise 14.11.   *Solve.*

k	ln k	T(K)	$1/T(\times 10^3)$
0.0521	−2.955	288	3.47
0.101	−2.293	298	3.36
0.184	−1.693	308	3.25
0.332	−1.103	318	3.14

The slope, -5.71×10^3, equals $-E_a/R$. Thus,

$E_a = 5.71 \times 10^3 \times 8.314$ J/mol $= 47.5$ kJ/mol.

14.57   *Analyze/Plan.* Given E_a, find the ratio of rates for a reaction at two temperatures. Assuming initial concentrations are the same at the two temperatures, the ratio of rates will be the ratio of rate constants, k_1/k_2. Use Equation [14.21] to calculate this ratio.   *Solve.*

$T_1 = 50°C + 273 = 323$ K; $T_2 = 0°C + 273 = 273$ K

$$\ln\left(\frac{k_1}{k_2}\right) = \frac{E_a}{R}\left[\frac{1}{T_2} - \frac{1}{T_1}\right] = \frac{65.7 \text{ kJ/mol}}{8.314 \text{ J/mol}} \times \frac{1000 \text{ J}}{1 \text{ kJ}}\left[\frac{1}{273} - \frac{1}{323}\right]$$

$\ln(k_1/k_2) = 7.902 \times 10^3 \ (5.670 \times 10^{-4}) = 4.481 = 4.5; \ k_1/k_2 = 88.3 = 9 \times 10^1$

The reaction will occur 90 times faster at 50°C, assuming equal initial concentrations.

Reaction Mechanisms

14.59   (a)   An *elementary reaction* is a process that occurs in a single event; the order is given by the coefficients in the balanced equation for the reaction.

     (b)   A *unimolecular* elementary reaction involves only one reactant molecule; the activated complex is derived from a single molecule. A *bimolecular* elementary reaction involves two reactant molecules in the activated complex and the overall process.

     (c)   A *reaction mechanism* is a series of elementary reactions that describe how an overall reaction occurs and explain the experimentally determined rate law.

14.61   *Analyze/Plan.* Elementary reactions occur as a single step, so the molecularity is determined by the number of reactant molecules; the rate law reflects reactant stoichiometry. *Solve.*

     (a)   unimolecular, rate $= k[Cl_2]$

     (b)   bimolecular, rate $= k[OCl^-][H_2O]$

     (c)   bimolecular, rate $= k[NO][Cl_2]$

14.63   *Analyze/Plan.* Use the definitions of the terms 'intermediate' and 'exothermic', along with the characteristics of reaction profiles, to answer the questions. *Solve.*

This is a three-step mechanism, $A \rightarrow B$, $B \rightarrow C$, and $C \rightarrow D$.

(a)   There are 2 intermediates, B and C.

(b)   There are 3 energy maxima in the reaction profile, so there are 3 transition states.

(c)   Step $C \rightarrow D$ has the lowest activation energy, so it is fastest.

(d)   The energy of D is slightly greater than the energy of A, so the overall reaction is endothermic.

14.65   (a) $H_2(g) + ICl(g) \rightarrow HI(g) + HCl(g)$

$\underline{HI(g) + ICl(g) \rightarrow I_2(g) + HCl(g)}$

$H_2(g) + 2ICl(g) \rightarrow I_2(g) + 2HCl(g)$

(b)   Intermediates are produced and consumed during reaction. HI is the intermediate.

(c)   Follow the logic in Sample Exercise 14.13.

First step: rate = $k_1[H_2][ICl]$

Second step: rate = $k_2[HI][ICl]$

(d)   The slow step determines the rate law for the overall reaction. If the first step is slow, the observed rate law is: rate = $k[H_2][HCl]$.

14.67   *Analyze/Plan.* Given a proposed mechanism and an observed rate law, determine which step is rate determining.   *Solve.*

(a)   If the first step is slow, the observed rate law is the rate law for this step.
rate = $k[NO][Cl_2]$

(b)   Since the observed rate law is second-order in [NO], the second step must be slow relative to the first step; the second step is rate determining.

Catalysis

14.69   (a)   A catalyst increases the rate of reaction by decreasing the activation energy, E_a, or increasing the frequency factor A. Lowering the activation energy is more common and more dramatic.

(b)   A homogeneous catalyst is in the same phase as the reactants; a heterogeneous catalyst is in a different phase and is usually a solid.

14.71   (a) $2[NO_2(g) + SO_2(g) \rightarrow NO(g) + SO_3(g)]$

$\underline{2NO(g) + O_2(g) \rightarrow 2NO_2(g)}$

$2SO_2(g) + O_2(g) \rightarrow 2SO_2(g)$

(b) $NO_2(g)$ is a catalyst because it is consumed and then reproduced in the reaction sequence. (NO(g) is an intermediate because it is produced and then consumed.)

(c)   Since NO_2 is in the same state as the other reactants, this is homogeneous catalysis.

14.73   Use of chemically stable supports such as alumina and silica makes it possible to obtain very large surface areas per unit mass of the precious metal catalyst. This is so because the metal can be deposited in a very thin, even monomolecular, layer on the surface of the support.

14.75   As illustrated in Figure 14.21, the two C–H bonds that exist on each carbon of the ethylene molecule before adsorption are retained in the process in which a D atom is added to each C (assuming we use D_2 rather than H_2). To put two deuteriums on a single carbon, it is necessary that one of the already existing C–H bonds in ethylene be broken while the molecule is adsorbed, so the H atom moves off as an adsorbed atom, and is replaced by a D. This requires a larger activation energy than simply adsorbing C_2H_4 and adding one D atom to each carbon.

14.77   (a)   Living organisms operate efficiently in a very narrow temperature range; heating to increase reaction rate is not an option. Therefore, the role of enzymes as homogeneous catalysts that speed up desirable reactions without heating and undesirable side-effects is crucial for biological systems.

        (b)   *catalase*: $2H_2O_2 \rightarrow 2H_2O + O_2$; *nitrogenase*: $N_2 \rightarrow 2NH_3$ (nitrogen fixation)

14.79   *Analyze/Plan.* Let k = the rate constant for the uncatalyzed reaction,

                    k_c = the rate constant for the catalyzed reaction

        According to Equation [14.20], $\ln k = -E_a / RT + \ln A$

        Subtracting $\ln k$ from $\ln k_c$,

        $$\ln k_c - \ln k = -\left[\frac{55 \text{ kJ/mol}}{RT} + \ln A\right] - \left[-\frac{95 \text{ kJ/mol}}{RT} + \ln A\right]. \quad Solve.$$

        (a)   RT = 8.314 J/K • mol × 298 K × 1 kJ/1000 J = 2.478 kJ/mol; ln A is the same for both reactions.

               $$\ln (k_c / k) = \frac{95 \text{ kJ/mol} - 55 \text{ kJ/mol}}{2.478 \text{ kJ/mol}}; \quad k_c/k = 1.024 \times 10^7 = 1 \times 10^7$$

               The catalyzed reaction is approximately 10,000,000 (ten million) times faster at 25°C.

        (b)   RT = 8.314 J/K • mol × 398 K × 1 kJ/1000 J = 3.309 kJ/mol

               $$\ln (k_c / k) = \frac{40 \text{ kJ/mol}}{3.309 \text{ kJ/mol}}; \quad k_c/k = 1.778 \times 10^5 = 2 \times 10^5$$

               The catalyzed reaction is 200,000 times faster at 125°C.

Additional Exercises

14.81 $$\text{rate} = \frac{-\Delta[H_2S]}{\Delta t} = \frac{\Delta[Cl^-]}{2\Delta t} = k[H_2S][Cl_2]$$

        $$\frac{-\Delta[H_2S]}{\Delta t} = (3.5 \times 10^{-2} \; M^{-1}s^{-1})(2.0 \times 10^{-4} \; M)(0.050 \; M) = 3.50 \times 10^{-7} = 3.5 \times 10^{-7} \; M/s$$

        $$\frac{\Delta[Cl^-]}{\Delta t} = \frac{-2\Delta[H_2S]}{\Delta t} = 2(3.50 \times 10^{-7} \; M/s) = 7.0 \times 10^{-7} \; M/s$$

14.84   (a)   The rate increases by a factor of nine when $[C_2O_4^{2-}]$ triples (compare experiments 1 and 2). The rate doubles when $[HgCl_2]$ doubles (compare experiments 2 and 3). The rate law is apparently: rate = $k[HgCl_2][C_2O_4^{2-}]^2$

       (b) $k = \dfrac{rate}{[HgCl_2][C_2O_4^{2-}]^2}$ Using the data for Experiment 1,

           $k = \dfrac{(3.2 \times 10^{-5} \ M/s)}{[0.164 \ M][0.15 \ M]^2} = 8.672 \times 10^{-3} = 8.7 \times 10^{-3} \ M^{-2}s^{-1}$

       (c)   rate = $(8.672 \times 10^{-3} \ M^{-2}s^{-1})(0.050 \ M)(0.10 \ M)^2 = 4.3 \times 10^{-6} \ M/s$

14.87   (a) $k = (8.56 \times 10^{-5} \ M/s)/(0.200 \ M) = 4.28 \times 10^{-4} \ s^{-1}$

       (b) $\ln [urea] = -(4.28 \times 10^{-4}s^{-1} \times 4.00 \times 10^3 \ s) + \ln (0.500)$

           $\ln [urea] = -1.712 - 0.693 = -2.405 = -2.41; \ [urea] = 0.0903 = 0.090 \ M$

       (c) $t_{1/2} = 0.693/k = 0.693/4.28 \times 10^{-4} \ s^{-1} = 1.62 \times 10^3 \ s$

14.91

ln k	1/T
−24.17	3.33×10^{-3}
−20.72	3.13×10^{-3}
−17.32	2.94×10^{-3}
−15.24	2.82×10^{-3}

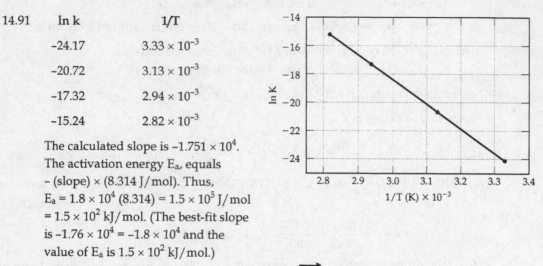

The calculated slope is -1.751×10^4. The activation energy E_a, equals $-$ (slope) $\times$ (8.314 J/mol). Thus, $E_a = 1.8 \times 10^4 \ (8.314) = 1.5 \times 10^5 \ J/mol = 1.5 \times 10^2 \ kJ/mol$. (The best-fit slope is $-1.76 \times 10^4 = -1.8 \times 10^4$ and the value of E_a is $1.5 \times 10^2 \ kJ/mol$.)

14.94   (a)

$$Cl_2(g) \rightleftharpoons 2Cl(g)$$
$$Cl(g) + CHCl_3(g) \rightarrow HCl(g) + CCl_3(g)$$
$$Cl(g) + CCl_3(g) \rightarrow CCl_4(g)$$

$$\overline{Cl_2(g) + 2Cl(g) + CHCl_3(g) + CCl_3(g) \rightarrow 2Cl(g) + HCl(g) + CCl_3(g) + CCl_4(g)}$$
$$Cl_2(g) + CHCl_3(g) \rightarrow HCl(g) + CCl_4(g)$$

       (b) $Cl(g), CCl_3(g)$

       (c)   Reaction 1 - unimolecular, Reaction 2 - bimolecular, Reaction 3 - bimolecular

       (d)   Reaction 2, the slow step, is rate determining.

       (e)   If Reaction 2 is rate determining, rate = $k_2[CHCl_3][Cl]$. Cl is an intermediate formed in reaction 1, an equilibrium. By definition, the rates of the forward and reverse processes are equal; $k_1 [Cl_2] = k_{-1} [Cl]^2$. Solving for [Cl] in terms of $[Cl_2]$,

$$[Cl]^2 = \frac{k_1}{k_{-1}}[Cl_2]; \; [Cl] = \left(\frac{k_1}{k_{-1}}[Cl_2]\right)^{1/2}$$

Substituting into the overall rate law

$$\text{rate} = k_2 \left(\frac{k_1}{k_{-1}}\right)^{1/2} [CHCl_3][Cl_2]^{1/2} = k[CHCl_3][Cl_2]^{1/2} \text{ (The overall order is 3/2.)}$$

14.96   *Enzyme*: carbonic anhydrase; *substrate*: carbonic acid (H_2CO_3); *turnover number*: 1×10^7 molecules/s.

Integrative Exercises

14.98   *Analyze/Plan.* $2N_2O_5 \rightarrow 4NO_2 + O_2$ rate = $k[N_2O_5] = 1.0 \times 10^{-5} \text{ s}^{-1} [N_2O_5]$

Use the integrated rate law for a first-order reaction, Equation [14.13], to calculate $k[N_2O_5]$ at 20.0 hr. Build a stoichiometry table to determine mol O_2 produced in 20.0 hr. Assuming that $O_2(g)$ is insoluble in chloroform, calculate the pressure of O_2 in the 10.0 L container. *Solve.*

$$20.0 \text{ hr} \times \frac{60 \text{ min}}{1 \text{ hr}} \times \frac{60 \text{ s}}{1 \text{ min}} \times 7.20 \times 10^4 \text{ s}; \; [N_2O_5]_0 = 0.600 \, M$$

$$\ln[A]_t - \ln[A]_0 = -kt; \; \ln[N_2O_5]_t = -kt + \ln[N_2O_5]_0$$

$$\ln[N_2O_5]_t = -1.0 \times 10^{-5} \text{ s}^{-1} (7.20 \times 10^4 \text{ s}) + \ln(0.600) = -0.720 - 0.511 = -1.231$$

$$[N_2O_5]_t = e^{-1.231} = 0.292 \, M$$

N_2O_5 was present initially as 1.00 L of 0.600 M solution.

mol $N_2O_5 = M \times L = 0.600$ mol N_2O_5 initial, 0.292 mol N_2O_5 at 20.0 hr

	$2N_2O_5$	$\rightarrow$	$4NO_2$	+	O_2
t = 0	0.600 mol		0		0
change	–0.308 mol		0.616 mol		0.154 mol
t = 20 hr	0.292 mol		0.616 mol		0.154 mol

[Note that the reaction stoichiometry is applied to the 'change' line.]

PV = nRT; P = nRT/V; V = 10.0 L, T = 45°C = 318 K, n = 0.154 mol

$$P = 0.154 \text{ mol} \times \frac{318 \text{ K}}{10.0 \text{ L}} \times \frac{0.08206 \text{ L} \bullet \text{atm}}{\text{mol} \bullet \text{K}} = 0.402 \text{ atm}$$

14.100   (a)   Use an apparatus such as the one pictured in Figure 10.3 (an open-end manometer), a clock, a ruler and a constant temperature bath. Since P = (n/V)RT, $\Delta P/\Delta t$ at constant temperature is an acceptable measure of reaction rate.

                Load the flask with HCl(aq) and read the height of the Hg in both arms of the manometer. Quickly add Zn(s) to the flask and record time = 0 when the Zn(s) contacts the acid. Record the height of the Hg in one arm of the manometer at convenient time intervals such as 5 sec. (The decrease in the short arm will be the same as the increase in the tall arm). Calculate the pressure of $H_2(g)$ at each time.

(b)   Keep the amount of Zn(s) constant and vary the concentration of HCl(aq) to determine the reaction order for H^+ and Cl^-. Keep the concentration of HCl(aq) constant and vary the amount of Zn(s) to determine the order for Zn(s). Combine this information to write the rate law.

(c) $-\Delta[H^+]/2\Delta t = \Delta[H_2]/\Delta t; \ -\Delta[H^+]/\Delta t = 2\Delta[H_2]/\Delta t$

     $[H_2]$ = mol H_2/L H_2 = n/V; $[H_2]$ = P (in atm)/RT

     Then, the rate of disappearance of H^+ is twice the rate of appearance of $H_2(g)$.

(d)   By changing the temperature of the constant temperature bath, measure the rate data at several (at least three) temperatures and calculate the rate constant k at these temperatures. Plot ln k vs 1/T. The slope of the line is $-E_a/R$ and E_a = –slope (R).

(e)   Measure rate data at constant temperature, HCl concentration and mass of Zn(s), varying only the form of the Zn(s). Compare the rate of reaction for metal strips and granules.

14.103   In the lock and key model of enzyme action, the active site is the specific location in the enzyme where reaction takes place. The precise geometry (size and shape) of the active site both accommodates and activates the substrate (reactant). Proteins are large biopolymers, with the same structural flexibility as synthetic polymers (Chapter 12). The three-dimensional shape of the protein in solution, including the geometry of the active site, is determined by many intermolecular forces of varying strengths.

Changes in temperature change the kinetic energy of the various groups on the enzyme and their tendency to form intermolecular associations or break free from them. Thus, changing the temperature changes the overall shape of the protein and specifically the shape of the active site. At the operating temperature of the enzyme, the competition between kinetic energy driving groups apart and intermolecular attraction pulling them together forms an active site that is optimum for a specific substrate. At temperatures above the temperature of maximum activity, sufficient kinetic energy has been imparted so that the forces driving groups apart win the competition, and the three-dimensional structure of the enzyme is destroyed. This is the process of *denaturation*. The activity of the enzyme is destroyed because the active site has collapsed. The protein or enzyme is denatured, because it is no longer capable of its "natural" activity.

15 Chemical Equilibrium

Visualizing Concepts

15.1   (a) $k_f > k_r$. According to the Arrhenius equation [14.19], $k = Ae^{-E_a/RT}$. As the magnitude of E_a increases, k decreases. On the energy profile, E_a is the difference in energy between the starting point and the energy at the top of the barrier. Clearly this difference is smaller for the forward reaction, so $k_f > k_r$.

          (b)   From the Equation [15.5], the equilibrium constant = k_f/k_r. Since $k_f > k_r$, the equilibrium constant for the process shown in the energy profile is greater than 1.

15.4   *Analyze/Plan.* Given that element A = red and element B = blue, evaluate the species in the reactant and product boxes, and write the reaction. Answer the remaining questions based on the balanced equation.   *Solve.*

          (a)   reactants: $4A_2 + 4B$; products: $4A_2B$

                balanced equation: $A_2 + B \rightarrow A_2B$

          (b) $K_c = \dfrac{[A_2B]}{[A_2][B]}$

          (c) $\Delta n = \Sigma n(prod) - \Sigma n(react) = 1 - 2 = -1$.

          (d) $K_p = K_c(RT)^{\Delta n}$, Equation [15.14].

                If you have a balanced equation, calculate Δn. Use Equation [15.14] to calculate K_p from K_c, or vice versa.

15.7   *Analyze.* Given the diagram and reaction type, calculate the equilibrium constant K_c.

        Plan. Analyze the contents of the cylinder. Express them as concentrations, using number of particles as a measure of moles, and V = 1 L. Write the equilibrium expression in terms of concentration and calculate K_c.   *Solve.*

          (a)   The mixture contains $2A_2$, 2B, 4AB. $[A_2] = 2$, $[B] = 2$, $[AB] = 4$.

$$K_c = \frac{[AB]^2}{[A_2][B]^2} = \frac{(4)^2}{(2)(2)^2} = 2$$

          (b)   A decrease in volume favors the reaction with fewer particles. This reaction has two particles in products and three in reactants, so a decrease in volume favors products. The number of AB (product) molecules will increase.

                Note that a change in volume does not change the value of K_c. If V decreases, the number of AB molecules must increase in order to maintain the equilibrium value of K_c.

15 Chemical Equilibrium   Solutions to Red Exercises

Equilibrium; the Equilibrium Constant

15.9   *Analyze/Plan.* Given the forward and reverse rate constants, calculate the equilibrium constant using Equation [15.5]. At equilibrium, the rates of the forward and reverse reactions are equal. Write the rate laws for the forward and reverse reactions and use their equality to answer part (b).   *Solve.*

(a) $K_c = \dfrac{k_f}{k_r}$, Equation [15.5]; $K_c = \dfrac{3.8 \times 10^{-2}\ s^{-1}}{3.1 \times 10^{-1}\ s^{-1}} = 0.12$

  For this reaction, $K_p = K_c = 0.12$.

(b) $\text{rate}_f = \text{rate}_r$; $k_f[A] = k_r[B]$

  Since $k_f < k_r$, in order for the two rates to be equal, $[A]$ must be greater than $[B]$.

15.11   (a)   The *law of mass action* expresses the relationship between the concentrations of reactants and products at equilibrium for any reaction. The law of mass action is a generic equilibrium expression.

$$K_c = \dfrac{[NOBr_2]}{[NO][Br_2]}$$

(b)   The *equilibrium-constant expression* is an algebraic equation where the variables are the equilibrium concentrations of the reactants and products for a specific chemical reaction. The *equilibrium constant* is a number; it is the ratio calculated from the equilibrium expression for a particular chemical reaction. For any reaction, there is an infinite number of sets of equilibrium concentrations, depending on initial concentrations, but there is only one equilibrium constant.

(c)   Introduce a known quantity of $NOBr_2(g)$ into a vessel of known volume at constant (known) temperature. After equilibrium has been established, measure the total pressure in the flask. Using an equilibrium table, such as the one in Sample Exercise 15.12, calculate equilibrium pressures and concentrations of $NO(g)$, $Br_2(g)$, and $NOBr_2(g)$ and calculate K_c.

15.13   *Analyze/Plan.* Follow the logic in Sample Exercises 15.1 and 15.5.   *Solve.*

(a) $K_c = \dfrac{[N_2O][NO_2]}{[NO]^3}$ (b) $K_c = \dfrac{[CS_2][H_2]^4}{[CH_4][H_2S]^2}$

(c) $K_c = \dfrac{[CO]^4}{[Ni(CO)_4]}$ (d) $K_c = \dfrac{[H^+][F^-]}{[HF]}$

(e) $K_c = \dfrac{[Ag^+]^2}{[Zn^{2+}]}$

homogeneous: (a), (b), (d); heterogeneous: (c), (e)

15.15   *Analyze.* Given the value of K_c, predict the contents of the equilibrium mixture.

Plan. If $K_c \gg 1$, products dominate; if $K_c \ll 1$, reactants dominate.   *Solve.*

(a)   mostly reactants ($K_c \ll 1$)

(b)   mostly products ($K_c \gg 1$)

15.17 *Analyze/Plan.* Follow the logic in Sample Exercise 15.2.   *Solve.*

$PCl_3(g) + Cl_2(g) \rightleftharpoons PCl_5(g)$, $K_c = 0.042$. $\Delta n = 1 - 2 = -1$

$K_p = K_c(RT)^{\Delta n} = 0.042(RT)^{-1} = 0.042/RT$

$K_p = \dfrac{0.042}{(0.08206)(500)} = 0.001024 = 1.0 \times 10^{-3}$

15.19 *Analyze.* Given K_c for a chemical reaction, calculate K_c for the reverse reaction.

Plan. The equilibrium expressions for the reaction and its reverse are the reciprocals of each other, and the values of K_c are also reciprocal. Evaluate which species are favored by examining the magnitude of K_c.   *Solve.*

(a) $K_c(\text{forward}) = \dfrac{[NOBr]^2}{[NO]^2[Br_2]} = 1.3 \times 10^{-2}$

       $K_c(\text{reverse}) = \dfrac{[NO]^2[Br_2]}{[NOBr]^2} = \dfrac{1}{1.3 \times 10^{-2}} = 76.92 = 77$

(b) $K_c < 1$ when NOBr is the product, and $K_c > 1$ when NOBr is the reactant. At this temperature, the equilibrium favors NO and Br_2.

15.21 *Analyze.* Given K_p for a reaction, calculate K_p for a related reaction.

Plan. The algebraic relationship between the K_p values is the same as the algebraic relationship between equilibrium expressions.

Solve. $K_p = \dfrac{P_{SO_3}}{P_{SO_2} \times P_{O_2}^{1/2}} = 1.85$

(a) $K_p = \dfrac{P_{SO_2} \times P_{O_2}^{1/2}}{P_{SO_3}} = \dfrac{1}{1.85} = 0.541$

(b) $K_p = \dfrac{P_{SO_3}^2}{P_{SO_2}^2 \times P_{O_2}} = (1.85)^2 = 3.4225 = 3.42$

(c) $K_p = K_c(RT)^{\Delta n}$; $\Delta n = 2 - 3 = -1$; $T = 1000\ K$

       $K_p = K_c(RT)^{-1} = K_c/RT$; $K_c = K_p(RT)$

       $K_c = 3.4225(0.08206)(1000) = 280.85 = 281$

15.23 *Analyze/Plan.* Follow the logic in Sample Exercise 15.5.   *Solve.*

$$A(aq) + B(aq) \rightleftharpoons C(aq) \qquad\qquad K_1 = 1.9 \times 10^{-4}$$
$$C(aq) + D(aq) \rightleftharpoons E(aq) + A(aq) \qquad K_2 = 8.5 \times 10^2$$
$$\overline{A(aq) + B(aq) + C(aq) + D(aq) \rightleftharpoons A(aq) + C(aq) + E(aq)}$$
$$B(aq) + D(aq) \rightleftharpoons E(aq) \qquad\qquad K_c = K_1 \times K_2 = 0.16$$

$K_c = (1.9 \times 10^{-4})(8.5 \times 10^2) = 0.162 = 0.16$

15.25   *Analyze/Plan.* Follow the logic in Sample Exercise 15.6.   *Solve.*

(a) $K_p = P_{O_2}$

(b)   The molar concentration, the ratio of moles of a substance to volume occupied by the substance, is a constant for pure solids and liquids.

Calculating Equilibrium Constants

15.27   *Analyze/Plan.* Follow the logic in Sample Exercise 15.8 using concentrations rather than pressures.   *Solve.*

$$K_c = \frac{[H_2][I_2]}{[HI]^2} = \frac{(4.79 \times 10^{-4})(4.79 \times 10^{-4})}{(3.53 \times 10^{-3})^2} = 0.018413 = 0.0184$$

15.29   *Analyze/Plan.* Follow the logic in Sample Exercise 15.8.   *Solve.*

$$2NO(g) + Cl_2(g) \rightleftharpoons 2NOCl(g)$$

$$K_p = \frac{P_{NOCl}^2}{P_{NO}^2 \times P_{Cl_2}} = \frac{(0.28)^2}{(0.095)^2(0.171)} = 50.80 = 51$$

15.31   *Analyze/Plan.* Follow the logic in Sample Exercise 15.9. Since the container volume is 1.0 L, mol = M.   *Solve.*

(a)   First calculate the change in [NO], 0.10 – 0.062 = 0.038 = 0.04 M. From the stoichiometry of the reaction, calculate the changes in the other pressures. Finally, calculate the equilibrium pressures.

	$2NO(g)$ +	$2H_2(g)$ $\rightleftharpoons$	$N_2(g)$ +	$2H_2O(g)$
initial	0.10 M	0.050 M	0 M	0.10 M
change	–0.038 M	–0.038 M	+0.019 M	+0.038 M
equil.	0.062 M	0.012 M	0.019 M	0.138 M

Strictly speaking, the change in [NO] has one decimal place and thus one sig fig. This limits equilibrium pressures to one decimal place for all but H_2O, and K_c to one sig fig. We compute the extra figures and then round.

(b) $K_c = \dfrac{[N_2][H_2O]^2}{[NO]^2[H_2]^2} = \dfrac{(0.019)(0.138)^2}{(0.062)^2(0.012)^2} = \dfrac{(0.02)(0.14)^2}{(0.06)^2(0.01)^2} = 653.7 = 7 \times 10^2$

15.33   *Analyze/Plan.* Follow the logic in Sample Exercise 15.9, using partial pressures, rather than concentrations.   *Solve.*

(a) $P = nRT/V$; $P_{CO_2} = 0.2000 \text{ mol} \times \dfrac{500 \text{ K}}{2.000 \text{L}} \times \dfrac{0.08206 \text{ L} \cdot \text{atm}}{\text{K} \cdot \text{mol}} = 4.1030 = 4.10 \text{ atm}$

$P_{H_2} = 0.1000 \text{ mol} \times \dfrac{500 \text{ K}}{2.000 \text{L}} \times \dfrac{0.08206 \text{ L} \cdot \text{atm}}{\text{K} \cdot \text{mol}} = 2.0515 = 2.05 \text{ atm}$

$P_{H_2O} = 0.1600 \times \dfrac{500 \text{ K}}{2.000 \text{L}} \times \dfrac{0.08206 \text{ L} \cdot \text{atm}}{\text{K} \cdot \text{mol}} = 3.2824 = 3.28 \text{ atm}$

(b)   The change in P_{H_2O} is 3.51 – 3.28 = 0.2276 = 0.23 atm. From the reaction stoichiometry, calculate the change in the other pressures and the equilibrium pressures.

	$CO_2(g)$ +	$H_2(g)$ ⇌	$CO(g)$ +	$H_2O(g)$
initial	4.10 atm	2.05 atm	0 atm	3.28 atm
change	–0.23 atm	–0.23 atm	+0.23	+0.23 atm
equil	3.87 atm	1.82 atm	0.23 atm	3.51 atm

(c) $K_p = \dfrac{P_{CO} \times P_{H_2O}}{P_{CO_2} \times P_{H_2}} = \dfrac{(0.23)(3.51)}{(3.87)(1.82)} = 0.1146 = 0.11$

Without intermediate rounding, equilibrium pressures are $P_{H_2O} = 3.51$, $P_{CO} = 0.2276$, $P_{H_2} = 1.8239$, $P_{CO_2} = 3.8754$ and $K_p = 0.1130 = 0.11$, in good agreement with the value above.

Applications of Equilibrium Constants

15.35   (a)   A reaction quotient is the result of the law of mass action for a general set of concentrations, whereas the equilibrium constant requires equilibrium concentrations.

(b)   In the direction of more products, to the right.

(c)   If $Q_c = K_c$, the system is at equilibrium; the concentrations used to calculate Q must be equilibrium concentrations.

15.37   *Analyze/Plan.* Follow the logic in Sample Exercise 15.10. We are given molarities, so we calculate Q directly and decide on the direction to equilibrium.   *Solve.*

$K_c = \dfrac{[CO][Cl_2]}{[COCl_2]} = 2.19 \times 10^{-10}$ at 100°C

(a) $Q = \dfrac{(3.3 \times 10^{-6})(6.62 \times 10^{-6})}{(2.00 \times 10^{-3})} = 1.1 \times 10^{-8}; Q > K$

The reaction will proceed left to attain equilibrium.

(b) $Q = \dfrac{(1.1 \times 10^{-7})(2.25 \times 10^{-6})}{(4.50 \times 10^{-2})} = 5.5 \times 10^{-12}; Q < K$

The reaction will proceed right to attain equilibrium.

(c) $Q = \dfrac{(1.48 \times 10^{-6})^2}{(0.0100)} = 2.19 \times 10^{-10}; Q = K$

The reaction is at equilibrium.

15.39   *Analyze/Plan.* Follow the logic in Sample Exercise 15.11. We are given concentrations, so write the K_c expression and solve for $[Cl_2]$. Change molarity to partial pressure using the ideal gas equation and the definition of molarity.   *Solve.*

$$K_c = \frac{[SO_2][Cl_2]}{[SO_2Cl_2]}; \quad [Cl_2] = \frac{K_c[SO_2Cl_2]}{[SO_2]} = \frac{(0.078)(0.108)}{0.052} = 0.16200 = 0.16 \; M$$

$$PV = nRT, \; P = \frac{n}{V}RT; \; \frac{n}{V} = M; \; P = M \; RT; \; T = 100°C + 273 = 373 \; K$$

$$P_{Cl_2} = \frac{0.16200 \; mol}{L} \times \frac{0.08206 \; L \bullet atm}{mol \bullet K} \times 373 \; K = 4.959 = 5.0 \; atm$$

Check. $K_c = \dfrac{(0.052)(0.162)}{(0.108)} = 0.078$. Our values are self-consistent.

15.41   *Analyze/Plan.* Follow the logic in Sample Exercise 15.11. In each case, change given masses to molarities solve for the equilibrium molarity of the desired component, and calculate mass of that substance present at equilibrium.   *Solve.*

(a) $K_c = \dfrac{[Br]^2}{[Br_2]} = 1.04 \times 10^{-3}$

$$[Br_2] = \frac{0.245 \; g \; Br_2}{0.200 \; L} \times \frac{1 \; mol \; Br_2}{159.8 \; g \; Br_2} = 0.007666 = 0.00767 \; M$$

$$[Br] = (K_c[Br_2])^{1/2} = [(1.04 \times 10^{-3})(0.007666)]^{1/2} = 0.002824 = 0.00282 \; M$$

$$\frac{0.002824 \; mol \; Br}{L} \times 0.200 \; L \times \frac{79.90 \; g \; Br}{mol} = 0.0451 \; g \; Br(g)$$

Check. $K_c = (0.002824)^2 / (0.007666) = 1.04 \times 10^{-3}$

(b) $K_c = \dfrac{[HI]^2}{[H_2][I_2]} = 55.3; \quad [HI] = (K_c[H_2][I_2])^{1/2}$

$$[H_2] = \frac{0.056 \; g \; H_2}{2.00 \; L} \times \frac{1 \; mol \; H_2}{2.016 \; g \; H_2} = 0.01389 = 0.014 \; M$$

$$[I_2] = \frac{4.36 \; g \; I_2}{2.00 \; L} \times \frac{1 \; mol \; I_2}{253.8 \; g I_2} = 0.008589 = 0.00859 \; M$$

$$[HI] = [(55.3)(0.01389)(0.008589)]^{1/2} = 0.08122 = 0.081 \; M$$

$$0.08122 \; mol \; HI \times 2.00 \; L \times \frac{127.9 \; g \; HI}{mol \; HI} = 20.78 = 21 \; g \; HI$$

Check. $K_c = \dfrac{(0.08122)^2}{(0.01389)(0.008589)} = 55.3$

15.43   *Analyze/Plan.* Follow the logic in Sample Exercise 15.12. Since molarity of NO is given directly, we can construct the equilibrium table straight away.   *Solve.*

	2NO(g)	$\rightleftharpoons$	N₂(g)	+	O₂(g)	$K_c = \dfrac{[N_2][O_2]}{[NO]^2} = 2.4 \times 10^3$
initial	0.200 *M*		0		0	
change	–2x		+x		+x	
equil.	0.200 – 2x		+x		+x	

$$2.4 \times 10^3 = \frac{x^2}{(0.200 - 2x)^2} \, ; (2.4 \times 10^3)^{1/2} = \frac{x}{0.200 - 2x}$$

$x = (2.4 \times 10^3)^{1/2} (0.200 - 2x); x = 9.798 - 97.98x; 98.98x = 9.798, x = 0.09899 = 0.099 \, M$

$[N_2] = [O_2] = 0.099 \, M; [NO] = 0.200 - 2(0.09899) = 0.00202 = 0.002 \, M$

Check. $K_c = (0.09899)^2 / (0.00202)^2 = 2.4 \times 10^3$

15.45   *Analyze/Plan.* Write the K_p expression, substitute the stated pressure relationship, and solve for P_{Br_2} .   *Solve.*

$$K_p = \frac{P_{NO}^2 \times P_{Br_2}}{P_{NOBr}^2}$$

When $P_{NOBr} = P_{NO}$, these terms cancel and $P_{Br_2} = K_p = 0.416$ atm.   This is true for all cases where $P_{NOBr} = P_{NO}$.

15.47   (a) $CaSO_4(s) \rightleftharpoons Ca^{2+}(aq) + SO_4^{2-}(aq)$ $K_c = [Ca^{2+}][SO_4^{2-}] = 2.4 \times 10^{-5}$

At equilibrium, $[Ca^{2+}] = [SO_4^{2-}] = x$

$K_c = 2.4 \times 10^{-5} = x^2; x = 4.9 \times 10^{-3} \, M \, Ca^{2+}$ and SO_4^{2-}

(b)   A saturated solution of $CaSO_4(aq)$ is $4.9 \times 10^{-3} \, M$.

3.0 L of this solution contain:

$$\frac{4.9 \times 10^{-3} \, mol}{L} \times 3.0 \, L \times \frac{136.14 \, g \, CaSO_4}{mol} = 2.001 = 2.0 \, g \, CaSO_4$$

A bit more than 2.0 g $CaSO_4$ is needed in order to have some undissolved $CaSO_4(s)$ in equilibrium with 3.0 L of saturated solution.

15.49   *Analyze/Plan.* Follow the approach in Solution 15.43. Calculate [IBr] from mol IBr and construct the equilibrium table.   *Solve.*

[IBr] = 0.500 mol/1.00 L = 0.500 M

Since no I_2 or Br_2 was present initially, the amounts present at equilibrium are produced by the reverse reaction and stoichiometrically equal. Let these amounts equal x. The amount of HBr that reacts is then 2x. Substitute the equilibrium molarities (in terms of x) into the equilibrium expression and solve for x.

	I_2	+	Br_2	$\rightleftharpoons$	2IBr	

$$K_c = \frac{[IBr]^2}{[I_2 Br_2]} = 280$$

	I_2	Br_2	2IBr
initial	0 M	0 M	0.500 M
change	+x M	+x M	–2x M
equil.	x M	x M	(0.500 – 2x) M

$K_c = 280 = \dfrac{(0.500 - 2x)^2}{x^2}$; taking the square root of both sides

$$16.733 = \frac{0.500 - 2x}{x}; \; 16.733x + 2x = 0.500; \;\; 18.733x = 0.500$$

$x = 0.02669 = 0.0267 \, M; \; [I_2] = [Br_2] = 0.0267 \, M$

$[IBr] = 0.500 - 2(0.02669) = 0.4466 = 0.447 \, M$

Check. $\dfrac{(0.447)^2}{(0.0267)^2} = 280.$ Our values are self-consistent.

LeChâtelier's Principle

15.51   *Analyze/Plan.* Follow the logic in Sample Exercise 15.13.   *Solve.*

(a)   Shift equilibrium to the right; more $SO_3(g)$ is formed, the amount of $SO_2(g)$ decreases.

(b)   Heating an exothermic reaction decreases the value of K. More SO_2 and O_2 will form, the amount of SO_3 will decrease.

(c)   Since, $\Delta n = -1$, a change in volume will affect the equilibrium position and favor the side with more moles of gas. The amounts of SO_2 and O_2 increase and the amount of SO_3 decreases; equilibrium shifts to the left.

(d)   No effect. Speeds up the forward and reverse reactions equally.

(e)   No effect. Does not appear in the equilibrium expression.

(f)   Shift equilibrium to the right; amounts of SO_2 and O_2 decrease.

15.53   *Analyze/Plan.* Given certain changes to a reaction system, determine the effect on K_p, if any. Only changes in temperature cause changes to the value of K_p.   *Solve.*

(a)   no effect   (b)   no effect   (c)   increase equilibrium constant   (d)   no effect

15.55   *Analyze/Plan.* Use Hess's Law, $\Delta H^\circ = \Sigma \Delta H_f^\circ$ products $- \Sigma \Delta H_f^\circ$ reactants, to calculate ΔH°. According to the sign of ΔH°, describe the effect of temperature on the value of K. According to the value of Δn, describe the effect of changes to container volume. *Solve.*

(a) $\Delta H^\circ = \Delta H_f^\circ \, NO_2(g) + \Delta H_f^\circ \, N_2O(g) - 3\Delta H_f^\circ \, NO(g)$

      $\Delta H^\circ = 33.84 \, kJ + 81.6 \, kJ - 3(90.37 \, kJ) = -155.7 \, kJ$

(b)   The reaction is exothermic because it has a negative value of ΔH°. The equilibrium constant will decrease with increasing temperature.

(c) Δn does not equal zero, so a change in volume at constant temperature will affect the fraction of products in the equilibrium mixture. An increase in container volume would favor reactants, while a decrease in volume would favor products.

Additional Exercises

15.57   (a)   Since both the forward and reverse processes are elementary steps, we can write the rate laws directly from the chemical equation.

      $\text{rate}_f = k_f \, [CO][Cl_2] = \text{rate}_r = k_r \, [COCl][Cl]$

$$\frac{k_f}{k_r} = \frac{[COCl][Cl]}{[CO][Cl_2]} = K$$

$$K_c = \frac{k_f}{k_r} = \frac{1.4 \times 10^{-28}\ M^{-1}\ s^{-1}}{9.3 \times 10^{10}\ M^{-1}\ s^{-1}} = 1.5 \times 10^{-39}$$

For a homogeneous equilibrium in the gas phase, we usually write K in terms of partial pressures. In this exercise, concentrations are more convenient because the rate constants are expressed in terms of molarity. For this reaction, the value of K is the same regardless of how it is expressed, because there is no change in the moles of gas in going from reactants to products.

(b) Since the K is quite small, reactants are much more plentiful than products at equilibrium.

15.59 $[SO_2Cl_2] = \dfrac{2.00\ mol}{2.00\ L} = 1.00\ M$

The change in $[SOCl_2] = 0.56(1.00\ M) = 0.56\ M$

	$SO_2Cl_2(g)$	$\rightleftharpoons$	$SO_2(g)$ +	$Cl_2(g)$	$K_c = \dfrac{[SO_2][Cl_2]}{[SO_2Cl_2]}$
initial	1.00 M		0	0	
change	−0.56 M		+0.56 M	+0.56 M	
equil.	0.44 M		+0.56 M	+0.56 M	

$$K_c = \frac{(0.56)^2}{0.44} = 0.7127 = 0.71$$

15.62   (a)

	A(g)	$\rightleftharpoons$	2B(g)
initial	0.55 atm		0
change	−0.19 atm		+0.38 atm
equil.	0.36 atm		0.38 atm

$P_t = P_A + P_B = 0.36\ atm + 0.38\ atm = 0.74\ atm$

(b) $K_p = \dfrac{(P_B)^2}{P_A} = \dfrac{(0.38)^2}{0.36} = 0.4011 = 0.40$

15.65   (a) $K_p = 0.052$; $K_p = K_c(RT)^{\Delta n}$; $\Delta n = 2 - 0 = 2$; $K_c = K_p/(RT)^2$

$K_c = 0.052/[0.08206)(333)]^2 = 6.964 \times 10^{-5} = 7.0 \times 10^{-5}$

(b) PH_3BCl_3 is a solid and its concentration is taken as a constant, C.

$[BCl_3] = 0.0128\ mol/0.500\ L = 0.0256\ M$

$$PH_3BCl_3 \rightleftharpoons PH_3 + BCl_3$$

initial	C	$0\,M$	$0.0256\,M$
change		$+x\,M$	$+x\,M$
equil.	C	$+x\,M$	$(0.0256 + x)\,M$

$K_c = [PH_3][BCl_3];\ 6.964 \times 10^{-5} = x(0.0256 + x);\ x^2 + 0.0256x - 6.964 \times 10^{-5} = 0$

$x = \dfrac{-0.0256 \pm [(0.0256)^2 - 4(1)(-6.964 \times 10^{-5})]^{1/2}}{2(1)} = 0.002480 = 2 \times 10^{-3}\,M\ PH_3$

Check. The numerator in the quadratic has 1 sig fig, which leads to $[PH_3]$ with 1 sig fig. $K_c = (2 \times 10^{-3})(0.0256 + 2 \times 10^{-3}) = 6 \times 10^{-5}$. Using 2 or 3 figures for $[PH_3]$ leads to closer agreement.

15.68   In general, the reaction quotient is of the form $Q = \dfrac{P_{NOCl}^2}{P_{NO}^2 \times P_{Cl_2}}$.

(a) $Q = \dfrac{(0.11)^2}{(0.15)^2\,(0.31)} = 1.7$

$Q > K_p$. Therefore, the reaction will shift toward reactants, to the left, in moving toward equilibrium.

(b) $Q = \dfrac{(0.050)^2}{(0.12)^2\,(0.10)} = 1.7$

$Q > K_p$. Therefore, the reaction will shift toward reactants, to the left, in moving toward equilibrium.

(c) $Q = \dfrac{(5.10 \times 10^{-3})^2}{(0.15)^2\,(0.20)} = 5.8 \times 10^{-3}$

$Q < K_p$. Therefore, the reaction mixture will shift in the direction of more product, to the right, in moving toward equilibrium.

15.71 $K_p = \dfrac{P_{CO_2}}{P_{CO}} = 6.0 \times 10^2$

If P_{CO} is 150 torr, P_{CO_2} can never exceed $760 - 150 = 610$ torr. Then $Q = 610/150 = 4.1$. Since this is far less than K, the reaction will shift in the direction of more product. Reduction will therefore occur.

15.74   *Analyze/Plan.* Equilibrium pressures of H_2, I_2, HI $\rightarrow K_p \rightarrow$ equilibrium table $\rightarrow$ new equilibrium pressures.   *Solve.*

$P_{H_2} = P_{I_2} = 0.112\ \text{mol} \times 11.997\ \dfrac{\text{atm}}{\text{mol}} = 1.344 = 1.34\ \text{atm}$

$P_{HI} = 0.775\ \text{mol} \times 11.997\ \dfrac{\text{atm}}{\text{mol}} = 9.298 = 9.30\ \text{atm}$

$$H_2(g) + I_2(g) \rightleftharpoons 2HI(g); \quad K_p = \frac{P_{HI}^2}{P_{H_2} \times P_{I_2}} = \frac{(9.298)^2}{(1.344)^2} = 47.861 = 47.9$$

$$P_{HI} \text{ (added)} = 0.100 \text{ mol} \times \frac{11.997 \text{ atm}}{\text{mol}} = 1.1997 = 1.20 \text{ atm}$$

	$H_2(g)$	+	$I_2(g)$	$\rightleftharpoons$	$2HI(g)$
initial	1.34 atm		1.34 atm		9.30 atm + 1.20 atm
change	+x atm		+x atm		–2x atm
equil.	(1.34+x) atm		(1.34+x) atm		(10.50–2x) atm

$$K_p = 47.86 = \frac{(10.50 - 2x)^2}{(1.34 + x)^2}. \quad \text{Take the square root of both sides :}$$

$$6.918 = \frac{10.50 - 2x}{1.34 + x} \, ; \, 9.270 + 6.918\,x = 10.50 - 2x; \, 8.918\,x = 1.230; \, x = 0.1379 = 0.138$$

$$P_{H_2} = P_{I_2} = 1.34 + 0.138 = 1.48 \text{ atm}; \, P_{HI} = 10.50 - 2(0.138) = 10.22 \text{ atm}$$

15.77 The patent claim is false. A catalyst does not alter the position of equilibrium in a system, only the rate of approach to the equilibrium condition.

Integrative Exercises

15.78 (a)   (i) $K_c = [Na^+]/[Ag^+]$

            (ii) $K_c = [Hg^{2+}]^3 / [Al^{3+}]^2$

            (iii) $K_c = [Zn^{2+}][H_2] / [H^+]^2$

      (b) According to Table 4.5, the activity series of the metals, a metal can be oxidized by any metal cation below it on the table.

            (i) Ag^+ is far below Na, so the reaction will proceed to the right and K_c will be large.

            (ii) Al^{3+} is above Hg, so the reaction will not proceed to the right and K_c will be small.

            (iii) H^+ is below Zn, so the reaction will proceed to the right and K_c will be large.

      (c) $K_c < 1$ for this reaction, so Fe^{2+} (and thus Fe) is above Cd on the table. In other words, Cd is below Fe. The value of K_c, 0.06, is small but not extremely small, so Cd will be only a few rows below Fe.

15.80 (a)   At equilibrium, the forward and reverse reactions occur at **equal** rates.

      (b) One expects the reactants to be favored at equilibrium since they are lower in energy.

(c)   A catalyst lowers the activation energy for both the forward and reverse reactions; the "hill" would be lower.

(d)   Since the activation energy is lowered for both processes, the new rates would be equal and the ratio of the rate constants, k_f/k_r, would remain unchanged.

(e)   Since the reaction is endothermic (the energy of the reactants is lower than that of the products, ΔE is positive), the value of K should increase with increasing temperature.

15.83   Mole % = pressure %. Since the total pressure is 1 atm, mol %/100 = mol fraction = partial pressure. $K_p = P_{CO}^2 / P_{CO_2}$.

Temp (K)	P_{CO_2} (atm)	P_{CO} (atm)	K_P
1123	0.0623	0.9377	14.1
1223	0.0132	0.9868	73.8
1323	0.0037	0.9963	2.7×10^2
1473	0.0006	0.9994	1.7×10^3 (2×10^3)

Because K grows larger with increasing temperature, the reaction must be endothermic in the forward direction.

16 Acid-Base Equilibria

Visualizing Concepts

16.1   *Analyze*. From the structures decide which reactant fits the description of a Brønsted-Lowry (B-L) acid, a B-L base, a Lewis acid, and a Lewis base. *Plan*. A B-L acid is an H^+ donor, and a B-L base is an H^+ acceptor. A Lewis acid is an electron pair acceptor and a Lewis base is an electron pair donor.   *Solve*.

(a)   H–X is a B-L acid, because it loses H^+ during reaction. NH_3 is a B-L base, because it gains H^+ during reaction.

(b)   By virtue of its unshared electron pair, NH_3 is the electron pair donor and Lewis base. HX is the electron pair acceptor and Lewis acid.

16.5   *Plan*. The stronger the acid, the greater the extent of ionization. The stronger the acid, the weaker its conjugate base. *Solve*.

(a)   HY has most H^+ and is strongest; HX has the fewest H^+ and is weakest. The order of base strength is the reverse of the order of acid strength. $Y^- < Z^- < X^-$

(b)   The strongest base, X^-, has the largest K_b value.

16.8   *Plan*. Evaluate the molecular structures to determine if the acids are binary acids or oxyacids. Consider the trends in acid strength for both classes of acids.   *Solve*.

(a)   The molecules are oxyacids; in both cases, the ionizable H atom is attached to O. The right molecule is a carboxylic acid; the ionizable H is part of a carboxyl,

$$-\overset{\overset{\text{O}}{\|}}{\text{C}}-\text{O}-\text{H} \, , \text{ group.}$$

(b)   Increasing the electronegativity of X increases the strength of both acids. As X becomes more electronegative and attracts more electron density, the O–H bond becomes weaker and more polar. This increases the likelihood of ionization and increases acid strength. An electronegative X group also stabilizes the anionic conjugate bases by delocalizing the negative charge. This causes the ionization equilibrium to favor products, and the values of K_a to increase.

Arrhenius and Brønsted-Lowry Acids and Bases

16.11   Solutions of HCl and H_2SO_4 taste sour, turn litmus paper red (are acidic), neutralize solutions of bases, react with active metals to form $H_2(g)$ and conduct electricity. The two solutions have these properties in common because both solutes are strong acids.

That is, they both ionize completely in H_2O to form $H^+(aq)$ and an anion. (The first ionization step for H_2SO_4 is complete, but the second is not.) The presence of ions enables the solutions to conduct electricity; the presence of $H^+(aq)$ in excess of $1 \times 10^{-7} M$ accounts for all the other listed properties.

16.13   (a)   According to the Arrhenius definition, an *acid* when dissolved in water increases $[H^+]$. According to the Brønsted-Lowry definition, an *acid* is capable of donating H^+, regardless of physical state. The Arrhenius definition of an acid is confined to an aqueous solution; the Brønsted-Lowry definition applies to any physical state.

(b) $HCl(g) + NH_3(g) \rightarrow NH_4^+Cl^-(s)$ HCl is the B-L (Brønsted-Lowry) acid; it donates an H^+ to NH_3 to form NH_4^+. NH_3 is the B-L base; it accepts the H^+ from HCl.

16.15   *Analyze/Plan.* Follow the logic in Sample Exercise 16.1. A conjugate base has one less H^+ than its conjugate acid.   *Solve.*

(a) IO_3^- (b) NH_3 (c) HPO_4^{2-} (d) $C_7H_5O_2^-$

16.17   *Analyze/Plan.* Use the definitions of B-L acids and bases, and conjugate acids and bases to make the designations. Evaluate the changes going from reactant to product to inform your choices.   *Solve.*

16.19   *Analyze/Plan.* Follow the logic in Sample Exercise 16.2.   *Solve.*

(a)   Acid: $HC_2O_4^-(aq) + H_2O(l) \rightleftharpoons C_2O_4^{2-}(aq) + H_3O^+(aq)$

                B-L acid   B-L base   conj. base   conj. acid

         Base: $HC_2O_4^-(aq) + H_2O(l) \rightleftharpoons H_2C_2O_4(aq) + OH^-(aq)$

                B-L base   B-L acid   conj. acid   conj. Base

(b) $H_2C_2O_4$ is the conjugate acid of $HC_2O_4^-$.

         $C_2O_4^{2-}$ is the conjugate base of $HC_2O_4^-$.

16.21   *Analyze/Plan.* Based on the chemical formula, decide whether the acid is strong, weak, or negligible. Is it one of the known seven strong acids (Section 16.5)? Also check Figure 16.4. Remove a single H and decrease the particle charge by one to write the formula of the conjugate base.   *Solve.*

(a) HNO_2, weak acid; NO_2^-, weak base

(b) H_2SO_4, strong acid; HSO_4^-, negligible base

(c) HPO_4^{2-}, weak acid; PO_4^{3-}, weak base

(d) CH_4, negligible acid; CH_3^-, strong base

(e) $CH_3NH_3^+$, weak acid; CH_3NH_2, weak base

16.23   *Analyze/Plan.* Given chemical formula, determine strength of acids and bases by checking the known strong acids (Section 16.5). Recall the paradigm "The stronger the acid, the weaker its conjugate base, and vice versa."   *Solve.*

(a)   HBr. It is one of the seven strong acids (Section 16.5).

(b) F^-. HCl is a stronger acid than HF, so F^- is the stronger conjugate base.

16.25   *Analyze/Plan.* Acid-base equilibria favor formation of the weaker acid and base. Compare the relative strengths of the substances acting as acids on opposite sides of the equation. (Bases can also be compared; the conclusion should be the same.)   *Solve.*

Base	+	Acid	⇌	Conjugate acid	+	Conjugate base

(a) $O^{2-}(aq)$ + $H_2O(l)$ ⇌ $OH^-(aq)$ + $OH^-(aq)$

H_2O is a stronger acid than OH^-, so the equilibrium lies to the right.

(b) $HS^-(aq)$ + $HC_2H_3O_2(aq)$ ⇌ $H_2S(aq)$ + $C_2H_3O_2^-(aq)$

$HC_2H_3O_2$ is a stronger acid than H_2S, so the equilibrium lies to the right.

(c) $NO_3^-(aq)$ + $H_2O(l)$ ⇌ $HNO_3(aq)$ + $OH^-(aq)$

HNO_3 is a stronger acid than H_2O (Solution 16.24), so the equilibrium lies to the left.

Autoionization of Water

16.27   (a)   *Autoionization* is the ionization of a neutral molecule (in the absence of any other reactant) into an anion and a cation. The equilibrium expression for the autoionization of water is $H_2O(l) \rightleftharpoons H^+(aq) + OH^-(aq)$.

(b)   Pure water is a poor conductor of electricity because it contains very few ions. Ions, mobile charged particles, are required for the conduction of electricity in liquids.

(c)   If a solution is acidic, it contains more H^+ than OH^- ($[H^+] > [OH^-]$).

16.29   *Analyze/Plan.* Follow the logic in Sample Exercise 16.5. In pure water at 25°C, $[H^+] = [OH^-] = 1 \times 10^{-7} M$. If $[H^+] > 1 \times 10^{-7} M$, the solution is acidic; if $[H^+] < 1 \times 10^{-7} M$, the solution is basic.   *Solve.*

(a) $[H^+] = \dfrac{K_w}{[OH^-]} = \dfrac{1.0 \times 10^{-14}}{4.5 \times 10^{-4} M} = 2.2 \times 10^{-11} M < 1 \times 10^{-7} M$; basic

(b) $[H^+] = \dfrac{K_w}{[OH^-]} = \dfrac{1.0 \times 10^{-14}}{8.8 \times 10^{-9} M} = 1.1 \times 10^{-6} M > 1 \times 10^{-7} M$; acidic

(c) $[OH^-] = 100[H^+]$; $K_w = [H^+] \times 100[H^+] = 100[H^+]^2$;

$[H^+] = (K_w/100)^{1/2} = 1.0 \times 10^{-8} M < 1 \times 10^{-7} M$; basic

16.31   *Analyze/Plan.* Follow the logic in Sample Exercise 16.4. Note that the value of the equilibrium constant (in this case, K_w) changes with temperature.   *Solve.*

At 0°C, $K_w = 1.2 \times 10^{-15} = [H^+][OH^-]$.

In pure water, $[H^+] = [OH^-]$; $2.4 \times 10^{-14} = [H^+]^2$; $[H^+] = (1.2 \times 10^{-15})^{1/2}$

$[H^+] = [OH^-] = 3.5 \times 10^{-9} M$

The pH Scale

16.33   *Analyze/Plan.* A change of one pH unit (in either direction) is:

$$\Delta pH = pH_2 - pH_1 = -(\log[H^+]_2 - \log[H^+]_1) = -\log\frac{[H^+]_2}{[H^+]_1} = \pm 1.$$ The antilog of +1 is 10;

the antilog of −1 is 1×10^{-1}. Thus, a ΔpH of one unit represents an increase or decrease in $[H^+]$ by a factor of 10.   *Solve.*

    (a) $\Delta pH = \pm 2.00$ is a change of $10^{2.00}$; $[H^+]$ changes by a factor of 100.

    (b) $\Delta pH = \pm 0.5$ is a change of $10^{0.50}$; $[H^+]$ changes by a factor of 3.2.

16.35   (a) $K_w = [H^+][OH^-]$. If NaOH is added to water, it dissociates into $Na^+(aq)$ and OH^- (aq). This increases $[OH^-]$ and necessarily decreases $[H^+]$. When $[H^+]$ decreases, pH increases.

    (b) $0.0006\ M = 6 \times 10^{-4}\ M$. On Figure 16.5, this is $[H^+] > 1 \times 10^{-4}$ but $< 1 \times 10^{-3}$. The pH is between 3 and 4, closer to 3. We estimate 3.3. If pH < 7, the solution is acidic.

    (c)   pH = 5.2 is between pH 5 and pH 6 on Figure 16.5, closer to pH = 5. At pH = 6, $[H^+] = 1 \times 10^{-6}$; at pH = 5, $[H^+] = 1 \times 10^{-5} = 10 \times 10^{-6}$. A good estimate is $7 \times 10^{-6}\ M\ H^+$.

By calculation: $[H^+] = 10^{-pOH} = 10^{-5.2} = 6 \times 10^{-6}\ M$

At pH = 5, $[OH^-] = 1 \times 10^{-9}$; at pH = 6, $[OH^-] = 1 \times 10^{-8} = 10 \times 10^{-9}$.

Since pH = 5.2 is closer to pH = 5, we estimate $3 \times 10^{-9}\ M\ OH^-$.

By calculation: pOH = 14.0 − 5.2 = 8.8

$[OH^-] = 10^{-pOH} = 10^{-8.8} = 2 \times 10^{-9}\ M\ OH^-$

16.37   *Analyze/Plan.* At 25°C, $[H^+][OH^-] = 1 \times 10^{-14}$; pH = pOH = 14. Use these relationships to complete the table. If pH < 7, the solution is acidic; if pH > 7, the solution is basic. *Solve.*

$[H^+]$	$[OH^-]$	pH	pOH	acidic or basic
$7.5 \times 10^{-3}\ M$	$1.3 \times 10^{-12}\ M$	2.12	11.88	acidic
$2.8 \times 10^{-5}\ M$	$3.6 \times 10^{-10}\ M$	4.56	9.44	acidic
$5.6 \times 10^{-9}\ M$	$1.8 \times 10^{-6}\ M$	8.25	5.75	basic
$5.0 \times 10^{-9}\ M$	$2.0 \times 10^{-6}\ M$	8.30	5.70	basic

16.39   *Analyze/Plan.* Given pH and a new value of the equilibrium constant K_w, calculate equilibrium concentrations of $H^+(aq)$ and $OH^-(aq)$. The definition of pH remains $pH = -\log[H^+]$.   *Solve.*

pH = 7.40; $[H^+] = 10^{-pH} = 10^{-7.40} = 4.0 \times 10^{-8}\ M$

$K_w = 2.4 \times 10^{-14} = [H^+][OH^-]$; $[OH^-] = 2.4 \times 10^{-14} / [H^+]$

$[OH^-] = 2.4 \times 10^{-14} / 4.0 \times 10^{-8} = 6.0 \times 10^{-7} M$, $pOH = -\log(6.0 \times 10^{-7}) = 6.22$

Alternately, $pH + pOH = pK_w$. At 37°C, $pH + pOH = -\log(2.4 \times 10^{-14})$

$pH + pOH = 13.62$; $pOH = 13.62 - 7.40 = 6.22$

$[OH^-] = 10^{-pOH} = 10^{-6.22} = 6.0 \times 10^{-7} M$

Strong Acids and Bases

16.41　(a)　A strong acid is completely ionized in aqueous solution; a strong acid is a strong electrolyte.

　　　　(b)　For a strong acid such as HCl, $[H^+]$ = initial acid concentration. $[H^+] = 0.500 M$

　　　　(c)　HCl, HBr, HI

16.43　*Analyze/Plan.* Follow the logic in Sample Exercise 16.8. Strong acids are completely ionized, so $[H^+]$ = original acid concentration, and $pH = -\log[H^+]$. For the solutions obtained by dilution, use the "dilution" formula, $M_1V_1 = M_2V_2$, to calculate molarity of the acid.　*Solve.*

　　　　(a)　$8.5 \times 10^{-3} M$ HBr $= 8.5 \times 10^{-3} M$ H$^+$; $pH = -\log(8.5 \times 10^{-3}) = 2.07$

　　　　(b)　$\dfrac{1.52 \text{ g HNO}_3}{0.575 \text{ L soln}} \times \dfrac{1 \text{ mol HNO}_3}{63.02 \text{ g HNO}_3} = 0.041947 = 0.0419 M \text{ HNO}_3$

　　　　　　$[H^+] = 0.0419 M$; $pH = -\log(0.041947) = 1.377$

　　　　(c)　$M_c \times V_c = M_d \times V_d$; $0.250 M \times 0.00500 \text{ L} = ? M \times 0.0500 \text{ L}$

　　　　　　$M_d = \dfrac{0.250 M \times 0.00500 \text{ L}}{0.0500 \text{ L}} = 0.0250 M \text{ HCl}$

　　　　　　$[H^+] = 0.0250 M$; $pH = -\log(0.0250) = 1.602$

　　　　(d)　$[H^+]_{total} = \dfrac{\text{mol H}^+ \text{from HBr} + \text{mol H}^+ \text{from HCl}}{\text{total L solution}}$

　　　　　　$[H^+]_{total} = \dfrac{(0.100 M \text{ HBr} \times 0.0100 \text{ L}) + (0.200 M \times 0.0200 \text{ L})}{0.0300 \text{ L}}$

　　　　　　$[H^+]_{total} = \dfrac{1.00 \times 10^{-3} \text{ mol H}^+ + 4.00 \times 10^{-3} \text{ mol H}^+}{0.0300 \text{ L}} = 0.1667 = 0.167 M$

　　　　　　$pH = -\log(0.1667 M) = 0.778$

16.45　*Analyze/Plan.* Follow the logic in Sample Exercise 16.9. Strong bases dissociate completely upon dissolving. $pOH = -\log[OH^-]$; $pH = 14 - pOH$.

　　　　(a)　Pay attention to the formula of the base to get $[OH^-]$.　*Solve.*

　　　　　　$[OH^-] = 2[Sr(OH)_2] = 2(1.5 \times 10^{-3} M) = 3.0 \times 10^{-3} M \text{ OH}^-$ (see Exercise 16.42(b))

　　　　　　$pOH = -\log(3.0 \times 10^{-3}) = 2.52$; $pH = 14 - pOH = 11.48$

　　　　(b)　mol/LiOH = g LiOH/molar mass LiOH. $[OH^-] = [LiOH]$.　*Solve.*

$$\frac{2.250 \text{ g LiOH}}{0.2500 \text{ L soln}} \times \frac{1 \text{ mol LiOH}}{23.948 \text{ g LiOH}} = 0.37581 = 0.3758 \text{ M LiOH} = [OH^-]$$

$$pOH = -\log(0.37581) = 0.4250; \ pH = 14 - pOH = 13.5750$$

(c)   Use the dilution formula to get the [NaOH] = [OH$^-$].   *Solve.*

$$M_c \times V_c = M_d \times V_d; \ 0.175 \ M \times 0.00100 \text{ L} = ? \ M \times 2.00 \text{ L}$$

$$M_d = \frac{0.0175 \ M \times 0.00100 \text{ L}}{2.00 \text{ L}} = 8.75 \times 10^{-5} \ M \text{ NaOH} = [OH^-]$$

$$pOH = -\log(8.75 \times 10^{-5}) = 4.058; \ pH = 14 - pOH = 9.942$$

(d)   Consider total mol OH$^-$ from KOH and Ca(OH)$_2$, as well as total solution volume. *Solve.*

$$[OH^-]_{total} = \frac{\text{mol OH}^- \text{ from KOH} + \text{mol OH}^- \text{ from Ca(OH)}_2}{\text{total L soln}}$$

$$[OH^-]_{total} = \frac{(0.105 \ M \times 0.00500 \text{ L}) + 2(9.5 \times 10^{-2} \times 0.0150 \text{ L})}{0.0200 \text{ L}}$$

$$[OH^-]_{total} = \frac{0.525 \times 10^{-3} \text{ mol OH}^- + 2.85 \times 10^{-3} \text{ mol OH}^-}{0.0200 \text{ L}} = 0.16875 = 0.17 \ M$$

$$pOH = -\log(0.16875) = 0.77; \ pH = 14 - pOH = 13.23$$

($9.5 \times 10^{-2} \ M$ has 2 sig figs, so the [OH$^-$] has 2 sig figs and pH and pOH have 2 decimal places.)

16.47   *Analyze/Plan.* pH $\rightarrow$ pOH $\rightarrow$ [OH$^-$] = [NaOH].   *Solve.*

pOH = 14 − pH = 14.00 − 11.50 = 2.50

$$pOH = 2.50 = -\log[OH^-]; \ [OH^-] = 10^{-2.50} = 3.2 \times 10^{-3} \ M$$

$$[OH^-] = [NaOH] = 3.2 \times 10^{-3} \ M$$

16.49   *Analyze/Plan.* NaH(aq) $\rightarrow$ Na$^+$(aq) + H$^-$(aq)

H$^-$(aq) + H$_2$O(l) $\rightarrow$ H$_2$(g) + OH$^-$(aq)

Thus, initial [NaH] = [OH$^-$]; [NaH] = g NaH/[M(NaH) × V].   *Solve.*

$$[NaH] = \frac{\text{mol NaH}}{\text{L solution}} = 15.00 \text{ g NaH} \times \frac{1 \text{ mol NaH}}{24.00 \text{ g NaH}} \times \frac{1}{2.50 \text{ L}} = 0.250 \ M$$

$$[OH^-] = 0.250 \ M; \ pOH = -\log(0.250) = 0.602, \ pH = 14 - pOH = 13.400$$

Weak Acids

16.51   *Analyze/Plan.* Remember that K_{eq} = [products]/[reactants]. If H$_2$O(l) appears in the equilibrium reaction, it will **not** appear in the K$_a$ expression, because it is a pure liquid. *Solve.*

(a) $HBrO_2(aq) \rightleftharpoons H^+(aq) + BrO_2^-(aq); \quad K_a = \dfrac{[H^+][BrO_2^-]}{[HBrO_2]}$

$HBrO_2(aq) + H_2O(l) \rightleftharpoons H_3O^+(aq) + BrO_2^-(aq); \quad K_a = \dfrac{[H_3O^+][BrO_2^-]}{[HBrO_2]}$

(b) $HC_3H_5O_2(aq) \rightleftharpoons H^+(aq) + C_3H_5O_2^-(aq); \quad K_a = \dfrac{[H^+][C_3H_5O_2^-]}{[HC_3H_5O_2]}$

$HC_3H_5O_2(aq) + H_2O(l) \rightleftharpoons H_3O^+(aq) + C_3H_5O_2^-(aq); \quad K_a = \dfrac{[H_3O^+][C_3H_5O_2^-]}{[HC_3H_5O_2]}$

16.53   *Analyze/Plan.* Follow the logic in Sample Exercise 16.10.   *Solve.*

$HC_3H_5O_3(aq) \; f \; H^+(aq) + C_3H_5O_3^-(aq); \quad K_a = \dfrac{[H^+][C_3H_5O_3^-]}{[HC_3H_5O_3]}$

$[H^+] = [C_3H_5O_3^-] = 10^{-2.44} = 3.63 \times 10^{-3} = 3.6 \times 10^{-3} \; M$

$[HC_3H_5O_3] = 0.10 - 3.63 \times 10^{-3} = 0.0964 = 0.096 \; M$

$K_a = \dfrac{(3.63 \times 10^{-3})^2}{(0.0964)} = 1.4 \times 10^{-1}$

16.55   *Analyze/Plan.* Write the equilibrium reaction and the K_a expression. Use % ionization to get equilibrium concentration of $[H^+]$, and by stoichiometry, $[X^-]$ and $[HX]$. Calculate K_a *Solve.*

$[H^+] = 0.110 \times [CH_2ClCOOH]_{initial} = 0.0110 \; M$

	$CH_2ClCOOH(aq)$	$\rightleftharpoons$	$H^+(aq)$	+	$CH_2ClCOO^-(aq)$
initial	0.100 M		0		0
equil.	0.089 M		0.0110 M		0.0110 M

$K_a = \dfrac{[H^+][CH_2ClCOO^-]}{[CH_2ClCOOH]} = \dfrac{(0.0110)^2}{0.089} = 1.4 \times 10^{-3}$

16.57   *Analyze/Plan.* Write the equilibrium reaction and the K_a expression.

$[H^+] = 10^{-pH} = [C_2H_3O_2^-] \quad [HC_2H_3O_2] = x - [H^+]$.

Substitute into the K_a expression and solve for x.   *Solve.*

$[H^+] = 10^{-pH} = 10^{-2.90} = 1.26 \times 10^{-3} = 1.3 \times 10^{-3} \; M$

$K_a = 1.8 \times 10^{-5} = \dfrac{[H^+][C_2H_3O_2^-]}{[HC_2H_3O_2]} = \dfrac{(1.26 \times 10^{-3})^2}{(x - 1.26 \times 10^{-3})}$

$1.8 \times 10^{-5}\,(x - 1.26 \times 10^{-3}) = (1.26 \times 10^{-3})^2;$

$1.8 \times 10^{-5}\,x = 1.585 \times 10^{-6} + 2.266 \times 10^{-8} = 1.608 \times 10^{-6};$

$x = 0.08931 = 0.089 \; M \; HC_2H_3O_2$

16.59   *Analyze/Plan.* Follow the logic in Sample Exercise 16.11. Write K_a, construct the equilibrium table, solve for $x = [H^+]$, then get equilibrium $[C_7H_5O_2^-]$ and $[HC_7H_5O_2]$ by substituting $[H^+]$ for x. *Solve.*

	$HC_7H_5O_2(aq)$	$\rightleftharpoons$	$H^+(aq)$	$+$	$C_7H_5O_2^-(aq)$
initial	0.050 M		0		0
equil.	$(0.050 - x)$ M		x M		x M

$$K_a = \frac{[H^+][C_7H_5O_2^-]}{[HC_7H_5O_2]} = \frac{x^2}{(0.050-x)} \approx \frac{x^2}{0.050} = 6.3 \times 10^{-5}$$

$x^2 = 0.050 (6.3 \times 10^{-5}); x = 1.8 \times 10^{-3} M = [H^+] = [H_3O^+] = [C_7H_6O_2^-]$

$[HC_7H_5O_2] = 0.050 - 0.0018 = 0.048 M$

Check. $\dfrac{1.8 \times 10^{-3} M H^+}{0.050 \, M \, HC_7H_5O_2} \times 100 = 3.6\%$ ionization; the assumption is valid

16.61   *Analyze/Plan.* Follow the logic in Sample Exercise 16.11.

(a)   *Solve.*

	$HC_3H_5O_2(aq)$	$\rightleftharpoons$	$H^+(aq)$	$+$	$C_3H_5O_2^-$ (aq)
initial	0.095 M		0		0
equil	$(0.095 - x)$ M		x M		x M

$$K_a = \frac{[H^+][C_3H_5O_2^-]}{[HC_3H_5O_2]} = \frac{x^2}{(0.095-x)} \approx \frac{x^2}{0.095} = 1.3 \times 10^{-5}$$

$x^2 = 0.095(1.3 \times 10^{-5}); x = 1.111 \times 10^{-3} = 1.1 \times 10^{-3} M H^+; pH = 2.95$

Check. $\dfrac{1.1 \times 10^{-3} M H^+}{0.095 \, M \, HC_3H_5O_2} \times 100 = 1.2\%$ ionization; the assumption is valid

(b)   *Solve.*

$$K_a = \frac{[H^+][CrO_4^{2-}]}{[HCrO_4^-]} = \frac{x^2}{(0.100-x)} \approx \frac{x^2}{0.100} = 3.0 \times 10^{-7}$$

$x^2 = 0.100(3.0 \times 10^{-7}); x = 1.732 \times 10^{-4} = 1.7 \times 10^{-4} M H^+$

$pH = -\log(1.732 \times 10^{-4}) = 3.7614 = 3.76$

Check. $\dfrac{1.7 \times 10^{-4} M H^+}{0.100 \, M \, HCrO_4^-} \times 100 = 0.17\%$ ionization; the assumption is valid

(c)   Follow the logic in Sample Exercise 16.14. $pOH = -\log[OH^-]$. $pH = 14 - pOH$ *Solve.*

	$C_5H_5N(aq) + H_2O(l)$	$\rightleftharpoons$	$C_5H_5NH^+(aq)$	$+$	$OH^-(aq)$
initial	0.120 M		0		0
equil	$(0.120 - x)$ M		x M		x M

$$K_b = \frac{[C_5H_5NH^+][OH^-]}{[C_5H_5N]} = \frac{x^2}{(0.120-x)} \approx \frac{x^2}{0.120} = 1.7 \times 10^{-9}$$

$$x^2 = 0.120(1.7 \times 10^{-9}); \; x = 1.428 \times 10^{-5} = 1.4 \times 10^{-5} \, M \, OH^-; \; pH = 9.15$$

$$Check. \frac{1.4 \times 10^{-5} \, M \, OH^-}{0.120 \, M \, C_5H_5N} \times 100 = 0.011\% \; \text{ionization; the assumption is valid}$$

16.63   *Analyze/Plan.* $K_a = 10^{-pK_a}$. Follow the logic in Sample Exercise 16.11.   *Solve.*

Let $[H^+] = [NC_7H_4SO_3^-] = z$. $K_a = $ antilog $(-2.32) = 4.79 \times 10^{-3} = 4.8 \times 10^{-3}$

$\dfrac{z^2}{0.10-z} = 4.79 \times 10^{-3}$.   Since K_a is relatively large, solve the quadratic.

$$z^2 = 4.79 \times 10^{-3}z - 4.79 \times 10^{-4} = 0$$

$$z = \frac{-4.79 \times 10^{-3} \pm \sqrt{(4.79 \times 10^{-3})^2 - 4(1)(-4.79 \times 10^{-4})}}{2(1)} = \frac{-4.79 \times 10^{-3} \pm \sqrt{1.937 \times 10^{-3}}}{2}$$

$$z = 1.96 \times 10^{-2} = 2.0 \times 10^{-2} \, M \, H^+; \; pH = -\log(1.96 \times 10^{-2}) = 1.71$$

16.65   *Analyze/Plan.* Follow the logic in Sample Exercise 16.12.   *Solve.*

(a)

	$HN_3(aq)$	$\rightleftharpoons$	$H^+(aq)$	$+$	$N_3^-(aq)$
initial	0.400 M		0		0
equil	(0.400 − x) M		x M		x M

$$K_a = \frac{[H^+][N_3^-]}{[HN_3]} = 1.9 \times 10^{-5}; \; \frac{x^2}{(0.400-x)} = \frac{x^2}{0.400} = 1.9 \times 10^{-5}$$

$$x = 0.00276 = 2.8 \times 10^{-3} \, M = [H^+]; \; \% \; \text{ionization} = \frac{2.76 \times 10^{-3}}{0.400} \times 100 = 0.69\%$$

(b) $1.9 \times 10^{-5} = \dfrac{x^2}{0.100}; \; x = 0.00138 = 1.4 \times 10^{-3} \, M \, H^+$

$$\% \; \text{ionization} = \frac{1.38 \times 10^{-3} \, M \, H^+}{0.100 \, M \, HN_3} \times 100 = 1.4\%$$

(c) $1.9 \times 10^{-5} = \dfrac{x^2}{0.0400}; \; x = 8.72 \times 10^{-4} = 8.7 \times 10^{-4} \, M \, H^+$

$$\% \; \text{ionization} = \frac{8.72 \times 10^{-4} \, M \, H^+}{0.0400 \, M \, HN_3} \times 100 = 2.2\%$$

Check. Notice that a tenfold dilution [part (a) versus part (c)] leads to a slightly more than threefold increase in percent ionization.

16.67   *Analyze/Plan.* Let the weak acid be HX. $HX(aq) \rightleftharpoons H^+(aq) + X^-(aq)$. Solve the K_a expression symbolically for $[H^+]$ in terms of $[HX]$. Substitute into the formula for % ionization, $([H^+]/[HX]) \times 100$.   *Solve.*

$$K_a = \frac{[H^+][X^-]}{[HX]}; [H^+] = [X^-] = y; K_a = \frac{y^2}{[HX]-y}; \text{assume that \% ionization is small}$$

$$K_a = \frac{y^2}{[HX]}; y = K_a^{1/2}[HX]^{1/2}$$

$$\% \text{ ionization} = \frac{y}{[HX]} \times 100 = \frac{K_a^{1/2}[HX]^{1/2}}{[HX]} \times 100 = \frac{K_a^{1/2}}{[HX]^{1/2}} \times 100$$

That is, percent ionization varies inversely as the square root of concentration HX.

16.69   Analyze/Plan. Follow the logic in Sample Exercise 16.13. Citric acid is a triprotic acid with three K_a values that do not differ by more than 10^3. We must consider all three steps. Also, $C_6H_5O_7^{3-}$ is only produced in step 3.   *Solve.*

$$H_3C_6H_5O_7(aq) \rightleftharpoons H^+(aq) + H_2C_6H_5O_7^-(aq) \qquad K_{a1} = 7.4 \times 10^{-4}$$

$$H_2C_6H_5O_7^-(aq) \rightleftharpoons H^+(aq) + HC_6H_5O_7^{2-}(aq) \qquad K_{a2} = 1.7 \times 10^{-5}$$

$$HC_6H_5O_7^{2-}(aq) \rightleftharpoons H^+(aq) + C_5H_5O_7^{3-}(aq) \qquad K_{a3} = 4.0 \times 10^{-7}$$

To calculate the pH of a 0.050 M solution, assume initially that only the first ionization is important:

	$H_3C_6H_5O_7(aq)$	$\rightleftharpoons$	$H^+(aq)$	+	$H_2C_6H_5O_7^-(aq)$
initial	0.050 M		0		0
equil.	(0.050 – x) M		x M		x M

$$K_{a1} = \frac{[H^+][H_2C_6H_5O_7^-]}{[H_3C_6H_5O_7]} = \frac{x^2}{(0.050-x)} = 7.4 \times 10^{-4}$$

$$x^2 = (0.050 - x)(7.4 \times 10^{-4}); x^2 = (0.050)(7.4 \times 10^{-4}); x = 0.00608 = 6.1 \times 10^{-3} M$$

Since this value for x is rather large in relation to 0.050, a better approximation for x can be obtained by substituting this first estimate into the expression for x^2, then solving again for x:

$$x^2 = (0.050 - x)(7.4 \times 10^{-4}) = (0.050 - 6.08 \times 10^{-3})(7.4 \times 10^{-4})$$

$$x^2 = 3.2 \times 10^{-5}; x = 5.7 \times 10^{-3} M$$

(This is the same result obtained from the quadratic formula.)

The correction to the value of x, though not large, is significant. Does the second ionization produce a significant additional concentration of H^+?

	$H_2C_6H_5O_7^-(aq)$	$\rightleftharpoons$	$H^+(aq)$	+	$HC_6H_5O_7^{2-}(aq)$
initial	$5.7 \times 10^{-5} M$		$5.7 \times 10^{-3} M$		0
equil.	$(5.7 \times 10^{-3} - y)$		$(5.7 \times 10^{-3} + y)$		y

$$K_{a2} = \frac{[H^+][HC_6H_5O_7^{2-}]}{[H_2C_6H_5O_7^-]} = 1.7 \times 10^{-5}; \frac{(5.7 \times 10^{-3} + y)(y)}{(5.7 \times 10^{-3} - y)} = 1.7 \times 10^{-5}$$

Assume that y is small relative to 5.7×10^{-3}; that is, that additional ionization of $H_2C_6H_5O_7^-$ is small, then

$$\frac{(5.7 \times 10^{-3})y}{(5.7 \times 10^{-3})} = 1.7 \times 10^{-5} \ M; \ y = 1.7 \times 10^{-5} \ M$$

This value is indeed small compared to $5.7 \times 10^{-3} \ M$; $[H^+]$ and pH are determined by the first ionization step. If we were only interested in pH, we could stop here. However, to calculate $[C_6H_5O_7^{3-}]$, we must consider the third ionization, with adjusted $[H^+] = 5.7 \times 10^{-3} + 1.7 \times 10^{-5} = 5.72 \times 10^{-3} \ M \ (= 5.7 \times 10^{-3})$

	$HC_6H_5O_7^{2-}$	$\rightleftharpoons$	$H^+ \ (aq)$	$+$	$HC_5H_5O_7^{3-} \ (aq)$
initial	$1.7 \times 10^{-5} \ M$		$5.72 \times 10^{-3} \ M$		0
equil.	$1.7 \times 10^{-5} - z$		$5.72 \times 10^{-3} + z$		z

$$K_{a3} = \frac{[H^+][C_6H_5O_7^{3-}]}{[HC_6H_5O_7^{2-}]} = \frac{(5.72 \times 10^{-3} + z)(z)}{(1.7 \times 10^{-5} - z)} = 4.0 \times 10^{-7}$$

Assume z is small relative to 5.72×10^{-3}, but not relative to 1.7×10^{-5}.

$(4.0 \times 10^{-7})(1.7 \times 10^{-5} - z) = 5.72 \times 10^{-3} \ z; \ 6.8 \times 10^{-12} - 4.0 \times 10^{-7} \ z = 5.72 \times 10^{-3} \ z;$

$6.8 \times 10^{-12} = 5.72 \times 10^{-3} \ z + 4.0 \times 10^{-7} \ z = 5.72 \times 10^{-3} \ z; \ z = 1.19 \times 10^{-9} = 1.2 \times 10^{-9} \ M$

$[C_6H_5O_7^{3-}] = 1.2 \times 10^{-9} \ M; \ [H^+] = 5.72 \times 10^{-3} \ M + 1.2 \times 10^{-9} \ M = 5.72 \times 10^{-3} \ M$

$pH = -\log(5.72 \times 10^{-3}) = 2.24$

Note that neither the second nor third ionizations contributed significantly to $[H^+]$ and pH.

Weak Bases

16.71   All Brønsted-Lowry bases contain at least one unshared (lone) pair of electrons to attract H^+.

16.73   *Analyze/Plan.* Remember that $K_{aq} = [products]/[reactants]$. If $H_2O(l)$ appears in the equilibrium reaction, it will not appear in the K_b expression, because it is a pure liquid. *Solve.*

(a) $(CH_3)_2NH(aq) + H_2O(l) \rightleftharpoons (CH_3)_2NH_2^+(aq) + OH^-(aq); \ K_b = \dfrac{[(CH_3)_2NH_2^+][OH^-]}{[(CH_3)_2NH]}$

(b) $CO_3^{2-}(aq) + H_2(l) \rightleftharpoons HCO_3^-(aq) + OH^-(aq); \ K_b = \dfrac{[HCO_3^-][OH^-]}{[CO_3^{2-}]}$

(c) $CHO_2^-(aq) + H_2O(l) \rightleftharpoons HCHO_2(aq) + OH^-(aq); \ K_b = \dfrac{[HCHO_2][OH^-]}{[CHO_2^-]}$

16.75   *Analyze/Plan.* Follow the logic in Sample Exercise 16.14.   *Solve.*

	$C_2H_5NH_2(aq)$ + $H_2O(l)$	$\rightleftharpoons$	$C_2H_5NH_3^+(aq)$	+	$OH^-(aq)$
initial	0.075 M		0		0
equil.	(0.075 – x) M		x M		x M

$$K_b = \frac{[C_2H_5NH_3^+][OH^-]}{[C_2H_5NH_2]} = \frac{(x)(x)}{(0.075-x)} = \frac{x^2}{0.075} = 6.4 \times 10^{-4}$$

$x^2 = 0.075 \,(6.4 \times 10^{-4})$; $x = [OH^-] = 6.9 \times 10^{-3}\ M$; pH = 11.84

Check. $\dfrac{6.9 \times 10^{-3}\ M\ OH^-}{0.075\ M\ C_2H_5NH_2} \times 100 = 9.2\%$ ionization; the assumption is not valid

To obtain a more precise result, the K_b expression is rewritten in standard quadratic form and solved via the quadratic formula.

$$\frac{x^2}{0.075-x} = 6.4 \times 10^{-4}; x^2 + 6.4 \times 10^{-4}\,x - 4.8 \times 10^{-5} = 0$$

$$x = \frac{b \pm \sqrt{b^2 - 4ac}}{2a} = \frac{-6.4 \times 10^{-4} \pm \sqrt{(6.4 \times 10^{-4})^2 - 4(1)(-4.8 \times 10^{-5})}}{2}$$

$x = 6.61 \times 10^{-3} = 6.6 \times 10^{-3}\ M\ OH^-$; pOH = 2.18, pH = 14.00 – pOH = 11.82

Note that the pH values obtained using the two algebraic techniques are very similar.

16.77   *Analyze/Plan.* Given pH and initial concentration of base, calculate all equilibrium concentrations. pH $\rightarrow$ pOH $\rightarrow$ [OH^-] at equilibrium. Construct the equilibrium table and calculate other equilibrium concentrations. Substitute into the K_b expression and calculate K_b.   *Solve.*

(a)   [OH^-] = 10^{-pOH}; pOH = 14 – pH = 14.00 – 11.33 = 2.67

[OH^-] = $10^{-2.67}$ = 2.138×10^{-3} = $2.1 \times 10^{-3}\ M$

	$C_{10}H_{15}ON(aq)$ + $H_2O(l)$	$\rightleftharpoons$	$C_{10}H_{15}ONH^+(aq)$	+	$OH^-(aq)$
initial	0.035 M		0		0
equil.	0.033 M		$2.1 \times 10^{-3}\ M$		$2.1 \times 10^{-3}\ M$

(b) $K_b = \dfrac{[C_{10}H_{15}ONH^+][OH^-]}{[C_{10}H_{15}ON]} = \dfrac{(2.138 \times 10^{-3})^2}{(0.03286)} = 1.4 \times 10^{-4}$

The K_a – K_b Relationship; Acid-Base Properties of Salts

16.79   (a)   For a conjugate acid/conjugate base pair such as $C_6H_5OH/C_6H_5O^-$, K_b for the conjugate base is always K_w/K_a for the conjugate acid. K_b for the conjugate base can always be calculated from K_a for the conjugate acid, so a separate list of K_b values is not necessary.

(b) $K_b = K_w/K_a = 1.0 \times 10^{-14} / 1.3 \times 10^{-10} = 7.7 \times 10^{-5}$

(c) K_b for phenolate (7.7×10^{-5}) > K_b for ammonia (1.8×10^{-5}).

Phenolate is a stronger base than NH_3.

16.81   *Analyze/Plan.* Given K_a, determine relative strengths of the acids and their conjugate bases. The greater the magnitude of K_a, the stronger the acid and the weaker the conjugate base.

K_b (conjugate base) = K_w/K_a.   *Solve.*

(a)   Acetic acid is stronger, because it has the larger K_a value.

(b)   Hypochlorite ion is the stronger base because the weaker acid, hypochlorous acid, has the stronger conjugate base.

(c) K_b for $C_2H_3O_2^-$ = K_w/K_a for $HC_2H_3O_2$ = $1.0 \times 10^{-14}/1.8 \times 10^{-5}$ = 5.6×10^{-10}

K_b for ClO^- = K_w/K_a for $HClO$ = $1 \times 10^{-14}/3.0 \times 10^{-8}$ = 3.3×10^{-7}

Note that K_b for ClO^- is greater than K_b for $C_2H_3O_2^-$.

16.83   *Analyze.* When the solute in an aqueous solution is a salt, evaluate the acid/base properties of the component ions.

(a)   *Plan.* NaCN is a soluble salt and thus a strong electrolyte. When it is dissolved in H_2O, it dissociates completely into Na^+ and CN^-. [NaCN] = [Na^+] = [CN^-] = 0.10 M. Na^+ is the conjugate acid of the strong base NaOH and thus does not influence the pH of the solution. CN^-, on the other hand, is the conjugate base of the weak acid HCN and **does** influence the pH of the solution. Like any other weak base, it hydrolyzes water to produce OH^-(aq). Solve the equilibrium problem to determine [OH^-].   *Solve.*

$$CN^-(aq) \ + \ H_2O(l) \ \rightleftharpoons \ HCN(aq) \ + \ OH^-(aq)$$

initial	0.10 M	0	0
equil.	(0.10 − x) M	x M	x M

$$K_b \text{ for } CN^- = \frac{[HCN][OH^-]}{[CN^-]} = \frac{K_w}{K_a \text{ for HCN}} - \frac{1 \times 10^{-14}}{4.9 \times 10^{-10}} = 2.04 \times 10^{-5} = 2.0 \times 10^{-5}$$

$2.04 \times 10^{-5} = \dfrac{(x)(x)}{(0.10-x)}$; assume the percent of CN^- that hydrolyzes is small

$x^2 = 0.10 \, (2.04 \times 10^{-5})$; x = [$OH^-$] = 0.00143 = 1.4×10^{-3} M

pOH = 2.85; pH = 14 − 2.85 = 11.15

(b)   *Plan.* $Na_2CO_3(aq) \ \rightarrow \ 2Na^+(aq) + CO_3^{2-}(aq)$

CO_3^{2-} is the conjugate base of HCO_3^- and its hydrolysis reaction will determine the [OH^-] and pH of the solution (see similar explanation for NaCN in part (a)). We will assume the process $HCO_3^-(aq) + H_2O(l) \rightleftharpoons H_2CO_3(aq) + OH^-$ will not add significantly to the [OH^-] in solution because [HCO_3^- (aq)] is so small. Solve the equilibrium problem for [OH^-].   *Solve.*

$$CO_3^{2-}(aq) \ + \ H_2O(l) \ \rightleftharpoons \ HCO_3^-(aq) + OH^-(aq)$$

initial	0.080 M	0	0
equil.	(0.080 – x) M	x	x

$$K_b = \frac{[HCO_3^-][OH^-]}{[CO_3^{2-}]} = \frac{K_w}{K_a \text{ for } HCO_3^-} = \frac{1.0 \times 10^{-14}}{5.6 \times 10^{-11}} = 1.79 \times 10^{-4} = 1.8 \times 10^{-4}$$

$$1.8 \times 10^{-4} = \frac{x^2}{(0.080 - x)} ; x^2 = 0.080 \, (1.79 \times 10^{-4}); x = 0.00378 = 3.8 \times 10^{-3} \, M \, OH^-$$

(Assume x is small compared to 0.080); pOH = 2.42; pH = 14 – 2.42 = 11.58

Check. $\dfrac{3.8 \times 10^{-3} \, M \, OH^-}{0.080 \, M \, CO_3^{2-}} \times 100 = 4.75\%$ hydrolysis; the assumption is valid

(c) *Plan.* For the two salts present, Na^+ and Ca^{2+} are negligible acids. NO_2^- is the conjugate base of HNO_2 and will determine the pH of the solution. *Solve.*

Calculate total $[NO_2^-]$ present initially.

$[NO_2^-]_{total} = [NO_2^-]$ from $NaNO_2 + [NO_2^-]$ from $Ca(NO_2)_2$

$[NO_2^-]_{total} = 0.10 \, M + 2(0.20 \, M) = 0.50 \, M$

The hydrolysis equilibrium is:

$$NO_2^-(aq) \ + \ H_2O(l) \ \rightleftharpoons \ HNO_2 \ + \ OH^-(aq)$$

initial	0.50 M	0	0
equil.	(0.50 – x) M	x M	x M

$$K_b = \frac{[HNO_2][OH^-]}{[NO_2^-]} = \frac{K_w}{K_a \text{ for } HNO_2} = \frac{1.0 \times 10^{-14}}{4.5 \times 10^{-4}} = 2.22 \times 10^{-11} = 2.2 \times 10^{-11}$$

$$2.2 \times 10^{-11} = \frac{x^2}{(0.50 - x)} \approx \frac{x^2}{0.50} ; x^2 = 0.50 \, (2.22 \times 10^{-11})$$

$x = 3.33 \times 10^{-6} = 3.3 \times 10^{-6} \, M \, OH^-$; pOH = 5.48; pH = 14 – 5.48 = 8.52

16.85 *Analyze/Plan.* Given the formula of a salt, predict whether an aqueous solution will be acidic, basic, or neutral. Evaluate the acid-base properties of both ions and determine the overall effect on solution pH. *Solve.*

(a) acidic; NH_4^+ is a weak acid, Br^- is negligible.

(b) acidic; Fe^{3+} is a highly charged metal cation and a Lewis acid; Cl^- is negligible.

(c) basic; CO_3^{2-} is the conjugate base of HCO_3^-; Na^+ is negligible.

(d) neutral; both K^+ and ClO_4^- are negligible.

(e) acidic; $HC_2O_4^-$ is amphoteric, but K_a for the acid dissociation (6.4×10^{-5}) is much greater than K_b for the base hydrolysis ($1.0 \times 10^{-14} / 5.9 \times 10^{-2} = 1.7 \times 10^{-13}$).

16.87   *Plan.* Estimate pH using relative base strength and then calculate to confirm prediction. NaCl is a neutral salt, so it is not the unknown. The unknown is a relatively weak base, because a pH of 8.08 is not very basic. Since F^- is a weaker base than OCl^-, the unknown is probably NaF. Calculate K_b for the unknown from the data provided. *Solve.*

$[OH^-] = 10^{-pOH}$; $pOH = 14.00 - pH = 14.00 - 8.08 = 5.92$

$[OH^-] = 10^{-6.92} = 1.202 \times 10^{-6} = 1.2 \times 10^{-6} M = [HX]$

$[NaX] = [X^-] = 0.050$ mol salt$/0.500$ L $= 0.10 M$

$$K_b = \frac{[OH^-][HX]}{[X^-]} = \frac{(1.202 \times 10^{-6})^2}{(0.10 - 1.2 \times 10^{-6})} = \frac{(1.202 \times 10^{-6})^2}{0.10} = 1.4 \times 10^{-11}$$

K_b for $F^- = K_w/K_a$ for $HF = 1.0 \times 10^{-14}/6.8 \times 10^{-4} = 1.5 \times 10^{-11}$

The unknown is NaF.

16.89   *Analyze/Plan.* The solution will be basic because of the hydrolysis of the sorbate anion, $C_6H_7O_2^-$. Calculate the initial molarity of $C_6H_7O_2^-$. Calculate K_b from K_w/K_a. Solve the K_b expression for $[OH^-]$.   *Solve.*

$$\frac{11.25 \text{ g } KC_6H_7O_2}{1.75 \text{ L}} \times \frac{1 \text{ mol } KC_6H_7O_2}{150.2 \text{ g } KC_6H_7O_2} = 0.04280 = 0.0428 M \text{ } KC_6H_7O_2$$

$[C_6H_7O_2^-] = [KC_6H_7O_2] = 0.0428 M$

	$C_6H_7O_2^-$(aq) $+$ H$_2$O(l) $\rightleftharpoons$	HC$_6$H$_7$O$_2$(aq)	$+$	OH$^-$(aq)
initial	$0.0428 M$	0		0
equil.	$(0.0428 - x) M$	$x M$		$x M$

$$K_b = \frac{[HC_6H_7O_2][OH^-]}{[C_6H_7O_2^-]} = \frac{K_w}{K_a \text{ for } HC_6H_7O_2} = \frac{1.0 \times 10^{-14}}{1.7 \times 10^{-5}} = 5.88 \times 10^{-10} = 5.9 \times 10^{-10}$$

$$5.88 \times 10^{-10} = \frac{x^2}{0.0428 - x} \approx \frac{x^2}{0.0428}; x^2 = 0.0428 (5.88 \times 10^{-10})$$

$x = [OH^-] = 5.018 \times 10^{-6} = 5.0 \times 10^{-6} M$; $pOH = 5.30$; $pH = 14 - pOH = 8.70$

Acid-Base Character and Chemical Structure

16.91   (a)   As the electronegativity of the central atom (X) increases, more electron density is withdrawn from the X–O and O–H bonds, respectively. In water, the O–H bond is ionized to a greater extent and the strength of the oxyacid increases.

           (b)   As the number of nonprotonated oxygen atoms in the molecule increases, they withdraw electron density from the other bonds in the molecule and the strength of the oxyacid increases.

16.93   (a) HNO_3 is a stronger acid than HNO_2 because it has one more nonprotonated oxygen atom, and thus a higher oxidation number on N.

(b)   For binary hydrides, acid strength increases going down a family, so H_2S is a stronger acid than H_2O.

(c) H_2SO_4 is a stronger acid because H^+ is much more tightly held by the anion HSO_4^-.

(d)   For oxyacids, the greater the electronegativity of the central atom, the stronger the acid, so H_2SO_4 is a stronger acid than H_2SeO_4.

(e) CCl_3COOH is stronger because the electronegative Cl atoms withdraw electron density from other parts of the molecule, which weakens the O–H bond and makes H^+ easier to remove. Also, the electronegative Cl delocalizes negative charge on the carboxylate anion. This stabilizes the conjugate base, favoring products in the ionization equilibrium and increasing K_a.

16.95   (a) BrO^- (HClO is the stronger acid due to a more electronegative central atom, so BrO^- is the stronger base.)

(b) BrO^- ($HBrO_2$ has more nonprotonated O atoms and is the stronger acid, so BrO^- is the stronger base.)

(c) HPO_4^{2-} (larger negative charge, greater attraction for H^+)

16.97   (a)   True.

(b)   False. In a series of acids that have the same central atom, acid strength increases with the number of nonprotonated oxygen atoms bonded to the central atom.

(c)   False. H_2Te is a stronger acid than H_2S because the H–Te bond is longer, weaker, and more easily dissociated than the H–S bond.

Lewis Acids and Bases

16.99   Yes. If a substance is an Arrhenius base, it must also be a Brønsted base and a Lewis base. The Arrhenius definition (hydroxide ion) is the most restrictive, the Brønsted (H^+ acceptor) more general and the Lewis (electron pair donor) most general. Since a hydroxide ion is both an H^+ acceptor and an electron pair donor, any substance that fits the narrow Arrhenius definition will fit the broader Brønsted and Lewis definitions.

16.101   *Analyze/Plan.* Identify each reactant as an electron pair donor (Lewis base) or electron pair acceptor (Lewis acid). Remember that a Brønsted acid is necessarily a Lewis acid, and a Brønsted base is necessarily a Lewis base (Solution 16.89).   *Solve.*

16.103   (a) Cu^{2+}, higher cation charge

(b) Fe^{3+}, higher cation charge

(c) Al^{3+}, smaller cation radius, same charge

Additional Exercises

16.105   (a) K_w is the equilibrium constant for the reaction of two water molecules to form hydronium ion and hydroxide ion.

(b) K_a is the equilibrium constant for the reaction of any acid, neutral or ionic, with water to form hydronium ion and the conjugate base of the acid.

(c)   pOH is the negative log of hydroxide ion concentration; pOH decreases as hydroxide ion concentration increases.

16.107   (a)   A higher O_2 concentration displaces protons from Hb, producing a more acidic solution, with lower pH in the lungs than in the tissues.

(b) $[H^+]$ = antilog $(-7.4) = 4.0 \times 10^{-8}$ M. At body temperature, 37°C, $K_w = 2.4 \times 10^{-14}$ (see Solution 16.39). At this temperature, a "neutral" solution has $[H^+] = 1.5 \times 10^{-7}$ and pH 6.81. Even though the frame of reference is a bit different at this temperature, blood at pH 7.4 is slightly basic.

(c)   The equilibrium indicates that a high $[H^+]$ shifts the equilibrium toward the proton-bound form HbH^+, which means a lower concentration of HbO_2 in the blood. Thus the ability of hemoglobin to transport oxygen is impeded.

16.110 $K_a = \dfrac{[H^+][C_6H_{11}O_2^-]}{[HC_6H_{11}O_2]}$; $[H^+] = [C_6H_{11}O^{2-}] = 10^{-pH} = 10^{-2.94} = 0.001148 = 1.1 \times 10^{-3}$ M

$$[HC_6H_{11}O_2] = \frac{11 \text{ g } HC_6H_{11}O_2}{1.} \times \frac{1 \text{ mol } HC_6H_{11}O_2}{116.16 \text{ g } HC_6H_{11}O_2} = 0.09470 = 0.095 \ M$$

$$K_a = \frac{[H^+][C_6H_{11}O_2^-]}{[HC_6H_{11}O_2]} = \frac{(0.001148)^2}{(0.09470 - 0.001148)} = 1.4092 \times 10^{-5} = 1.41 \times 10^{-5}$$

16.113   Considering the stepwise dissociation of H_3PO_4:

$$H_3PO_4(aq) \ \rightleftharpoons \ H^+(aq) \ + \ H_2PO_4^-(aq)$$

initial   0.025   0   0

equil.   (0.025 − x) M   x   x

$$K_{a1} = \frac{[H^+][H_2PO_4^-]}{[H_3PO_4]} = \frac{x^2}{(0.025 - x)} = 7.5 \times 10^{-3}; x^2 + 7.5 \times 10^{-3}\,x - 1.875 \times 10^{-4} = 0$$

$$x = \frac{-7.5 \times 10^{-3} \pm \sqrt{(7.5 \times 10^{-3})^2 - 4(1)(-1.875 \times 10^{-4})}}{2(1)} = \frac{-7.5 \times 10^{-3} \pm \sqrt{8.06 \times 10^{-4}}}{2}$$

x = 0.01045 = 0.010 M H^+, 0.010 M $H_2PO_4^-$ available for further ionization

$$H_2PO_4^-(aq) \ \rightleftharpoons \ H^+(aq) \ + \ HPO_4^{2-}(aq)$$

initial   0.010 M   0.010 M   0 M

equil.   (0.010 − y) M   (0.010 + y) M   y M

$$K_{a2} = \frac{[H^+][HPO_4^{2-}]}{[H_2PO_4^-]} = \frac{(y)(0.010 + y)}{(0.010 - y)} = 6.2 \times 10^{-8}$$

Since K_{a2} is very small, assume y is small compared to 0.010 M.

$$6.2 \times 10^{-8} = \frac{0.010 \, y}{0.010}; y = [HPO_4^{2-}] = 6.2 \times 10^{-8} \ M$$

$[H_2PO_4^-] = (0.010 \ M + 6.2 \times 10^{-8} \ M) = 0.010 \ M$

$[H^+] = (0.010 \ M - 6.2 \times 10^{-8} \ M) = 0.010 \ M$

$[HPO_4^{2-}]$ available for further ionization $= 6.2 \times 10^{-8} \ M$

	HPO_4^{2-}	$\rightleftharpoons$	$H^+(aq)$	$+$	$PO_4^{3-}(aq)$
initial	$6.2 \times 10^{-8} \ M$		$0.010 \ M$		$0 \ M$
equil.	$(6.2 \times 10^{-8} - z) \ M$		$(0.010 + z) \ M$		$z \ M$

$$K_{a3} = \frac{[H^+][PO_4^{3-}]}{[HPO_4^{2-}]} = \frac{(0.010 + z)(z)}{(6.2 \times 10^{-8} - z)} = 4.2 \times 10^{-13}$$

Assuming z is small compared to 6.2×10^{-8} (and 0.010),

$$\frac{(0.010)(z)}{(6.2 \times 10^{-8})} = 4.2 \times 10^{-13}; z = 2.6 \times 10^{-18} \ M \ PO_4^{3-}$$

The contribution of z to $[HPO_4^{2-}]$ and $[H^+]$ is negligible. In summary, after all dissociation steps have reached equilibrium:

$[H^+] = 0.010 \ M$, $[H_2PO_4^-] = 0.010 \ M$, $[HPO_4^{2-}] = 6.2 \times 10^{-8} \ M$, $[PO_4^{3-}] = 2.6 \times 10^{-18} \ M$

Note that the first ionization step is the major source of H^+ and the others are important as sources of HPO_4^{2-} and PO_4^{3-}. The $[PO_4^{3-}]$ is very small at equilibrium.

Integrative Exercises

16.117 At 25°C, $[H^+] = [OH^-] = 1.0 \times 10^{-7} \ M$

$$\frac{1.0 \times 10^{-7} \ mol \ H^+}{1L \ H_2O} \times 0.0010 \ L \times \frac{6.022 \times 10^{23} \ H^+ \ ions}{mol \ H^+} = 6.0 \times 10^{13} \ H^+ \ ions$$

16.120 *Analyze.* If pH were directly related to CO_2 concentration, this exercise would be simple. Unfortunately, we must solve the equilibrium problem for the diprotic acid H_2CO_3 to calculate $[H^+]$ and pH. We are given ppm CO_2 in the atmosphere at two different times, and the pH that corresponds to one of these CO_2 levels. We are asked to find pH at the other atmospheric CO_2 level.

Plan. Assume all dissolved CO_2 is present as H_2CO_3 (aq) (Sample Exercise 16.13).

pH $\rightarrow$ $[H^+]$ $\rightarrow$ $[H_2CO_3]$. While H_2CO_3 is a diprotic acid, the two K_a values differ by more than 10^3, so we can ignore the second ionization when calculating $[H_2CO_3]$. Change 375 ppm CO_2 to pressure and calculate the Henry's law constant for CO_2. Calculate the dissolved $[CO_2] = [H_2CO_3]$ at 315 ppm, then solve the K_{a1} expression for $[H^+]$ and pH.

(a)   *Solve.* $H_2CO_3(aq) \rightleftharpoons H^+(aq) + HCO_3^- (aq)$

$$K_{a1} = 4.3 \times 10^{-7} = \frac{[H^+][HCO_3^-]}{[H_2CO_3]}; [H^+] = 10^{-5.4} = 3.98 \times 10^{-6} = 4 \times 10^{-6} \ M$$

$[H^+] = [HCO_3^-]; [H_2CO_3] = x - 4 \times 10^{-5}$

$$4.3 \times 10^{-7} = \frac{(3.98 \times 10^{-6})^2}{(x - 3.98 \times 10^{-6})}; \; 4.3 \times 10^{-7} x = 1.585 \times 10^{-11} + 1.712 \times 10^{-12}$$

$x = 1.756 \times 10^{-11} / 4.3 \times 10^{-7} = 4.084 \times 10^{-5} = 4 \times 10^{-5} \, M \, H_2CO_3$

375 ppm = 375 mol CO_2/1 × 10^6 mol air = 0.000375 mol % CO_2

Because of the properties of gases, mol % = pressure %. $P_{CO_2} = 0.000375$ atm. According to Equation [13.4], $S_{CO_2} = kP_{CO_2}$;

4.084×10^{-5} mol/L = k(3.75×10^{-4} atm) k = 0.1089 = 0.1 mol/L •atm.

Forty years ago, $S_{CO_2} = 0.1089 \dfrac{\text{mol}}{\text{L} \bullet \text{atm}} \times 3.15 \times 10^{-4}$ atm = 3.4305×10^{-5}

$$= 3.4 \times 10^{-5} \, M$$

Now solve K_{a1} for [H^+] at this [H_2CO_3]. [H^+] = x

We cannot assume x is small, because [H_2CO_3] is so low.

$4.3 \times 10^{-7} = x^2/(3.4305 \times 10^{-5} - x); \; x^2 + 4.3 \times 10^{-7} x - 1.475 \times 10^{-11} = 0$

$$x = \frac{-4.3 \times 10^{-7} \pm \sqrt{(4.3 \times 10^{-7})^2 - 4(-1.475 \times 10^{-11})}}{2} = \frac{-4.3 \times 10^{-7} + 7.693 \times 10^{-6}}{2}$$

$$= 3.632 \times 10^{-6} = 3.6 \times 10^{-6} \, M \, H^+$$

[H^+] = 3.6×10^{-6} M, pH = 5.440 = 5.4

(Note that, to the precision that the pH data is reported, the change in atmospheric CO_2 leads to no change in pH.)

(b)   From part (a), [H_2CO_3] today = 4.084×10^{-5} M

$$\frac{4.084 \times 10^{-5} \text{ mol } H_2CO_3}{1 \, L} \times 20.0 \, L = 8.168 \times 10^{-4} = 8 \times 10^{-4} \text{ mol } CO_2$$

$$V = \frac{nRT}{P} = 8.168 \times 10^{-5} \text{ mol} \times \frac{298 \, K}{1.0 \text{ atm}} \times \frac{0.08206 \, L \bullet \text{atm}}{\text{mol} \bullet K} = 0.01997 = 0.02 \, L = 20 \text{ mL}$$

16.123   Calculate M of the solution from osmotic pressure, and K_b using the equilibrium expression for the hydrolysis of cocaine. Let Coc = cocaine and CocH$^+$ be the conjugate acid of cocaine.

$$\pi = M\,RT; \; M = \pi/RT = \frac{52.7 \text{ torr}}{288 \, K} \times \frac{1 \text{ atm}}{760 \text{ torr}} \times \frac{\text{mol} \bullet K}{0.08206 \, L \bullet \text{atm}}$$

$$= 0.002934 = 2.93 \times 10^{-3} \, M \text{ Coc}$$

pH = 8.53; pOH = 14 – pH = 5.47; [OH^-] = $10^{-5.47}$ = 3.39×10^{-6} = 3.4×10^{-6} M

	Coc(aq) + H$_2$O(l)	$\rightleftharpoons$	CocH$^+$(aq)	+	OH$^-$(aq)
initial	2.93×10^{-3} M		0		0
equil.	$(2.93 \times 10^{-3} - 3.4 \times 10^{-6})$ M		3.4×10^{-6} M		3.4×10^{-6} M

$$K_b = \frac{[CocH^+][OH^-]}{[Coc]} = \frac{(3.39 \times 10^{-6})^2}{(2.934 \times 10^{-3} - 3.39 \times 10^{-6})} = 3.9 \times 10^{-9}$$

Note that % hydrolysis is small in this solution, so "x," 3.4×10^{-6} M, is small compared to 2.93×10^{-3} M and could be ignored in the denominator of the calculation.

16.125   (a)   (i)

$HCO_3^-(aq) \rightleftharpoons H^+(aq) + CO_3^{2-}(aq)$ $K_1 = K_{a2}$ for $H_2CO_3 = 5.6 \times 10^{-11}$

$H^+(aq) + OH^-(aq) \, f \, H_2O(l)$ $K_2 = 1/K_w = 1 \times 10^{14}$

$\rule{9cm}{0.4pt}$

$HCO_3^-(aq) + OH^-(aq) \rightleftharpoons CO_3^{2-}(aq) + H_2O(l)$ $K = K_1 \times K_2 = 5.6 \times 10^3$

        (ii)

$NH_4^+(aq) \rightleftharpoons H^+(aq) + NH_3(aq)$ $K_1 = K_a$ for $NH_4^+ = 5.6 \times 10^{-10}$

$CO_3^{2-}(aq) + H^+(aq) \rightleftharpoons HCO_3^-(aq)$ $K_2 = 1/K_{a2}$ for $H_2CO_3 = 1.8 \times 10^{10}$

$\rule{9cm}{0.4pt}$

$NH_4^+(aq) + CO_3^{2-}(aq) \rightleftharpoons HCO_3^-(aq) + NH_3(aq)$ $K = K_1 \times K_2 = 10$

  (b)   Both (i) and (ii) have K > 1, although K = 10 is not **much** greater than 1. Both could be written with a single arrow. (This is true in general when a strong acid or strong base, $H^+(aq)$ or $OH^-(aq)$, is a reactant.)

17 Additional Aspects of Aqueous Equilibria

Visualizing Concepts

17.1   *Analyze.* Given diagrams showing equilibrium mixtures of HX and X^- with different compositions, decide which has the highest pH. HX is a weak acid and X^- is its conjugate base. *Plan.* Evaluate the contents of the boxes. Use acid-base equilibrium principles to relate $[H^+]$ to box composition.   *Solve.*

Use the following acid ionization equilibrium to describe the mixtures:
$HX(aq) \rightleftharpoons H^+(aq) + X^-(aq)$. Each box has 4 HX molecules, but differing amounts of X^- ions. The greater the amount of X^- (conjugate base), for the same amount of HX (weak acid), the lower the amount of H^+ and the higher the pH. The middle box, with most X^-, has least H^+ and highest pH.

17.3   *Analyze/Plan.* When strong acid is added to a buffer, it reacts with conjugate base (CB) to produce conjugate acid (CA). [CA] increases and [CB] decreases. The opposite happens when strong base is added to a buffer, [CB] increases and [CA] decreases. Match these situations to the drawings.   *Solve.*

The buffer begins with equal concentrations of HX and X^-.

(a)   After addition of strong acid, [HX] will increase and $[X^-]$ will decrease. Drawing (3) fits this description.

(b)   Adding of strong base causes [HX] to decrease and $[X^-]$ to increase. Drawing (1) matches the description.

(c)   Drawing (2) shows both [HX] and $[X^-]$ to be smaller than the initial concentrations shown on the left. This situation cannot be achieved by adding strong acid or strong base to the original buffer.

17.6   *Analyze/Plan.* The product of the ion concentrations in a saturated solution equals K_{sp}. Use numbers of anions and cations as a measure of concentration to calculate relative "K_{sp}" values. Counting cations is not adequate, because excess anions in some of the boxes drive down the cation concentrations. Ion-products must be considered. *Solve.*

AgX: $(4\ Ag^+)(4X^-) = 16$

AgY: $(1\ Ag^+)(9\ Y^-) = 9$

AgZ: $(3\ Ag^+)(6\ Y^-) = 18$

AgY has the smallest K_{sp}.

17 Additional Aspects of Aqueous Equilibria

Common-Ion Effect

17.9 (a) The extent of ionization of a weak electrolyte is decreased when a strong electrolyte containing an ion in common with the weak electrolyte is added to it.

(b) $NaNO_2$

17.11 *Analyze/Plan.* Given the formula of two substances, determine the effect on pH when one is added to the other. In general, when an acid is added to a solution, pH decreases; when a base is added to a solution, pH increases. Based on its formula, determine whether the substance being added is an acid or a base and predict the change in pH of the solution. *Solve.*

(a) pH increases; NO_2^- decreases the ionization of HNO_2 and decreases $[H^+]$.

(b) pH decreases; $CH_3NH_3^+$ decreases the ionization (hydrolysis) of CH_3NH_2 and decreases $[OH^-]$.

(c) pH increases; CHO_2^- decreases the ionization of $HCHO_2$ and decreases $[H^+]$.

d) No change; Br^- is a negligible base and does not affect the 100% ionization of the strong acid HBr.

(e) pH decreases; the pertinent equilibrium is

$$C_2H_3O_2^-(aq) + H_2O(l) \rightleftharpoons HC_2H_3O_2 + OH^-(aq).$$

HCl reacts with $OH^-(aq)$, decreasing $[OH^-]$ and pH.

17.13 *Analyze/Plan.* Follow the logic in Sample Exercise 17.1.

(a)

	$HC_3H_5O_2$ (aq)	$\rightleftharpoons$	H^+ (aq)	+	$C_3H_5O_2^-$ (aq)
i	0.085 M				0.060 M
c	$-x$		$+x$		$+x$
e	$(0.085 - x)\,M$		$+x\,M$		$(0.060 + x)\,M$

$$K_a = 1.3 \times 10^{-5} = \frac{[H^+][C_3H_5O_2^-]}{[HC_3H_5O_2]} = \frac{(x)(0.060 + x)}{(0.085 - x)}$$

Assume x is small compared to 0.060 and 0.085.

$$1.3 \times 10^{-5} = \frac{0.060\,x}{0.085} ; x = 1.8 \times 10^{-5} = [H^+], pH = 4.73$$

Check. Since the extent of ionization of a weak acid or base is suppressed by the presence of a conjugate salt, the 5% rule usually holds true in buffer solutions.

(b)

	$(CH_3)_3N$(aq) + H_2O(l)	$\rightleftharpoons$	$(CH_3)_3NH^+$(aq)	+	OH^- (aq)
i	0.075 M		0.10 M		
c	$-x$		$+x$		$+x$
e	$(0.075 - x)\,M$		$(0.10 + x)\,M$		$+x\,M$

$$K_b = 6.4 \times 10^{-5} = \frac{[OH^-][(CH_3)_3NH^+]}{[(CH_3)_3N]} = \frac{(x)(0.10+x)}{(0.075-x)} \approx \frac{0.10\,x}{0.075}$$

$x = 4.8 \times 10^{-5} = [OH^-]$, pOH = 4.32, pH = 14.00 – 4.32 = 9.68

Check. In a buffer, if [conj. acid] > [conj. base], pH < pK_a of the conj. acid.
If [conj. acid] < [conj. base], pH > pK_a of the conj. acid. In this buffer, pK_a of
$(CH_3)_3NH^+$ is 9.81. $[(CH_3)_3N]$ and pH = 9.61, less than 9.81.

(c)   mol = $M \times$ L; mol $HC_2H_3O_2$ = 0.15 M $\times$ 0.0500 L = 7.5×10^{-3} mol

mol $C_2H_3O_2^-$ = 0.20 $M \times$ 0.0500 L = 0.010 mol

	$HC_2H_3O_2(aq)$	$\rightleftharpoons$	$H^+(aq)$	+	$C_2H_3O_2^-(aq)$
i	7.5×10^{-3} mol		0		0.010 mol
c	$-x$		$+x$		$+x$
e	$(7.5 \times 10^{-3}-x)$ mol		x mol		$(0.010 + x)$ mol

$[HC_2H_3O_2] = (7.5 \times 10^{-3} - x)$ mol/0.1000 L; $[C_2H_3O_2^-] = (0.010 + x)$ mol/0.1000 L

$$K_a = 1.8 \times 10^{-5} = \frac{[H+][C_2H_3O_2]}{[HC_2H_3O_2]} = \frac{(x)(0.010+x)/0.1000\,L}{(0.0075-x)/0.1000\,L} \approx \frac{x(0.010)}{0.0075}$$

$x = 1.35 \times 10^{-5}\,M = 1.4 \times 10^{-5}\,M\,H^+$; pH = 4.87

Check. pK_a for $HC_2H_3O_2$ = 4.74. $[C_2H_3O_2^-] > [HC_2H_3O_2]$, pH of buffer = 4.87,
greater than 4.74.

$$K_a = 1.8 \times 10^{-4} = \frac{[H^+][CHO_2^-]}{[HCHO_2]} = \frac{(x)(0.160+x)}{(0.260-x)} \approx \frac{0.160\,x}{0.260}$$

$x = 2.93 \times 10^{-4} = 2.9 \times 10^{-4}\,M = [H^+]$, pH = 3.53

Check. Since the extent of ionization of a weak acid or base is suppressed by the
presence of a conjugate salt, the 5% rule usually holds true in buffer solutions.

17.15   *Analyze/Plan.* We are asked to calculate % ionization of (a) a weak acid and (b) a weak
acid in a solution containing a common ion, its conjugate base. Calculate % ionization
as in Sample Exercise 16.12. In part (b), the concentration of the common ion is 0.085 M,
not x, as in part (a).   *Solve.*

	HBu(aq)	$\rightleftharpoons$	H^+ (aq)	+	Bu^-(aq)	$K_a = \dfrac{[H^+][Bu^-]}{[HBu]} = 1.5 \times 10^{-5}$
equil (a)	0.0075– x M		x M		x M	
equil (b)	0.0075– x M		x M		0.085 + x M	

(a) $K_a = 1.5 \times 10^{-5} = \dfrac{x^2}{0.0075-x} \approx \dfrac{x^2}{0.0075}$; $x = [H^+] = 3.354 \times 10^{-4} = 3.4 \times 10^{-4}\,M\,H^+$

      % ionization = $\dfrac{3.4 \times 10^{-4}\,M\,H^+}{0.0075\,M\,HBu} \times 100 = 4.5\%$ ionization

(b) $K_a = 1.5 \times 10^{-5} = \dfrac{(x)(0.085+x)}{0.0075-x} \approx \dfrac{0.085\,x}{0.0075}; x = 1.3 \times 10^{-6}\ M\,H^+$

% ionization $= \dfrac{1.3 \times 10^{-6}\ M\,H^+}{0.0075\ M\ HBu} \times 100 = 0.018\%$ ionization

Check. Percent ionization is much smaller when the "common ion" is present.

Buffers

17.17 $HC_2H_3O_2$ and $NaC_2H_3O_2$ are a weak conjugate acid/conjugate base pair which acts as a buffer because unionized $HC_2H_3O_2$ reacts with added base, while $C_2H_3O_2^-$ combines with added acid, leaving $[H^+]$ relatively unchanged. Although HCl and NaCl are a conjugate acid/conjugate base pair, Cl^- is a negligible base. That is, it has no tendency to combine with added acid to form molecular HCl. Any added acid simply increases $[H^+]$ in an HCl/NaCl mixture. In general, the conjugate bases of strong acids are negligible and mixtures of strong acids and their conjugate salts do not act as buffers.

17.19   *Analyze/Plan.* Follow the logic in Sample Exercise 17.3. Assume that % ionization is small in these buffers (Solutions 17.15 and 17.16).   *Solve.*

(a) $K_a = \dfrac{[H^+][Lac^-]}{[HLac]}; [H^+] = \dfrac{[K_a][HLac]}{[Lac^-]} = \dfrac{1.4 \times 10^{-4}\,(0.12)}{(0.11)}$

$[H^+] = 1.53 \times 10^{-4} = 1.5 \times 10^{-4}$; pH = 3.82

(b)   mol $= M \times L$; total volume = 85 mL + 95 mL = 180 mL

$[H^+] = \dfrac{K_a[HLac]}{[Lac^-]} = \dfrac{1.4 \times 10^{-4}\,(0.13\ M \times 0.085\ L)/0.180\ L}{(0.15\ M \times 0.095\ L)/0.180\ L} = \dfrac{1.4 \times 10^{-4}\,(0.13 \times 0.085)}{(0.15 \times 0.095)}$

$[H^+] = 1.086 \times 10^{-4} = 1.1\ M\,H^+$; pH = 3.96

17.21   (a)   *Analyze/Plan.* Follow the logic in Sample Exercises 17.1 and 17.3. As in Sample Exercise 17.1, start by calculating concentrations of the components.   *Solve.*

$HC_2H_3O_2(aq) \rightleftharpoons H^+(aq) + C_2H_3O_2^-(aq);\ K_a = 1.8 \times 10^{-5} = \dfrac{[H^+][C_2H_3O_2^-]}{[HC_2H_3O_2]}$

$[HC_2H_3O_2] = \dfrac{20.0\ g\ HC_2H_3O_2}{2.00\ L\ soln} \times \dfrac{1\ mol\ HC_2H_3O_2}{60.05\ g\ HC_2H_3O_2} = 0.167\ M$

$[C_2H_3O_2^-] = \dfrac{20.0\ g\ NaC_2H_3O_2}{2.00\ L\ soln} \times \dfrac{1\ mol\ NaC_2H_3O_2}{82.04\ g\ NaC_2H_3O_2} = 0.122\ M$

$[H^+] = \dfrac{K_a[HC_2H_3O_2]}{[C_2H_3O_2^-]} = \dfrac{1.8 \times 10^{-5}\,(0.167-x)}{(0.122+x)} \approx \dfrac{1.8 \times 10^{-5}\,(0.167)}{(0.122)}$

$[H^+] = 2.4843 \times 10^{-5} = 2.5 \times 10^{-5}\ M$, pH = 4.60

(b)   *Plan.* On the left side of the equation, write all ions present in solution after HCl or NaOH is added to the buffer. Using acid-base properties and relative strengths, decide which ions will combine to form new products.   *Solve.*

$$Na^+(aq) + C_2H_3O_2^-(aq) + H^+(aq) + Cl^-(aq) \rightarrow HC_2H_3O_2(aq) + Na^+(aq) + Cl^-(aq)$$

(c)　$HC_2H_3O_2(aq) + Na^+(aq) + OH^-(aq) \rightarrow C_2H_3O_2^-(aq) + H_2O(l) + Na^+(aq)$

17.23　*Analyze/Plan.* Follow the logic in Sample Exercise 17.4.　*Solve.*

In this problem, $[BrO^-]$ is the unknown.

pH = 9.15, $[H^+] = 10^{-9.15} = 7.0795 \times 10^{-10} = 7.1 \times 10^{-10}\ M$

$[HBrO] = 0.050 - 7.1 \times 10^{-10} \approx 0.050\ M$

$K_a = 2.5 \times 10^{-9} = \dfrac{7.0795 \times 10^{-10}\ [BrO^-]}{0.050}$; $[BrO] = 0.1766 = 0.18\ M$

For 1.00 L, 0.18 mol NaBrO are needed.

17.25　*Analyze/Plan.* Follow the logic in Sample Exercise 17.3 and 17.5.　*Solve.*

(a)　$K_a = \dfrac{[H^+][C_2H_3O_2^-]}{[HC_2H_3O_2]}$; $[H^+] = \dfrac{K_a[HC_2H_3O_2]}{[C_2H_3O_2^-]}$

$[H^+] \approx \dfrac{1.8 \times 10^{-5}\ (0.10)}{(0.13)} = 1.385 \times 10^{-5} = 1.4 \times 10^{-5}\ M$; pH = 4.86

(b)　$HC_2H_3O_2(aq)$　+　$KOH(aq)$　$\rightarrow$　$C_2H_3O_2^-(aq) + H_2O(l) + K^+(aq)$

	0.10 mol	0.02 mol	0.13 mol
	–0.02 mol	–0.02 mol	+0.02 mol
	0.08 mol	0 mol	0.15 mol

$[H^+] = \dfrac{1.8 \times 10^{-5}\ (0.08\ mol/0.100\ L)}{(0.15\ mol/0.100\ L)} = 9.60 \times 10^{-6} = 1 \times 10^{-5}\ M$; pH = 5.02 = 5.0

(c)　$C_2H_3O_2^-(aq)$　+　$HNO_3(aq)$　$\rightarrow$　$HC_2H_3O_2(aq) + NO_3^-(aq)$

	0.13 mol	0.02 mol	0.10 mol
	–0.02 mol	–0.02 mol	+0.02 mol
	0.11 mol	0 mol	0.12 mol

$[H^+] = \dfrac{1.8 \times 10^{-5}\ (0.12\ mol/0.100\ L)}{(0.11\ mol/0.100\ L)} = 1.96 \times 10^{-5} = 2.0 \times 10^{-5}\ M$; pH = 4.71

17.27　*Analyze/Plan.* Calculate the [conj. base]/[conj. acid] ratio in the H_2CO_3/HCO_3^- blood buffer. Write the acid dissociation equilibrium and K_a expression. Find K_a for H_2CO_3 in Appendix D. Calculate $[H^+]$ from the pH and solve for the ratio.　*Solve.*

$H_2CO_3(aq) \rightleftharpoons H^+(aq) + HCO_3^-(aq)$　$K_a = \dfrac{[H^+][HCO_3^-]}{[H_2CO_3]}$; $\dfrac{[HCO_3^-]}{[H_2CO_3]} = \dfrac{K_a}{[H^+]}$

(a)　at pH = 7.4, $[H^+] = 10^{-7.4} = 4.0 \times 10^{-8}\ M$; $\dfrac{[HCO_3^-]}{[H_2CO_3]} = \dfrac{4.3 \times 10^{-7}}{4.0 \times 10^{-8}} = 11$

(b)   at pH $= 7.1$, $[H^+] = 7.9 \times 10^{-8}$ M; $\dfrac{[HCO_3^-]}{[H_2CO_3]} = 5.4$

17.29   *Analyze.* Given six solutions, decide which two should be used to prepare a pH 3.50 buffer. Calculate the volumes of the two 0.10 M solutions needed to make approximately 1 L of buffer.

Plan. A buffer must contain a conjugate acid/conjugate base (CA/CB) pair. By examining the chemical formulas, decide which pairs of solutions could be used to make a buffer. If there is more than one possible pair, calculate pK_a for the acids. A buffer is most effective when its pH is within 1 pH unit of pK_a for the conjugate acid component. Select the pair with pK_a nearest to 3.50. Use Equation [17.9] to calculate the [CB]/[CA] ratio and the volumes of 0.10 M solutions needed to prepare 1 L of buffer. *Solve.*

There are three CA/CB pairs:

$HCHO_2/NaCHO_2$, $pK_a = 3.74$

$HC_2H_3O_2/NaC_2H_3O_2$, $pK_a = 4.74$

H_3PO_4/NaH_2PO_4, $pK_a = 2.12$

The most appropriate solutions are $HCHO_2/NaCHO_2$, because pK_a for $HCHO_2$ is nearest to 3.50.

$$pH = pKa + \log\frac{[CB]}{[CA]} \quad 3.50 = 3.7447 + \log\frac{[NaCHO_2]}{[HCHO_2]}$$

$$\log\frac{[NaCHO_2]}{[HCHO_2]} = -0.2447, \frac{[NaCHO_2]}{[HCHO_2]} = 0.5692 = 0.57$$

Since we are making a total of 1 L of buffer, let y = vol $NaCHO_2$ and
(1 – y) = vol $HCHO_2$.

$$0.5692 = \frac{[NaCHO_2]}{[HCHO_2]} = \frac{(0.1\,M \times y)/1L}{[0.10\,M \times (1-y)]/1\,L}; 0.5692[0.10(1-y)] = 0.10\,y;$$

0.05692 = 0.15692 y; y = 0.3627 = 0.36 L

360 mL of 0.10 M $NaCHO_2$, 640 mL of 0.10 M $HCHO_2$.

Check. The pH of the buffer is less than pK_a for the conjugate acid, indicating that the amount of CA in the buffer is greater than the amount of CB. This agrees with our result.

Acid-Base Titrations

17.31   (a)   Curve B. The initial pH is lower and the equivalence point region is steeper.

(b)   pH at the approximate equivalence point of curve A = 8.0

pH at the approximate equivalence point of curve B = 7.0

(c)   Volume of base required to reach the equivalence point depends only on moles of acid present; it is independent of acid strength. Since acid B requires 40 mL and acid A requires only 30 mL, more moles of acid B are being titrated. For equal volumes of A and B, the concentration of acid B is greater.

17.33   *Analyze.* Given reactants, predict whether pH at the equivalence point of a titration is less than, equal to or greater than 7.

Plan. At the equivalence point of a titration, only product is present in solution; there is no excess of either reactant. Determine the product of each reaction and whether a solution of it is acidic, basic or neutral.   *Solve.*

(a) $NaHCO_3(aq) + NaOH(aq) \rightarrow Na_2CO_3(aq) + H_2O(l)$

At the equivalence point, the major species in solution are Na^+ and CO_3^{2-}. Na^+ is negligible and CO_3^{2-} is the CB of HCO_3^-. The solution is basic, above pH 7.

(b) $NH_3(aq) + HCl(aq) \rightarrow NH_4Cl(aq)$

At the equivalence point, the major species are NH_4^+ and Cl^-. Cl^- is negligible, NH_4^+ is the CA of NH_3. The solution is acidic, below pH 7

(c) $KOH(aq) + HBr(aq) \rightarrow KBr(aq) + H_2O(l)$

At the equivalence point, the major species are K^+ and Br^-; both are negligible. The solution is at pH 7.

17.35   (a)   HX is weaker. The pH at the equivalence point is determined by the identity and concentration of the conjugate base, X^- or Y^-. The higher the pH at the equivalence point, the stronger the conjugate base (X^-) and the weaker the conjugate acid (HX).

(b)   Phenolphthalein, which changes color in the pH 8-10 range, is perfect for HX and probably appropriate for HY. Bromthymol blue changes from 6-7.5, and thymol blue between from 8-9.5, but these are two-color indicators. One-color indicators such as phenolphthalein are preferred because detection of the color change is more reproducible.

17.37   *Analyze/Plan.* We are asked to calculate the volume of 0.0850 *M* NaOH required to titrate various acid solutions to their equivalence point. At the equivalence point, moles base added equals moles acid initially present. Solve the stoichiometry problem, recalling that mol = *M* × L. In part (c) calculate molarity of HCl from g/L and proceed as outlined above.   *Solve.*

(a) $40.0 \text{ mL HNO}_3 \times \dfrac{0.0900 \text{ mol HNO}_3}{1000 \text{ mL soln}} \times \dfrac{1 \text{ mol NaOH}}{1 \text{ mol HNO}_3} \times \dfrac{1000 \text{ mL soln}}{0.0850 \text{ mol NaOH}}$

$= 42.353 = 42.4$ mL NaOH soln

(b) $35.0 \text{ mL HC}_2\text{H}_3\text{O}_2 \times \dfrac{0.0850 \text{ } M \text{ HC}_2\text{H}_3\text{O}_2}{1000 \text{ mL soln}} \times \dfrac{1 \text{ mol NaOH}}{1 \text{ mol HC}_2\text{H}_3\text{O}_2} \times \dfrac{1000 \text{ mL soln}}{0.0850 \text{ mol NaOH}}$

$= 35.0$ mL NaOH soln

(c) $\dfrac{1.85 \text{ g HCl}}{1 \text{ L soln}} \times \dfrac{1 \text{ mol HCl}}{36.46 \text{ g HCl}} = 0.05074 = 0.0507 \ M \text{ HCl}$

$50.0 \text{ mL HCl} \times \dfrac{0.05074 \text{ mol HCl}}{1000 \text{ mL}} \times \dfrac{1 \text{ mol NaOH}}{1 \text{ mol HCl}} \times \dfrac{1000 \text{ mL soln}}{0.0850 \text{ mol NaOH}}$

$= 29.847 = 29.8 \text{ mL NaOH soln}$

17.39   *Analyze/Plan.* Follow the logic in Sample Exercise 17.6 for the titration of a strong acid with a strong base.   *Solve.*

moles $H^+ = M_{HBr} \times L_{HBr} = 0.200 \ M \times 0.0200 \text{ L} = 4.00 \times 10^{-3} \text{ mol}$

moles $OH^- = M_{NaOH} \times L_{NaOH} = 0.200 \ M \times L_{NaOH}$

	mL_{HBr}	mL_{NaOH}	Total Volume	Moles H^+	Moles OH^-	Molarity Excess Ion	pH
(a)	20.0	15.0	35.0	4.00×10^{-3}	3.00×10^{-3}	$0.0286(H^+)$	1.544
(b)	20.0	19.9	39.9	4.00×10^{-3}	3.98×10^{-3}	$5 \times 10^{-4}(H^+)$	3.3
(c)	20.0	20.0	40.0	4.00×10^{-3}	4.00×10^{-3}	$1 \times 10^{-7}(H^+)$	7.0
(d)	20.0	20.1	40.1	4.00×10^{-3}	4.02×10^{-3}	$5 \times 10^{-4}(H^+)$	10.7
(e)	20.0	35.0	55.0	4.00×10^{-3}	7.00×10^{-3}	$0.0545(OH^-)$	12.737

molarity of excess ion = moles ion / total vol in L

(a) $\dfrac{4.00 \times 10^{-3} \text{ mol } H^+ - 3.00 \times 10^{-3} \text{ mol } OH^-}{0.0350 \text{ L}} = 0.0286 \ M \ H^+$

(b) $\dfrac{4.00 \times 10^{-3} \text{ mol } H^+ - 3.98 \times 10^{-3} \text{ mol } OH^-}{0.0339 \text{ L}} = 5.01 \times 10^{-4} = 5 \times 10^{-4} \ M \ H^+$

(c)   equivalence point, mol H^+ = mol OH^-

NaBr does not hydrolyze, so $[H^+] = [OH^-] = 1 \times 10^{-7} \ M$

(d) $\dfrac{4.02 \times 10^{-3} \text{ mol } H^+ - 4.00 \times 10^{-3} \text{ mol } OH^-}{0.041 \text{ L}} = 4.88 \times 10^{-4} = 5 \times 10^{-4} \ M \ OH^-$

(e) $\dfrac{7.00 \times 10^{-3} \text{ mol } H^+ - 4.00 \times 10^{-3} \text{ mol } OH^-}{0.0550 \text{ L}} = 0.054545 = 0.0545 \ M \ OH^-$

17.41   *Analyze/Plan.* Follow the logic in Sample Exercise 17.7 for the titration of a weak acid with a strong base.   *Solve.*

(a)   At 0 mL, only weak acid, $HC_2H_3O_2$, is present in solution. Using the acid ionization equilibrium

$$HC_2H_3O_2(aq) \rightleftharpoons H^+(aq) \ + \ C_2H_3O_2^-(aq)$$

Initial	0.150 M	0	0
equil	0.150 – x M	x M	x M

$$K_a = \frac{[H^+][C_2H_3O_2^-]}{[HC_2H_3O_2]} = 1.8 \times 10^{-5} \text{ (Appendix D)}$$

$$1.8 \times 10^{-5} = \frac{x^2}{(0.150-x)} \approx \frac{x^2}{0.150}; x^2 = 2.7 \times 10^{-6}; x = [H^+] = 0.001643$$

$$= 1.6 \times 10^{-3} \ M; pH = 2.78$$

(b)–(f) Calculate the moles of each component after the acid-base reaction takes place. Moles $HC_2H_3O_2$ originally present = $M \times L$ = 0.150 M × 0.0350 L = 5.25 × 10⁻³ mol. Moles NaOH added = $M \times L$ = 0.150 M × y mL.

		NaOH(aq) +	HC₂H₃O₂(aq) →	Na⁺C₂H₃O₂⁻(aq)+H₂O(l)
		(0.150 M × 0.0175 L) =		
(b)	before rx	2.625 × 10⁻³ mol	5.25 × 10⁻³ mol	
	after rx	0	**2.625 × 10⁻³ mol**	**2.63 × 10⁻³ mol**
		(0.150 M × 0.0345 L) =		
(c)	before rx	5.175 × 10⁻³ mol	5.25 × 10⁻³ mol	
	after rx	0	**0.075 × 10⁻³ mol**	**5.18 × 10⁻³ mol**
		(0.150 M × 0.0350 L) =		
(d)	before rx	5.25 × 10⁻³ mol	5.25 × 10⁻³ mol	
	after rx	0	0	**5.25 × 10⁻³ mol**
		(0.150 M × 0.0355 L) =		
(e)	before rx	5.325 × 10⁻³ mol	5.25 × 10⁻³ mol	
	after rx	**0.075 × 10⁻³ mol**	0	5.25 × 10⁻³ mol
		(0.150 M × 0.0500 L) =		
(f)	before rx	7.50 × 10⁻³ mol	5.25 × 10⁻³ mol	
	after rx	**2.25 × 10⁻³ mol**	0	5.25 × 10⁻³ mol

Calculate the molarity of each species (M = mol/L) and solve the appropriate equilibrium problem in each part.

(b) V_T = 35.0 mL $HC_2H_3O_2$ + 17.5 mL NaOH = 52.5 mL = 0.0525 L

$$[HC_2H_3O_2] = \frac{2.625 \times 10^{-3} \text{ mol}}{0.0525} = 0.0500 \ M$$

$$[C_2H_3O_2^-] = \frac{2.625 \times 10^{-3} \text{ mol}}{0.0525} = 0.0500 \ M$$

$$HC_2H_3O_2(aq) \rightleftharpoons H^+(aq) + C_2H_3O_2^-(aq)$$

equil   0.0500 – x M x M 0.0500 + x M

$$K_a = \frac{[H^+][C_2H_3O_2^-]}{[HC_2H_3O_2]} ; [H^+] = \frac{K_a[HC_2H_3O_2]}{[C_2H_3O_2^-]}$$

$$[H^+] = \frac{1.8 \times 10^{-5}(0.0500-x)}{(0.0500+x)} = 1.8 \times 10^{-5} \ M \ H^+; pH = 4.74$$

(c) $[HC_2H_3O_2] = \dfrac{7.5 \times 10^{-5} \ \text{mol}}{0.0695 \ L} = 0.001079 = 1.1 \times 10^{-3} \ M$

$$[C_2H_3O_2^-] = \frac{5.175 \times 10^{-3} \ \text{mol}}{0.0695 \ L} = 0.07446 = 0.074 \ M$$

$$[H^+] = \frac{1.8 \times 10^{-5}(1.079 \times 10^{-3} - x)}{(0.07446 + x)} \approx 2.6 \times 10^{-7} \ M \ H^+; pH = 6.58$$

(d)   At the equivalence point, only $C_3H_5O_2^-$ is present.

$$[C_2H_3O_2^-] = \frac{5.25 \times 10^{-3} \ \text{mol}}{0.0700 \ L} = 0.0750 \ M$$

The pertinent equilibrium is the base hydrolysis of $C_2H_3O_2^-$.

	$C_2H_3O_2^-(aq)$ + $H_2O(l)$	$\rightleftharpoons$	$HC_2H_3O_2(aq)$ +	$OH^-(aq)$
initial	0.0750 M		0	0
equil	0.0750 – x M		x	x

$$K_b = \frac{K_w}{K_a \text{ for } HC_2H_3O_2} = \frac{1.0 \times 10^{-14}}{1.8 \times 10^{-5}} = 5.56 \times 10^{-10} = 5.6 \times 10^{-10} = \frac{[HC_2H_3O_2][OH^-]}{[C_2H_3O_2^-]}$$

$$5.56 \times 10^{-10} = \frac{x^2}{0.0750 - x}; x^2 \approx 5.56 \times 10^{-10}(0.0750); x = 6.458 \times 10^{-6}$$

$$= 6.5 \times 10^{-6} \ M \ OH^-$$

$$pOH = -\log(6.458 \times 10^{-6}) = 5.19; pH = 14.00 - pOH = 8.81$$

(e)   After the equivalence point, the excess strong base determines the pOH and pH. The $[OH^-]$ from the hydrolysis of $C_2H_3O_2^-$ is small and can be ignored.

$$[OH^-] = \frac{0.075 \times 10^{-3} \ \text{mol}}{0.0705 \ L} = 1.064 \times 10^{-3} = 1.1 \times 10^{-3} \ M; pOH = 2.97$$

$$pH = 14.00 - 2.97 = 11.03$$

(f) $[OH^-] = \dfrac{2.25 \times 10^{-3} \ \text{mol}}{0.0850 \ L} = 0.0265 \ M \ OH^-; pOH = 1.577; pH = 14.00 - 1.577 = 12.423$

17.43   *Analyze/Plan.* Calculate the pH at the equivalence point for the titration of several bases with 0.200 M HBr. The volume of 0.200 M HBr required in all cases equals the volume of base and the final volume = $2V_{base}$. The concentration of the salt produced at the equivalence point is $\dfrac{0.200 \ M \times V_{base}}{2V_{base}} = 0.100 \ M.$

In each case, identify the salt present at the equivalence point, determine its acid-base properties (Section 16.9), and solve the pH problem. *Solve.*

(a)   NaOH is a strong base; the salt present at the equivalence point, NaBr, does not affect the pH of the solution. 0.100 M NaBr, pH = 7.00

(b) $HONH_2$ is a weak base, so the salt present at the equivalence point is $HONH_3^+Br^-$. This is the salt of a strong acid and a weak base, so it produces an acidic solution.

0.100 M $HONH_3^+Br^-$; $HONH_3^+(aq)$ ⇌ $H^+(aq)$ + $HONH_2$

                                [equil]   0.100 – x   x   x

$$K_a = \frac{[H^+][HONH_2]}{[HONH_3^+]} = \frac{K_w}{K_b} = \frac{1.0 \times 10^{-14}}{1.1 \times 10^{-8}} = 9.09 \times 10^{-7} = 9.1 \times 10^{-7}$$

Assume x is small with respect to [salt].

$K_a = x^2 / 0.100$; x = $[H^+]$ = 3.02×10^{-4} = 3.0×10^{-4} M, pH = 3.52

(c) $C_6H_5NH_2$ is a weak base and $C_6H_5NH_3^+Br^-$ is an acidic salt.

0.100 M $C_6H_5NH_3^+Br^-$.   Proceeding as in (b):

$$K_a = \frac{[H^+][C_6H_5NH_2]}{[C_6H_5NH_3^+]} = \frac{K_w}{K_b} = 2.33 \times 10^{-5} = 2.3 \times 10^{-5}$$

$[H^+]^2 = 0.100(2.33 \times 10^{-5})$; $[H^+]$ = 1.52×10^{-3} = 1.5×10^{-3} M, pH = 2.82

Solubility Equilibria and Factors Affecting Solubility

17.45   (a)   The concentration of undissolved solid does not appear in the solubility product expression because it is constant as long as there is solid present. Concentration is a ratio of moles solid to volume of the solid; solids occupy a specific volume not dependent on the solution volume. As the amount (moles) of solid changes, the volume changes proportionally, so that the ratio of moles solid to volume solid is constant.

(b)   *Analyze/Plan.* Follow the example in Sample Exercise 17.9.   *Solve.*

$K_{sp} = [Ag^+][I^-]$; $K_{sp} = [Sr^{2+}][SO_4^{2-}]$; $K_{sp} = [Fe^{2+}][OH^-]^2$; $K_{sp} = [Hg_2^{2+}][Br^-]^2$

17.47   *Analyze/Plan.* Follow the logic in Sample Exercise 17.10.   *Solve.*

(a) $CaF_2(s)$ ⇌ $Ca^{2+}(aq)$ + $2F^-(aq)$; $K_{sp} = [Ca^{2+}][F^-]^2$

The molar solubility is the moles of CaF_2 that dissolve per liter of solution. Each mole of CaF_2 produces 1 mol $Ca^{2+}(aq)$ and 2 mol $F^-(aq)$.

$[Ca^{2+}]$ = 1.24×10^{-3} M; $[F^-]$ = $2 \times 1.24 \times 10^{-3}$ M = 2.48×10^{-3} M

K_{sp} = $(1.24 \times 10^{-3})(2.48 \times 10^{-3})^2$ = 7.63×10^{-9}

(b) $SrF_2(s) \rightleftharpoons Sr^{2+}(aq) + 2F^-(aq)$; $K_{sp} = [Sr^{2+}][F^-]^2$

Transform the gram solubility to molar solubility.

$$\frac{1.1 \times 10^{-2} \text{ g } SrF_2}{0.100 \text{ L}} \times \frac{1 \text{ mol } SrF_2}{125.6 \text{ g } SrF_2} = 8.76 \times 10^{-4} = 8.8 \times 10^{-4} \text{ mol } SrF_2 / L$$

$[Sr^{2+}] = 8.76 \times 10^{-4} M$; $[F^-] = 2(8.76 \times 10^{-4} M)$

$K_{sp} = (8.76 \times 10^{-4})(2(8.76 \times 10^{-4}))^2 = 2.7 \times 10^{-9}$

(c) $Ba(IO_3)_2(s) \rightleftharpoons Ba^{2+}(aq) + 2IO_3^-(aq)$; $K_{sp} = [Ba^{2+}][IO_3^-]^2$

Since 1 mole of dissolved $Ba(IO_3)_2$ produces 1 mole of Ba^{2+}, the molar solubility of

$Ba(IO_3)_2 = [Ba^{2+}]$. Let $x = [Ba^{2+}]$; $[IO_3^-] = 2x$

$K_{sp} = 6.0 \times 10^{-10} = (x)(2x)^2$; $4x^3 = 6.0 \times 10^{-10}$; $x^3 = 1.5 \times 10^{-10}$; $x = 5.3 \times 10^{-4} M$

The molar solubility of $Ba(IO_3)_2$ is 5.3×10^{-4} mol/L.

17.49   *Analyze/Plan.* Given gram solubility of a compound, calculate K_{sp}. Write the dissociation equilibrium and K_{sp} expression. Change gram solubility to molarity of the individual ions, taking the stoichiometry of the compound into account. Calculate K_{sp}. *Solve.*

$CaC_2O_4(s) \rightleftharpoons Ca^{2+}(aq) + C_2O_4^{2-}(aq)$; $K_{sp} = [Ca^{2+}][C_2O_4^{2-}]$

$$[Ca^{2+}] = [C_2O_4^{2-}] = \frac{0.0061 \text{ g } CaC_2O_4}{1.00 \text{ L soln}} \times \frac{1 \text{ mol } CaC_2O_4}{128.1 \text{ g } CaC_2O_4} = 4.76 \times 10^{-5} = 4.8 \times 10^{-5} M$$

$K_{sp} = (4.76 \times 10^{-5} M)(4.76 \times 10^{-5} M) = 2.3 \times 10^{-9}$

17.51   *Analyze/Plan.* Follow the logic in Sample Exercises 17.11 and 17.12.   *Solve.*

(a) $AgBr(s) \rightleftharpoons Ag^+(aq) + Br^-(aq)$; $K_{sp} = [Ag^+][Br^-] = 5.0 \times 10^{-13}$

molar solubility $= x = [Ag^+] = [Br^-]$; $K_{sp} = x^2$

$x = (5.0 \times 10^{-13})^{1/2}$; $x = 7.1 \times 10^{-7}$ mol AgBr/L

(b)   Molar solubility $= x = [Br^-]$; $[Ag^+] = 0.030 M + x$

$K_{sp} = (0.030 + x)(x) \approx 0.030(x)$

$5.0 \times 10^{-13} = 0.030(x)$; $x = 1.7 \times 10^{-11}$ mol AgBr/L

(c)   Molar solubility $= x = [Ag^+]$

There are two sources of Br^-: NaBr(0.10 M) and AgBr(x M)

$K_{sp} = (x)(0.10 + x)$; Assuming x is small compared to 0.10 M

$5.0 \times 10^{-13} = 0.10 (x)$; $x \approx 5.0 \times 10^{-12}$ mol AgBr/L

17.53   *Analyze/Plan.* We are asked to calculate the solubility of a slightly-soluble hydroxide salt at various pH values. This is a common ion problem; pH tells us not only $[H^+]$ but also $[OH^-]$, which is an ion common to the salt. Use pH to calculate $[OH^-]$, then proceed as in Sample Exercise 17.12.   *Solve.*

$Mn(OH)_2(s) \rightleftharpoons Mn^{2+}(aq) + 2OH^-(aq); \quad K_{sp} = 1.6 \times 10^{-13}$

Since $[OH^-]$ is set by the pH of the solution, the solubility of $Mn(OH)_2$ is just $[Mn^{2+}]$.

(a)   pH = 7.0, pOH = 14 – pH = 7.0, $[OH^-] = 10^{-pOH} = 1.0 \times 10^{-7} M$

$$K_{sp} = 1.6 \times 10^{-13} = [Mn^{2+}](1.0 \times 10^{-7})^2; \quad [Mn^{2+}] = \frac{1.6 \times 10^{-13}}{1.0 \times 10^{-14}} = 16 M$$

$$\frac{16 \text{ mol } Mn(OH)_2}{1 L} \times \frac{88.95 \text{ g } Mn(OH)_2}{1 \text{ mol } Mn(OH)_2} = 1423 = 1.4 \times 10^3 \text{ g } Mn(OH)_2 / L$$

Check. Note that the solubility of $Mn(OH)_2$ in pure water is $3.6 \times 10^{-5} M$, and the pH of the resulting solution is 9.0. The relatively low pH of a solution buffered to pH 7.0 actually increases the solubility of $Mn(OH)_2$.

(b)   pH = 9.5, pOH = 4.5, $[OH^-] = 3.16 \times 10^{-5} = 3.2 \times 10^{-5} M$

$$K_{sp} = 1.6 \times 10^{-13} = [Mn^{2+}](3.16 \times 10^{-5})^2; \quad [Mn^{2+}] = \frac{1.6 \times 10^{-13}}{1.0 \times 10^{-9}} = 1.6 \times 10^{-4} M$$

$1.6 \times 10^{-4} M \, Mn(OH)_2 \times 88.95 \text{ g/mol} = 0.0142 = 0.014 \text{ g/L}$

(c)   pH = 11.8, pOH = 2.2, $[OH^-] = 6.31 \times 10^{-3} = 6.3 \times 10^{-3} M$

$$K_{sp} = 1.6 \times 10^{-13} = [Mn^{2+}](6.31 \times 10^{-3})^2; \quad [Mn^{2+}] = \frac{1.6 \times 10^{-13}}{3.98 \times 10^{-5}} = 4.0 \times 10^{-9} M$$

$4.02 \times 10^{-9} M \, Mn(OH)_2 \times 88.95 \text{ g/mol} = 3.575 \times 10^{-7} = 3.6 \times 10^{-7} \text{ g/L}$

17.55   *Analyze/Plan.* Follow the logic in Sample Exercise 17.13.   *Solve.*

If the anion of the salt is the conjugate base of a weak acid, it will combine with H^+, reducing the concentration of the free anion in solution, thereby causing more salt to dissolve. More soluble in acid: (a) $ZnCO_3$, (b) ZnS, (d) AgCN, (e) $Ba_3(PO_4)_2$

17.57   *Analyze/Plan.* Follow the logic in Sample Exercise 17.14.   *Solve.*

The formation equilibrium is

$$Cu^{2+}(aq) + 4NH_3(aq) \rightleftharpoons Cu(NH_3)_4{}^{2+}(aq) \quad K_f = \frac{[Cu(NH_3)_4{}^{2+}]}{[Cu^{2+}][NH_3]^4} = 5 \times 10^{12}$$

Assuming that nearly all the Cu^{2+} is in the form $Cu(NH_3)_4{}^{2+}$,

$[Cu(NH_3)_4{}^{2+}] = 1 \times 10^{-3} M; \; [Cu^{2+}] = x; \; [NH_3] = 0.10 M$

$$5 \times 10^{12} = \frac{(1 \times 10^{-3})}{x(0.10)^4}; \; x = 2 \times 10^{-12} M = [Cu^{2+}]$$

17.59   *Analyze/Plan.* We are asked to calculate K_{eq} for a particular reaction, making use of pertinent K_{sp} and K_f values from Appendix D and Table 17.1. Write the dissociation equilibrium for AgI and the formation reaction for $Ag(CN)_2{}^-$. Use algebra to manipulate these equations and their associated equilibrium constants to obtain the desired reaction and its equilibrium constant. *Solve.*

$$AgI(s) \rightleftharpoons Ag^+(aq) + I^-(aq)$$

$$\underline{Ag^+(aq) + 2CN^-(aq) \rightleftharpoons Ag(CN)_2^-(aq)}$$

$$AgI(s) + 2CN^-(aq) \rightleftharpoons Ag(CN)_2^-(aq) + I^-(aq)$$

$$K = K_{sp} \times K_f = [Ag^+][I^-] \times \frac{[Ag(CN)_2^-]}{[Ag^+][CN^-]^2} = (8.3 \times 10^{-17})(1 \times 10^{21}) = 8 \times 10^4$$

Precipitation; Qualitative Analysis

17.61 *Analyze/Plan.* Follow the logic in Sample Exercise 17.15. Precipitation conditions: will Q (see Chapter 15) exceed K_{sp} for the compound? *Solve.*

    (a)   In base, Ca^{2+} can form $Ca(OH)_2(s)$.

$$Ca(OH)_2(s) \rightleftharpoons Ca^{2+}(aq) + 2OH^-(aq); \quad K_{sp} = [Ca^{2+}][OH^-]^2$$

$$Q = [Ca^{2+}][OH^-]^2; [Ca^{2+}] = 0.050\ M; pOH = 14 - 8.0 = 6.0; [OH^-] = 1.0 \times 10^{-6}\ M$$

$$Q = (0.050)(1.0 \times 10^{-6})^2 = 5.0 \times 10^{-14}; K_{sp} = 6.5 \times 10^{-6} \text{ (Appendix D)}$$

$$Q < K_{sp}, \text{ no } Ca(OH)_2 \text{ precipitates.}$$

    (b) $Ag_2SO_4(s) \rightleftharpoons 2Ag^+(aq) + SO_4^{2-}(aq); \quad K_{sp} = [Ag+]^2[SO_4^{2-}]$

$$[Ag^+] = \frac{0.050\ M \times 100\ mL}{110\ mL} = 4.545 \times 10^{-2} = 4.5 \times 10^{-2}\ M$$

$$[SO_4^{2-}] = \frac{0.050\ M \times 10\ mL}{110\ mL} = 4.545 \times 10^{-3} = 4.5 \times 10^{-3}\ M$$

$$Q = (4.545 \times 10^{-2})^2 (4.545 \times 10^{-3}) = 9.4 \times 10^{-6}; K_{sp} = 1.5 \times 10^{-5}$$

$$Q < K_{sp}, \text{ no } Ag_2SO_4 \text{ precipitates.}$$

17.63 *Analyze/Plan.* We are asked to calculate pH necessary to precipitate $Mn(OH)_2(s)$ if the resulting Mn^{2+} concentration is $\leq 1\ \mu g/L$.

$$Mn(OH)_2(s) \rightleftharpoons Mn^{2+}(aq) + 2OH^-(aq); K_{sp} = [Mn^{2+}][OH^-]^2 = 1.6 \times 10^{-13}$$

At equilibrium, $[Mn^{2+}][OH^-]^2 = 1.6 \times 10^{-13}$. Change concentration $Mn^{2+}(aq)$ to mol/L and solve for $[OH^-]$. *Solve.*

$$\frac{1\ \mu g\ Mn^{2+}}{1\ L} \times \frac{1 \times 10^{-6}\ g}{1\ \mu g} \times \frac{1\ mol\ Mn^2 +}{54.94\ g\ Mn^{2+}} = 1.82 \times 10^{-8} = 2 \times 10^{-8}\ M\ Mn^{2+}$$

$$1.6 \times 10^{-13} = (1.82 \times 10^{-8})[OH^-]^2; [OH^-]^2 = 8.79 \times 10^{-6}; [OH^-] = 2.96 \times 10^{-3} = 3 \times 10^{-3}\ M$$

$$pOH = 2.53; pH = 14 - 2.53 = 11.47 = 11.5$$

17.65 *Analyze/Plan.* We are asked which ion will precipitate first from a solution containing $Pb^{2+}(aq)$ and $Ag^+(aq)$ when $I^-(aq)$ is added. Follow the logic in Sample Exercise 17.16. Calculate $[I^-]$ needed to initiate precipitation of each ion. The cation that requires lower $[I^-]$ will precipitate first. *Solve.*

Ag^+: $K_{sp} = [Ag^+][I^-]$; $8.3 \times 10^{-17} = (2.0 \times 10^{-4})[I^-]$; $[I^-] = \dfrac{8.3 \times 10^{-17}}{2.0 \times 10^{-4}} = 4.2 \times 10^{-13} \ M \ I^-$

Pb^{2+}: $K_{sp} = [Pb^{2+}][I^-]^2$; $7.9 \times 10^{-9} = (1.5 \times 10^{-3})[I^-]^2$; $[I^-] = \left(\dfrac{7.9 \times 10^{-9}}{1.5 \times 10^{-3}} \right)^{1/2} = 2.3 \times 10^{-3} \ M \ I^-$

AgI will precipitate first, at $[I^-] = 4.2 \times 10^{-13} \ M$.

17.67   *Analyze/Plan.* Use Figure 17.22 and the description of the five qualitative analysis "groups" in Section 17.7 to analyze the given data.   *Solve.*

The first two experiments eliminate Group 1 and 2 ions (Figure 17.22). The fact that no insoluble phosphates form in the filtrate from the third experiment rules out Group 4 ions. The ions which might be in the sample are those of Group 3, that is, Al^{3+}, Fe^{3+}, Zn^{2+}, Cr^{3+}, Ni^{2+}, Co^{2+}, or Mn^{2+}, and those of Group 5, NH_4^+, Na^+ or K^+.

17.69   *Analyze/Plan.* We are asked to devise a procedure to separate various pairs of ions in aqueous solutions. In each case, refer to Figure 17.22 to find a set of conditions where the solubility of the two ions differs. Construct a procedure to generate these conditions.   *Solve.*

(a) Cd^{2+} is in Gp. 2, but Zn^{2+} is not. Make the solution acidic using 0.5 M HCl; saturate with H_2S. CdS will precipitate, ZnS will not.

(b) $Cr(OH)_3$ is amphoteric but $Fe(OH)_3$ is not. Add excess base; $Fe(OH)_3(s)$ precipitates, but Cr^{3+} forms the soluble complex $Cr(OH)_4^-$.

(c) Mg^{2+} is a member of Gp. 4, but K^+ is not. Add $(NH_4)_2HPO_4$; Mg^{2+} precipitates as $MgNH_4PO_4$, K^+ remains in solution.

(d) Ag^+ is a member of Gp. 1, but Mn^{2+} is not. Add 6 M HCl, precipitate Ag^+ as $AgCl(s)$.

17.71   (a)   Because phosphoric acid is a weak acid, the concentration of free $PO_4^{3-}(aq)$ in an aqueous phosphate solution is low except in strongly basic media. In less basic media, the solubility product of the phosphates that one wishes to precipitate is not exceeded.

(b) K_{sp} for those cations in Group 3 is much larger. Thus, to exceed K_{sp} a higher $[S^{2-}]$ is required. This is achieved by making the solution more basic.

(c)   They should all redissolve in strongly acidic solution, e.g., in 12 M HCl (all the chlorides of Group 3 metals are soluble).

Additional Exercises

17.73   The equilibrium of interest is

$HC_5H_3O_3(aq) \rightleftharpoons H^+(aq) + C_5H_3O_3^-(aq)$; $K_a = 6.76 \times 10^{-4} = \dfrac{[H^+][C_5H_3O_3^-]}{[HC_5H_3O_3]}$

Begin by calculating $[HC_5H_3O_3]$ and $[C_5H_3O_3^-]$ for each case.

(a) $\dfrac{35.0 \text{ g HC}_5\text{H}_3\text{O}_3}{0.250 \text{ L soln}} \times \dfrac{1 \text{ mol HC}_5\text{H}_3\text{O}_3}{112.1 \text{ g HC}_5\text{H}_3\text{O}_3} = 1.249 = 1.25 \ M \text{ HC}_5\text{H}_3\text{O}_3$

$\dfrac{30.0 \text{ g NaC}_5\text{H}_3\text{O}_3}{0.250 \text{ L soln}} \times \dfrac{1 \text{ mol NaC}_5\text{H}_3\text{O}_3}{134.1 \text{ g NaC}_5\text{H}_3\text{O}_3} = 0.8949 = 0.895 \ M \text{ C}_5\text{H}_3\text{O}_3^-$

$[H^+] = \dfrac{K_a[HC_5H_3O_3]}{[C_5H_3O_3^-]} = \dfrac{6.76 \times 10^{-4} \,(1.249 - x)}{(0.8949 + x)} \approx \dfrac{6.76 \times 10^{-4}(1.249)}{(0.8949)}$

$[H^+] = 9.43 \times 10^{-4} \ M, \text{ pH} = 3.025$

(b) For dilution, $M_1V_1 = M_2V_2$

$[HC_5H_3O_3] = \dfrac{0.250 \ M \times 30.0 \text{ mL}}{125 \text{ mL}} = 0.0600 \ M$

$[C_5H_3O_3^-] = \dfrac{0.220 \ M \times 20.0 \text{ mL}}{125 \text{ mL}} = 0.0352 \ M$

$[H^+] \approx \dfrac{6.76 \times 10^{-4} \,(0.0600)}{0.0352} = 1.15 \times 10^{-3} \ M, \text{ pH} = 2.938$

(yes, $[H^+]$ is < 5% of 0.0352 M)

(c) $0.0850 \ M \times 0.500 \text{ L} = 0.0425 \text{ mol HC}_5\text{H}_3\text{O}_3$

$1.65 \ M \times 0.0500 \text{ L} = 0.0825 \text{ mol NaOH}$

	$HC_5H_3O_3$(aq)	+	NaOH(aq)	→	$NaC_5H_3O_3$(aq) + H_2O(l)
initial	0.0425 mol		0.0825 mol		
reaction	−0.0425 mol		−0.0425 mol		+0.0425 mol
after	0 mol		0.0400 mol		0.0425 mol

The strong base NaOH dominates the pH; the contribution of $C_5H_3O_3^-$ is negligible. This combination would be "after the equivalence point" of a titration. The total volume is 0.550 L.

$[OH^-] = \dfrac{0.0400 \text{ mol}}{0.550 \text{ L}} = 0.0727 \ M; \text{ pOH} = 1.138, \text{ pH} = 12.862$

17.76 (a) $K_a = \dfrac{[H^+][CHO_2]}{[HCHO_2]}; [H^+] = \dfrac{K_a[HCHO_2]}{[CHO_2^-]}$

Buffer A : $[HCHO_2] = [CHO_2^-] = \dfrac{1.00 \text{ mol}}{1.00 \text{ L}} = 1.00 \ M$

$[H^+] = \dfrac{1.8 \times 10^{-4} \,(1.00 \ M)}{(1.00 \ M)} = 1.8 \times 10^{-4} \ M, \text{ pH} = 3.74$

Buffer B : $[HCHO_2] = [CHO_2^-] = \dfrac{0.010 \text{ mol}}{1.00 \text{ L}} = 0.010 \ M$

$$[H^+] = \frac{1.8 \times 10^{-4}\,(0.010\,M)}{(0.010\,M)} = 1.8 \times 10^{-4}\,M,\ pH = 3.74$$

The pH of a buffer is determined by the identity of the conjugate acid/conjugate base pair (that is, the relevant K_a value) and the ratio of concentrations of the conjugate acid and conjugate base. The absolute concentrations of the components is not relevant. The pH values of the two buffers are equal because they both contain $HCHO_2$ and $NaCHO_2$ and the $[HCHO_2]$ / $[CHO_2^-]$ ratio is the same in both solutions.

(b)   Buffer capacity is determined by the absolute amount of conjugate acid and conjugate base available to absorb strong acid (H^+) or strong base (OH^-) that is added to the buffer. Buffer A has the greater capacity because it contains the greater absolute concentrations of $HCHO_2$ and CHO_2^-.

(c)   Buffer A:

CHO_2^-	+	HCl	$\rightarrow$	$HCHO_2$	+	Cl^-
1.00 mol		0.001 mol		1.00 mol		
0.999 mol		0		1.001 mol		

$$[H^+] = \frac{1.8 \times 10^{-4}\,(1.001)}{(0.999)} = 1.8 \times 10^{-4}\,M,\ pH = 3.74$$

(In a buffer calculation, volumes cancel and we can substitute moles directly into the K_a expression.)

Buffer B:

CHO_2^-	+	HCl	$\rightarrow$	$HCHO_2$	+	Cl^-
0.010 mol		0.001 mol		0.010 mol		
0.009 mol		0		0.011 mol		

$$[H^+] = \frac{1.8 \times 10^{-4}\,(0.011)}{(0.009)} = 2.2 \times 10^{-4}\,M,\ pH = 3.66$$

(d)   Buffer A: $1.00\,M\ HCl \times 0.010\,L = 0.010$ mol H^+ added

        mol $HCHO_2$ = 1.00 + 0.010 = 1.01 mol

        mol CHO_2^- = 1.00 – 0.010 = 0.99 mol

$$[H^+] = \frac{1.8 \times 10^{-4}\,(1.01)}{(0.99)} = 1.8 \times 10^{-4}\,M,\ pH = 3.74$$

Buffer B: mol $HCHO_2$ = 0.010 + 0.010 = 0.020 mol = 0.020 M

        mol CHO_2^- = 0.010 – 0.010 = 0.000 mol

The solution is no longer a buffer; the only source of CHO_2^- is the dissociation of $HCHO_2$.

$$K_a = \frac{[H^+][CHO_2^-]}{[HCHO_2]} = \frac{x^2}{(0.020-x)\,M}$$

The extent of ionization is greater than 5%; from the quadratic formula,
$x = [H^+] = 1.8 \times 10^{-3}$, pH = 2.74.

(e) Adding 10 mL of 1.00 *M* HCl to buffer B exceeded its capacity, while the pH of buffer A was unaffected. This is quantitative confirmation that buffer A has a significantly greater capacity than buffer B. In fact, 1.0 L of 1.0 *M* HCl would be required to exceed the capacity of buffer A. Buffer A, with 100 times more $HCHO_2$ and CHO_2^- has 100 times the capacity of buffer B.

17.78 (a) For a monoprotic acid (one H^+ per mole of acid), at the equivalence point moles OH^- added = moles H^+ originally present

$M_B \times V_B = $ g acid/molar mass

$$MM = \frac{\text{g acid}}{M_B \times V_B} = \frac{0.2140\text{ g}}{0.0950\,M \times 0.0274\text{ L}} = 82.21 = 82.2 \text{ g/mol}$$

(b) initial mol HA $= \dfrac{0.2140\text{g}}{82.21\text{ g/mol}} = 2.603 \times 10^{-3} = 2.60 \times 10^{-3}$ mol HA

mol OH^- added to pH 6.50 = $0.0950\,M \times 0.0150$ L = 1.425×10^{-3}

$\qquad\qquad\qquad\qquad\qquad\qquad\qquad = 1.43 \times 10^{-3}$ mol OH^-

	HA(aq)	+	NaOH(aq)	→	NaA(aq) + H$_2$O
before rx	2.603×10^{-3} mol		1.425×10^{-3} mol		0
change	-1.425×10^{-3} mol		-1.425×10^{-3} mol		$+1.425 \times 10^{-3}$ mol
after rx	1.178×10^{-3} mol		0		1.425×10^{-3} mol

$$[HA] = \frac{1.178 \times 10^{-3}\text{ mol}}{0.0400\text{ L}} = 0.02945 = 0.0295 \; M$$

$$[A^-] = \frac{1.425 \times 10^{-3}\text{ mol}}{0.0400\text{ L}} = 0.03563 = 0.0356 \; M; [H^+] = 10^{-6.50} = 3.162 \times 10^{-7}$$

$$= 3.2 \times 10^{-7} \; M$$

The mixture after reaction (a buffer) can be described by the acid dissociation equilibrium.

	HA(aq)	⇌	H$^+$(aq)	+	A$^-$(aq)
initial	0.0295 *M*		0		0.0356 *M*
equil	$(0.0295 - 3.2 \times 10^{-7}\,M)$		3.2×10^{-7} *M*		$(0.0356 + 3.2 \times 10^{-7})$ *M*

$$K_a = \frac{[H^+][A^-]}{[HA]} \approx \frac{(3.162 \times 10^{-7})(0.03563)}{(0.02945)} = 3.8 \times 10^{-7}$$

(Although we have carried 3 figures through the calculation to avoid rounding errors, the data dictate an answer with 2 significant figures.)

17.81 (a) Initially, the solution is 0.100 M in CO_3^{2-}.

$$CO_3^{2-}(aq) + H_2O(l) \rightleftharpoons HCO_3^-(aq) + OH^-(aq)$$

$$K_b = \frac{[HCO_3^-][OH^-]}{[CO_3^{2-}]} = \frac{K_w}{K_a[HCO_3^-]} = 1.79 \times 10^{-4} = 1.8 \times 10^{-4}$$

Proceeding in the usual way for a weak base,

calculate $[OH^-] = 4.23 \times 10^{-3} = 4.2 \times 10^{-3}$ M, pH = 11.63.

(b) It will require 40.00 mL of 0.100 M HCl to reach the first equivalence point, at which point HCO_3^- is the predominant species.

(c) An additional 40.00 mL are required to react with HCO_3^- to form H_2CO_3, the predominant species at the second equivalence point.

(d) At the second equivalence point there is a 0.0333 M solution of H_2CO_3. By the usual procedure for a weak acid:

$$H_2CO_3(aq) \rightleftharpoons H^+(aq) + HCO_3^-(aq)$$

$$K_a = \frac{[H^+][HCO_3^-]}{[H_2CO_3]} = 4.3 \times 10^{-7}; \frac{(x)^2}{(0.0333-x)} \approx \frac{x^2}{0.0333} \approx 4.3 \times 10^{-7}$$

$x = 1.20 \times 10^{-4} = 1.2 \times 10^{-4}$ M H^+; pH = 3.92

17.84 The pH of a buffer system is centered around pK_a for the conjugate acid component. For a diprotic acid, two conjugate acid/conjugate base pairs are possible.

$$H_2X(aq) \rightleftharpoons H^+(aq) + HX^-(aq); \quad K_{a1} = 2 \times 10^{-2}; \quad pK_{a1} = 1.70$$

$$HX^-(aq) \rightleftharpoons H^+(aq) + X^{2-}(aq); \quad K_{a2} = 5.0 \times 10^{-7}; \quad pK_{a2} = 6.30$$

Clearly HX^- / X^{2-} is the more appropriate combination for preparing a buffer with pH = 6.50. The $[H^+]$ in this buffer = $10^{-6.50} = 3.16 \times 10^{-7} = 3.2 \times 10^{-7}$ M. Using the K_{a2} expression to calculate the $[X^{2-}]$ / $[HX^-]$ ratio:

$$K_{a2} = \frac{[H^+][X^{2-}]}{[HX^-]}; \frac{K_{a2}}{[H^+]} = \frac{[X^{2-}]}{[HX^-]} = \frac{5.0 \times 10^{-7}}{3.16 \times 10^{-7}} = 1.58 = 1.6$$

Since X^{2-} and HX^- are present in the same solution, the ratio of concentrations is also a ratio of moles.

$$\frac{[X^{2-}]}{[HX^-]} = \left(\frac{mol\ X^{2-} / L\ soln}{mol\ HX^- / L\ soln}\right) = \frac{mol\ X^{2-}}{mol\ HX^-} = 1.58; mol\ X^{2-} = (1.58)\ mol\ HX^-$$

In the 1.0 L of 1.0 M H_2X, there is 1.0 mol of X^{2-} containing material.

Thus, mol HX^- + 1.58 (mol HX^-) = 1.0 mol. 2.58 (mol HX^-) = 1.0;

mol HX^- = 1.0 / 2.58 = 0.39 mol HX^-; mol X^{2-} = 1.0 − 0.39 = 0.61 mol X^{2-}.

Thus enough 1.0 M NaOH must be added to produce 0.39 mol HX^- and 0.61 mol X^{2-}.

Considering the neutralization in a step-wise fashion (see discussion of titrations of polyprotic acids in Section 17.3).

	$H_2X(aq)$	+	$NaOH(aq)$	$\rightarrow$	$HX^-(aq) + H_2O(l)$
before	1.0 mol		1 mol		0
after	0		0		1.0 mol

	$HX^-(aq)$	+	$NaOH(aq)$	$\rightarrow$	$X^{2-}(aq) + H_2O(l)$
before	1.0				0.61
change	−0.61		−0.61		+0.61
after	0.39		0		0.61

Starting with 1.0 mol of H_2X, 1.0 mol of NaOH is added to completely convert it to 1.0 mol of HX^-. Of that 1.0 mol of HX^-, 0.61 mol must be converted to 0.61 mol X^{2-}. The total moles of NaOH added is (1.00 + 0.61) = 1.61 mol NaOH.

$$L\,NaOH = \frac{mol\,NaOH}{M\,NaOH} = \frac{1.61\,mol}{1.0\,M} = 1.6\,L\,of\,1.0\,M\,NaOH$$

17.87   (a)   CdS: 8.0×10^{-28}; CuS: 6×10^{-37} CdS has greater molar solubility.

       (b) $PbCO_3$: 7.4×10^{-14}; $BaCrO_4$: 2.1×10^{-10} $BaCrO_4$ has greater molar solubility.

       (c)   Since the stoichiometry of the two complexes is not the same, K_{sp} values can't be compared directly; molar solubilities must be calculated from K_{sp} values.

$Ni(OH)_2$: $K_{sp} = 6.0 \times 10^{-16} = [Ni^{2+}][OH^-]^2$; $[Ni^{2+}] = x$, $[OH^-] = 2x$

$6.0 \times 10^{-16} = (x)(2x)^2 = 4x^3$; $x = 5.3 \times 10^{-6}\,M\,Ni^{2+}$

Note that $[OH^-]$ from the autoionization of water is less than 1% of $[OH^-]$ from $Ni(OH)_2$ and can be neglected.

$NiCO_3$: $K_{sp} = 1.3 \times 10^{-7} = [Ni^{2+}][CO_3^{2-}]$; $[Ni^{2+}] = [CO_3^{2-}] = x$

$1.3 \times 10^{-7} = x^2$; $x = 3.6 \times 10^{-4}\,M\,Ni^{2+}$

$NiCO_3$ has greater molar solubility than $Ni(OH)_2$, but the values are much closer than expected from inspection of K_{sp} values alone.

       (d)   Again, molar solubilities must be calculated for comparison.

Ag_2SO_4: $K_{sp} = 1.5 \times 10^{-5} = [Ag^+]^2[SO_4^{2-}]$; $[SO_4^{2-}] = x$, $[Ag^+] = 2x$

$1.5 \times 10^{-5} = (2x)^2(x) = 4x^3$; $x = 1.6 \times 10^{-2}\,M\,SO_4^{2-}$

AgI: $K_{sp} = 8.3 \times 10^{-17} = [Ag^+][I^-]$; $[Ag^+] = [I^-] = x$

$8.3 \times 10^{-17} = x^2$; $x = 9.1 \times 10^{-9}\,M\,Ag^+$

Ag_2SO_4 has greater molar solubility than AgI.

17.90 $K_{sp} = [Ba^{2+}][MnO_4^-]^2 = 2.5 \times 10^{-10}$

$[MnO_4^-]^2 = 2.5 \times 10^{-10} / 2.0 \times 10^{-8} = 0.0125$; $[MnO_4^-] = \sqrt{0.0125} = 0.11\,M$

17.93 $MgC_2O_4(s) \rightleftharpoons Mg^{2+}(aq) + C_2O_4^{2-}(aq)$

$K_{sp} = [Mg^{2+}][C_2O_4^{2-}] = 8.6 \times 10^{-5}$

If $[Mg^{2+}]$ is to be $3.0 \times 10^{-2}\,M$, $[C_2O_4^{2-}] = 8.6 \times 10^{-5}/3.0 \times 10^{-2} = 2.87 \times 10^{-3} = 2.9 \times 10^{-3}\,M$

The oxalate ion undergoes hydrolysis:

$C_2O_4^{2-}(aq) + H_2O(l) \rightleftharpoons HC_2O_4^-(aq) + OH^-(aq)$

$K_b = \dfrac{[HC_2O_4^-][OH^-]}{[C_2O_4^{2-}]} = 1.0 \times 10^{-14}/6.4 \times 10^{-5} = 1.56 \times 10^{-10} = 1.6 \times 10^{-10}$

$[Mg^{2+}] = 3.0 \times 10^{-2}\,M$, $[C_2O_4^{2-}] = 2.87 \times 10^{-3} = 2.9 \times 10^{-3}\,M$

$[HC_2O_4^-] = (3.0 \times 10^{-2} - 2.87 \times 10^{-3})\,M = 2.71 \times 10^{-2} = 2.7 \times 10^{-2}\,M$

$[OH^-] = 1.56 \times 10^{-10} \times \dfrac{[C_2O_4^{2-}]}{[HC_2O_4^-]} = 1.56 \times 10^{-10} \times \dfrac{(2.87 \times 10^{-3})}{(2.71 \times 10^{-2})} = 1.652 \times 10^{-11}$

$[OH^-] = 1.7 \times 10^{-11}$; pOH = 10.78, pH = 3.22

Integrative Exercises

17.96   (a)   Complete ionic:

$H^+(aq) + Cl^-(aq) + Na^+(aq) + CHO_2^-(aq) \rightarrow HCHO_2(aq) + Na^+(aq) + Cl^-(aq)$

Na^+ and Cl^- are spectator ions.

Net ionic: $H^+(aq) + CHO_2^-(aq) \rightleftharpoons HCHO_2(aq)$

(b)   The net ionic equation in part (a) is the reverse of the dissociation of $HCHO_2$.

$K = \dfrac{1}{K_a} = \dfrac{1}{1.8 \times 10^{-4}} = 5.55 \times 10^3 = 5.6 \times 10^3$

(c)   For Na^+ and Cl^-, this is just a dilution problem.

$M_1V_1 = M_2V_2$; V_2 is 50.0 mL + 50.0 mL = 100.0 mL

Cl^-: $\dfrac{0.15\,M \times 50.0\,mL}{100.0\,mL} = 0.075\,M$; Na^+: $\dfrac{0.15\,M \times 50.0\,mL}{100.0\,mL} = 0.075\,M$

H^+ and CHO_2^- react to form $HCHO_2$. Since K >> 1, the reaction essentially goes to completion.

$0.15\,M \times 0.0500\,mL = 7.5 \times 10^{-3}$ mol H^+

$\dfrac{0.15\,M \times 0.0500\,mL = 7.5 \times 10^{-3}\ \text{mol}\ CHO_2^-}{= 7.5 \times 10^{-3}\ \text{mol}\ HCHO_2}$

Solve the weak acid problem to determine $[H^+]$, $[CHO_2^-]$ and $[HCHO_2]$ at equilibrium.

$$K_a = \frac{[H^+][CHO_2^-]}{[HCHO_2]}; [H^+] = [CHO_2^-] = x\,M; [HCHO_2] = \frac{(7.5 \times 10^{-3} - x)\,mol}{0.100\,L}$$

$$= (0.075 - x)\,M$$

$$1.8 \times 10^{-4} = \frac{x^2}{(0.075 - x)} \approx \frac{x^2}{0.075}; x = 3.7 \times 10^{-3}\,M\,H^+ \text{ and } HCHO_2^-$$

$$[HCHO_2] = (0.075 - 0.0037) = 0.071\,M$$

$$\frac{[H^+]}{[HNO_2]} \times 100 = \frac{3.7 \times 10^{-3}}{0.075} \times 100 = 4.9\% \text{ dissociation}$$

In summary:

$$[Na^+] = [Cl^-] = 0.075\,M, [HCHO_2] = 0.071\,M, [H^+] = [CHO_2^-] = 0.0037\,M$$

17.98 $n = \dfrac{PV}{RT} = 735\,torr \times \dfrac{1\,atm}{760\,torr} \times \dfrac{7.5\,L}{295\,K} \times \dfrac{K \bullet mol}{0.08206\,L \bullet atm} = 0.300 = 0.30\,mol\,NH_3$

$0.40\,M \times 0.50\,L = 0.20\,mol\,HCl$

	HCl(aq)	+	NH$_3$(g)	→	NH$_4^+$(aq)	+	Cl$^-$(aq)
before	0.20 mol		0.30 mol				
after	0		0.10 mol		0.20 mol		0.20 mol

The solution will be a buffer because of the substantial concentrations of NH_3 and NH_4^+ present. Use K_a for NH_4^+ to describe the equilibrium.

$$NH_4^+(aq) \rightleftharpoons NH_3(aq) + H^+(aq)$$

equil. 0.20 – x   0.10 + x   x

$$K_a = \frac{1.0 \times 10^{-14}}{1.8 \times 10^{-5}} = 5.56 \times 10^{-10} = 5.6 \times 10^{-10}; K_a = \frac{[NH_3][H^+]}{[NH_4^+]}; [H^+] = \frac{K_a[NH_4^+]}{[NH_3]}$$

Since this expression contains a ratio of concentrations, volume will cancel and we can substitute moles directly. Assume x is small compared to 0.10 and 0.20.

$$[H^+] = \frac{5.56 \times 10^{-10}\,(0.20)}{(0.10)} = 1.111 \times 10^{-9} = 1.1 \times 10^{-9}\,M, pH = 8.95$$

17.101 $\pi = MRT, M = \dfrac{\pi}{RT} = \dfrac{21\,torr}{298\,K} \times \dfrac{1\,atm}{760\,torr} \times \dfrac{K \bullet mol}{0.08206\,L \bullet atm} = 1.13 \times 10^{-3} = 1.1 \times 10^{-3}\,M$

$$SrSO_4(s) \rightleftharpoons Sr^{2+}(aq) + SO_4^{2-}(aq); K_{sp} = [Sr^{2+}][SO_4^{2-}]$$

The total particle concentration is $1.13 \times 10^{-3}\,M$. Each mole of $SrSO_4$ that dissolves produces 2 mol of ions, so $[Sr^{2+}] = [SO_4^{2-}] = 1.13 \times 10^{-3}\,M/2 = 5.65 \times 10^{-4} = 5.7 \times 10^{-4}\,M$.

$$K_{sp} = (5.65 \times 10^{-4})^2 = 3.2 \times 10^{-7}$$

18 Chemistry of the Environment

Visualizing Concepts

18.1   *Analyze.* Given that one mole of an ideal gas at 1 atm and 298 K occupies 22.4 L, is the volume of one mole of ideal gas in the middle of the stratosphere greater than, equal to, or less than 22.4 L?

Plan. Consider the relationship between pressure, temperature, and volume of an ideal gas. Use Figure 18.1 to estimate the pressure and temperature in the middle of the stratosphere, and compare the two sets of temperature and pressure.

Solve. According to the ideal gas law, $PV = nRT$, so $V = nRT/P$. Since n and R are constant for this exercise, V is proportional to T/P.

(a)   The stratosphere ranges from 10 to 50 km, so the middle is at approximately 30 km. At this altitude, $T \approx 230$ K, $P \approx 40$ torr (from Figure 18.1). Since we are comparing T/P ratios, either atm or torr can be used as pressure units; we will use torr.

At sea level: T/P = 298 K/760 torr = 0.39

At 30 km: T/P = 230 K/40 torr = 5.75

The proportionality constant (T/P) is much greater at 30 km than sea level, so the volume of 1 mol of an ideal gas is greater at this altitude. The decrease in temperature at 30 km is more than offset by the substantial decrease in pressure.

(b)   Volume is proportional to T/P, not simply T. The relative volumes of one mole of an ideal gas at 50 km and 85 km depend on the temperature and pressure at the two altitudes. From Figure 18.1,

50 km: $T \approx 270$ K, $P \approx 20$ torr, T/P = 270 K/20 torr = 13.5

85 km: $T \approx 190$ K, $P < 0.01$ torr, T/P = 190 K/0.01 torr = 19,000

Again, the slightly lower temperature at 85 km is more than offset by a much lower pressure. One mole of an ideal gas will occupy a much larger volume at 85 km than 50 km.

18.3   Ozone concentration varies with altitude because conditions favorable to ozone formation (and unfavorable to decomposition) vary with altitude. Formation and persistence of O_3 require the presence of O atoms, O_2 molecules, and energy carriers (M^*, usually N_2 or O_2). Above 60 km, there are too few O_2 molecules for significant O_3 formation. Below 30 km, there are too few O atoms. Between 30 km and 60 km, O_3 concentration varies depending on the concentrations of O, O_2 and M^*.

18.5 $CO_2(g)$ dissolves in seawater to form $H_2CO_3(aq)$. The basic pH of the ocean encourages ionization of $H_2CO_3(aq)$ fo form $HCO_3^-(aq)$ and $CO_3^{2-}(aq)$. Under the correct conditions, carbon is removed from the ocean as $CaCO_3(s)$ (sea shells, coral, chalk cliffs). As carbon is removed, more $CO_2(g)$ dissolves to maintain the balance of complex and interacting acid-base and precipitation equilibria.

18.7   The guiding principle of green chemistry is that "an ounce of prevention is worth a pound of cure." Processes should be designed to minimize or eliminate solvents and waste, generate nontoxic waste, be energy efficient, employ renewable starting materials, and take advantage of catalysts that enable the use of safe and common reagents.

Earth's Atmosphere

18.9   (a)   The temperature profile of the atmosphere (Figure 18.1) is the basis of its division into regions. The center of each peak or trough in the temperature profile corresponds to a new region.

    (b)   Troposphere, 0–12 km; stratosphere, 12–50 km; mesosphere, 50–85 km; thermosphere, 85–110 km.

18.11   *Analyze/Plan.* Given O_3 concentration in ppm, calculate partial pressure. Use the definition of ppm to get mol fraction O_3. For gases mole fraction = pressure fraction;

$$P_{O_3} = \chi_{O_3} \cdot P_{atm}$$

$$0.441 \text{ ppm } O_3 = \frac{0.441 \text{ mol } O_3}{1 \times 10^6 \text{ mol air}} = 4.41 \times 10^{-7} = \chi_{O_3}$$

Solve. $P_{O_3} = \chi_{O_3} \cdot P_{atm} = 4.41 \times 10^{-7} (0.67 \text{ atm}) = 2.955 \times 10^{-7} = 3.0 \times 10^{-7} \text{ atm}$

18.13   *Analyze/Plan.*   Given CO concentration in ppm, calculate number of CO molecules in 1.0 L air at given conditions. ppm $CO \rightarrow \chi_{O_3} \rightarrow$ atm $CO \rightarrow$ mol $CO \rightarrow$ molecules CO. Use the ideal gas law to change atm CO to mol CO, then Avogadro's number to get molecules.   *Solve.*

$$3.4 \text{ ppm } CO = \frac{3.4 \text{ mol } CO}{1 \times 10^6 \text{ mol air}} = 3.4 \times 10^{-6} = \chi_{CO}$$

$$P_{CO} = \chi_{CO} \cdot P_{atm} = 3.4 \times 10^{-6} \times 755 \text{ torr} \times \frac{1 \text{ atm}}{760 \text{ torr}} = 3.378 \times 10^{-6} = 3.4 \times 10^{-6} \text{ atm}$$

$$n_{CO} = \frac{P_{CO}V}{RT} = \frac{3.378 \times 10^{-6} \text{ atm} \times 1.0 \text{ L}}{295 \text{ K}} \times \frac{K \cdot mol}{0.08206 \text{ L} \cdot atm} = 1.395 \times 10^{-7} = 1.4 \times 10^{-7} \text{ mol CO}$$

$$1.395 \times 10^{-7} \text{ mol CO} \times \frac{6.022 \times 10^{23} \text{ molecules}}{mol} = 8.402 \times 10^{16} = 8.4 \times 10^{16} \text{ molecules CO}$$

The Upper Atmosphere; Ozone

18.15   *Analyze/Plan.* Given bond dissociation energy in kJ/mol, calculate the wavelength of a single photon that will rupture a C–Br bond. kJ/mol $\rightarrow$ J/molecule. $\lambda = hc/E$.
    ($\lambda = hc/E$ describes the energy/wavelength relationship of a single photon.)   *Solve.*

$$\frac{210 \times 10^3 \text{ J}}{1 \text{ mol}} \times \frac{1 \text{ mol}}{6.022 \times 10^{23} \text{ molecules}} = 3.487 \times 10^{-19} = 3.49 \times 10^{-19} \text{ J/molecule}$$

$\lambda = c/\nu$ We also have that $E = h\nu$, so $\nu = E/h$. Thus,

$$\lambda = \frac{hc}{E} = \frac{(6.626 \times 10^{-34} \text{ J} \cdot \text{sec})(3.00 \times 10^8 \text{ m/sec})}{3.487 \times 10^{-19} \text{ J}} = 5.70 \times 10^{-7} \text{ m} = 570 \text{ nm}$$

18.17   (a)   *Photodissociation* is cleavage of the $O{=}O$ bond such that two neutral O atoms are produced: $O_2(g) \rightarrow 2O(g)$

Photoionization is absorption of a photon with sufficient energy to eject an electron from an O_2 molecule: $O_2(g) + h\nu \rightarrow O_2^+ + e^-$

(b)   Photoionization of O_2 requires 1205 kJ/mol.   Photodissociation requires only 495 kJ/mol. At lower elevations, solar radiation with wavelengths corresponding to 1205 kJ/mol or shorter has already been absorbed, while the longer wavelength radiation has passed through relatively well. Below 90 km, the increased concentration of O_2 and the availability of longer wavelength radiation cause the photodissociation process to dominate.

18.19   A *hydrofluorocarbon* is a compound that contains hydrogen, fluorine, and carbon; it contains hydrogen in place of chlorine. HFCs are potentially less harmful than CFCs because photodissociation does not produce Cl atoms, which catalyze the destruction of ozone.

18.21   (a)   In order to catalyze ozone depletion, the halogen must be present as single halogen atoms. These halogen atoms are produced in the stratosphere by photo-dissociation of a carbon halogen bond. According to Table 8.4, the C–F average bond dissociation energy is 485 kJ/mol, while that of C–Cl is 328 kJ/mol.   The C–F bond requires more energy for dissociation and is not readily cleaved by the available wavelengths of UV light.

(b)   Chlorine is present as chlorine atoms and chlorine oxide molecules, Cl and ClO.

Chemistry of the Troposphere

18.23   (a)   Methane, CH_4, arises from decomposition of organic matter by certain microorganisms; it also escapes from underground gas deposits.

(b) SO_2 is released in volcanic gases, and also is produced by bacterial action on decomposing vegetable and animal matter.

(c)   Nitric oxide, NO, results from oxidation of decomposing organic matter, and is formed in lightning flashes.

(d)   CO is a possible product of some vegetable matter decay.

18.25   (a)   Acid rain is primarily $H_2SO_4(aq)$.

$H_2SO_4(aq) + CaCO_3(s) \rightarrow CaSO_4(s) + H_2O(l) + CO_2(g)$

(b)   The $CaSO_4(s)$ would be much less reactive with acidic solution, since it would require a strongly acidic solution to shift the relevant equilibrium to the right.

$$CaSO_4(s) + 2H^+(aq) \rightleftharpoons Ca^{2+}(aq) + 2HSO_4^-(aq)$$

Note, however, that $CaSO_4(s)$ is brittle and easily dislodged; it provides none of the structural strength of limestone.

18.27   *Analyze/Plan.* Given wavelength of a photon, place it in the electromagnetic spectrum, calculate its energy in kJ/mol, and compare it to an average bond dissociation energy. Use Figure 6.4; $E(J/photon) = hc/\lambda$. $J/photon \rightarrow kJ/mol$.   *Solve.*

(a)   Ultraviolet (Figure 6.4)

(b) $E_{photon} = hc/\lambda = \dfrac{6.626 \times 10^{-34} \text{ J} \cdot \text{s} \times 3.00 \times 10^8 \text{ m/s}}{335 \times 10^{-9} \text{ m}} = 5.934 \times 10^{-19}$

$$= 5.93 \times 10^{-19} \text{ J/photon}$$

$$\frac{5.934 \times 10^{-19} \text{ J}}{1 \text{ photon}} \times \frac{6.022 \times 10^{23} \text{ photons}}{1 \text{ mol}} \times \frac{1 \text{kJ}}{1000 \text{ J}} = 357 \text{ kJ/mol}$$

(c)   The average C–H bond energy from Table 8.4 is 413 kJ/mol. The energy calculated in part (b), 357 kJ/mol, is the energy required to break 1 mol of C–H bonds in formaldehyde, CH_2O. The C–H bond energy in CH_2O must be less than the "average" C–H bond energy.

(d)
$$\text{H} - \overset{\displaystyle :\text{O}:}{\underset{\|}{\text{C}}} - \text{H} + h\nu \longrightarrow \text{H} - \overset{\displaystyle :\text{O}:}{\underset{\|}{\text{C}}}\cdot + \text{H}$$

18.29   Most of the energy entering the atmosphere from the sun is in the form of visible radiation, while most of the energy leaving the earth is in the form of infrared radiation. CO_2 is transparent to the incoming visible radiation, but absorbs the outgoing infrared radiation.

The World Ocean

18.31   *Analyze/Plan.* Given salinity and density, calculate molarity. A salinity of 5.6 denotes that there are 5.6 g of dry salt per kg of water.   *Solve.*

$$\frac{5.6 \text{ g NaCl}}{1 \text{ kg soln}} \times \frac{1.03 \text{ kg soln}}{1 \text{ L soln}} \times \frac{1 \text{ mol NaCl}}{58.44 \text{ g NaCl}} \times \frac{1 \text{ mol Na}^+}{1 \text{ mol NaCl}} = 0.0987 = 0.099 \ M \text{ Na}^+$$

18.33   *Analyze/Plan.* g $Mg(OH)_2 \rightarrow$ mol $Mg(OH)_2 \rightarrow$ mol ratio $\rightarrow$ mol CaO $\rightarrow$ g CaO.   *Solve.*

$$1000 \text{ lb Mg(OH)}_2 \times \frac{453.6 \text{ g}}{\text{lb}} \times \frac{1 \text{ mol Mg(OH)}_2}{58.33 \text{ g Mg(OH)}_2} \times \frac{1 \text{ mol CaO}}{1 \text{ mol Mg(OH)}_2} \times \frac{56.08 \text{ g CaO}}{1 \text{ mol CaO}}$$

$$= 4.361 \times 10^5 \text{ g CaO}$$

18.35   *Analyze/Plan.* Given temperature and the concentration difference between the two solutions, ($\Delta M = 0.22 - 0.01 = 0.21 \ M$), calculate the minimum pressure for reverse osmosis. Use the relationship $\pi = MRT$ from Section 13.5. This is the pressure required to halt osmosis from the more dilute (0.01 M) to the more concentrated (0.22 M) solution. Slightly more pressure will initiate reverse osmosis.   *Solve.*

$$\pi = \Delta MRT = \frac{0.21 \text{ mol}}{\text{L}} \times \frac{0.08206 \text{ L} \cdot \text{atm}}{\text{mol} \cdot \text{K}} \times 298 \text{ K} = 5.135 = 5.1 \text{ atm}$$

The minimum pressure required to initiate reverse osmosis is greater than 5.1 atm.

Fresh Water

18.37   *Analyze/Plan.* Under aerobic conditions, excess oxygen is present and decomposition leads to oxidized products, the element in its maximum oxidation state combined with oxygen. Under anaerobic conditions, little or no oxygen is present so decomposition leads to reduced products, the element in its minimum oxidation state combined with hydrogen.   *Solve.*

(a) CO_2, HCO_3^-, H_2O, SO_4^{2-}, NO_3^-, HPO_4^{2-}, $H_2PO_4^-$.

(b) $CH_4(g)$, $H_2S(g)$, $NH_3(g)$, $PH_3(g)$

18.39   *Analyze/Plan.* Given the balanced equation, calculate the amount of one reactant required to react exactly with a certain amount of the other reactants. Solve the stoichiometry problem. $g\ C_{18}H_{29}O_3S^- \to mol \to mol\ ratio \to mol\ O_2 \to g\ O_2$.   *Solve.*

$$1.0\ g\ C_{18}H_{29}O_3S^- \times \frac{1\ mol\ C_{18}H_{29}O_3S^-}{325\ g\ C_{18}H_{29}O_3S^-} \times \frac{51\ mol\ O_2}{2\ mol\ C_{18}H_{29}O_3S^-} \times \frac{32.0\ g\ O_2}{1\ mol\ O_2} = 2.5\ g\ O_2$$

Notice that the mass of O_2 required is 2.5 times greater than the mass of biodegradable material.

18.41   *Analyze/Plan.* Slaked lime is $Ca(OH)_2(s)$. The reaction is metathesis.   *Solve.*

$$Mg^{2+}(aq) + Ca(OH)_2(s) \to Mg(OH)_2(s) + Ca^{2+}(aq)$$

The excess $Ca^{2+}(aq)$ is removed as $CaCO_3$ by naturally occurring bicarbonate or added Na_2CO_3.

18.43   *Analyze/Plan.* Given $[Ca^{2+}]$ and $[HCO_3^-]$ calculate mole $Ca(OH)_2$ and Na_2CO_3 needed to remove the Ca^{2+} and HCO_3^-. Consider the chemical equations and reaction stoichiometry in the stepwise process.   *Solve.*

$Ca(OH)_2$ is added to remove Ca^{2+} as $CaCO_3(s)$, and Na_2CO_3 removes the remaining Ca^{2+}.

$$Ca^{2+}(aq) + 2HCO_3^-(aq) + [Ca^{2+}(aq) + 2OH^-(aq)] \to 2CaCO_3(s) + 2H_2O(l).$$

One mole $Ca(OH)_2$ is needed for each 2 moles of $HCO_3^-(aq)$ present.

$$\frac{7.0 \times 10^{-4}\ mol\ HCO_3^-}{L} \times \frac{1\ mol\ Ca(OH)_2}{2\ mol\ HCO_3^-} \times 1.200 \times 10^3\ L\ H_2O = 0.42\ mol\ Ca(OH)_2$$

$$1.200 \times 10^3\ L\ H_2O \times \frac{5.0 \times 10^{-4}\ mol\ Ca^{2+}}{L} = 0.60\ mol\ Ca^{2+}(aq)\ total$$

0.42 mol $Ca(OH)_2$ removes 0.42 mol of the 0.60 mol $Ca^{2+}(aq)$ in the sample. This leaves 0.18 mol $Ca^{2+}(aq)$ to be removed by Na_2CO_3.

$$Ca^{2+}(aq) + Na_2CO_3(aq) \to CaCO_3(s) + 2Na^+(aq)$$

0.18 mol of Na_2CO_3 is needed to remove the remaining $Ca^{2+}(aq)$.

18.45 $4FeSO_4(aq) + O_2(aq) + 2H_2O(l) \to 4Fe^{3+}(aq) + 4OH^-(aq) + 4SO_4^{2-}(aq)$

SO_4^{2-} is a spectator, so the net ionic equation is

$4Fe^{2+}(aq) + O_2(aq) + 2H_2O(l) \to 4Fe^{3+}(aq) + 4OH^-(aq)$.

$Fe^{3+}(aq) + 3HCO_3^-(aq) \rightarrow Fe(OH)_3(s) + 3CO_2(g)$

In this reaction, Fe^{3+} acts as a Lewis acid, and HCO_3^- acts as a Lewis base.

Green Chemistry

18.47   The fewer steps in a process, the less waste (solvents as well as unusable by-products) is generated. It is probably true that a process with fewer steps requires less energy at the site of the process, and it is certainly true that the less waste the process generates, the less energy is required to clean or dispose of the waste.

18.49   (a)

(b)   • It is better to prevent waste than to treat it. The alternative process eliminates production of 3-chlorobenzoic acid by-product, chlorine-containing waste that must be treated.

   • Produce as little, nontoxic waste as possible. The by-product of the alternative process is nontoxic water. The low molar mass of water means that a small amount of "waste" is generated.

   • Chemical processes should be efficient. The alternative process is catalyzed, which could mean that the process will be more energy efficient than the Baeyer-Villiger reaction (see Solution 18.44).

   • Raw materials should be renewable. The catalyst can be recovered from the reaction mixture and reused. We don't have information about solvents or other auxiliary substances.

Additional Exercises

18.51   (a) *Acid rain* is rain with a larger $[H^+]$ and thus a lower pH than expected. The additional H^+ is produced by the dissolution of sulfur and nitrogen oxides such as $SO_3(g)$ and $NO_2(g)$ in rain droplets to form sulfuric and nitric acid, $H_2SO_4(aq)$ and $HNO_3(aq)$.

   (b) A *greenhouse gas* absorbs infrared or "heat" radiation emitted from the earth's surface and serves to maintain a relatively constant temperature on the surface. A significant increase in the amount of atmospheric CO_2 (from burning fossil fuels and other sources) could cause a corresponding increase in the average surface temperature and drastically change the global climate.

   (c) *Photochemical smog* is an unpleasant collection of atmospheric pollutants initiated by photochemical dissociation of NO_2 to form NO and O atoms. The major components are $NO(g)$, $NO_2(g)$, $CO(g)$, and unburned hydrocarbons, all produced by automobile engines, and $O_3(g)$, ozone.

   (d) *Ozone depletion* is the reduction of O_3 concentration in the stratosphere, most notably over Antarctica. It is caused by reactions between O_3 and Cl atoms originating from chlorofluorocarbons (CFC's), CF_xCl_{4-x}. Depletion of the ozone layer allows damaging ultraviolet radiation disruptive to the plant and animal life in our ecosystem to reach earth.

18.55   (a)   The production of Cl atoms in the stratosphere is the result of the photodissociation of a C—H bond in the chlorofluorocarbon molecule.

$$CF_2Cl_2(g) \xrightarrow{h\nu} CF_2Cl(g) + Cl(g)$$

According to Table 8.4, the bond dissociation energy of a C—Br bond is 276 kJ/mol, while the value for a C—H bond is 328 kJ/mol. Photodissociation of $CBrF_3$ to form Br atoms requires less energy than the production of Cl atoms and should occur readily in the stratosphere.

(b) $CBrF_3(g) \xrightarrow{h\nu} CF_3(g) + Br(g)$

$Br(g) + O_3(g) \rightarrow BrO(g) + O_2(g)$

Also, under certain conditions

$BrO(g) + BrO(g) \rightarrow Br_2O_2(g)$

$Br_2O_2(g) + h\nu \rightarrow O_2(g) + 2Br(g)$

18.57   From section 18.4:

$N_2(g) + O_2(g) \rightleftharpoons 2NO(g)$ $\Delta H = +180.8$ kJ   [18.11]

$2 NO(g) + O_2(g) \rightleftharpoons 2NO_2(g)$ $\Delta H = -113.1$ kJ   [18.12]

In an endothermic reaction, heat is a reactant. As the temperature of the reaction increases, the addition of heat favors formation of products and the value of K increases. The reverse is true for exothermic reactions; as temperature increases, the value of K decreases. Thus, K for reaction [18.11], which is endothermic, increases with increasing temperature and K for reaction [18.12], which is exothermic, decreases with increasing temperature.

18.59   (a) $CH_4(g) + 2O_2(g) \rightarrow CO_2(g) + 2H_2O(g)$

(b) $2CH_4(g) + 3O_2(g) \rightarrow 2CO(g) + 4H_2O(g)$

(c) $vol\ CH_4 \rightarrow vol\ O_2 \rightarrow$ volume air ($\chi_{O_2} = 0.20948$)

Equal volumes of gases at the same temperature and pressure contain equal numbers of moles (Avogadro's law). If 2 moles of O_2 are required for 1 mole of CH_4, 2.0 L of pure O_2 are needed to burn 1.0 L of CH_4.

$$vol\ O_2 = \chi_{O_2} \times vol_{air} = \frac{vol\ O_2}{\chi_{O_2}} = \frac{2.0\ L}{0.20948} = 9.5\ L\ air$$

18.61   Most of the 390 watts/m^2 radiated from Earth's surface is in the infrared region of the spectrum. Tropospheric gases, particularly $H_2O(g)$ and $CO_2(g)$, absorb much of this radiation and prevent it from escaping into space (Figure 18.11). The energy absorbed by these so-called "greenhouse gases" warms the atmosphere close to Earth's surface and makes the planet livable.

18.63   (a) $NO(g) + h\nu \rightarrow N(g) + O(g)$

        (b) $NO(g) + h\nu \rightarrow NO^+(g) + e^-$

        (c) $NO(g) + O_3(g) \rightarrow NO_2(g) + O_2(g)$

        (d) $3NO_2(g) + H_2O(l) \rightarrow 2HNO_3(aq) + NO(g)$

18.65   Because NO has an odd electron, like Cl(g), it could act as a catalyst for decomposition of ozone in the stratosphere. The increased destruction of ozone by NO would result in less absorption of short wavelength UV radiation now being screened out primarily by the ozone. Radiation in this wavelength range is known to be harmful to humans; it causes skin cancer. There is evidence that many plants don't tolerate it very well either, though more research is needed to test this idea.

       In Chapter 22 the oxidation of NO to NO_2 by oxygen is described. On dissolving in water, NO_2 disproportionates into $NO_3^-(aq)$ and NO(g). Thus, over time the NO in the troposphere will be converted into NO_3^-, which is in turn incorporated into soils.

Integrative Exercises

18.67   (a) $0.019 \text{ ppm } NO_2 = \dfrac{0.019 \text{ mol } NO_2}{1 \times 10^6 \text{ mol air}} = 1.9 \times 10^{-8} = \chi_{NO_2}$

           $P_{NO_2} = \chi_{NO_2} \cdot P_{atm} = 1.9 \times 10^{-8} (755 \text{ torr}) = 1.4345 \times 10^{-5} = 1.4 \times 10^{-5} \text{ torr}$

        (b) $n = \dfrac{PV}{RT}$; molecules $= n \times \dfrac{6.022 \times 10^{23} \text{ molecules}}{mol} = \dfrac{PV}{RT} \times \dfrac{6.022 \times 10^{23} \text{ molecules}}{mol}$

           $V = 15 \text{ ft} \times 14 \text{ ft} \times 8 \text{ ft} \times \dfrac{12^3 \text{ in}^3}{ft^3} \times \dfrac{2.54^3 \text{ cm}^3}{in^3} \times \dfrac{1 L}{1000 \text{ cm}^3} = 4.757 \times 10^4 = 5 \times 10^4 \text{ L}$

           $1.4345 \times 10^{-5} \text{ torr} \times \dfrac{1 \text{ atm}}{760 \text{ torr}} \times \dfrac{4.757 \times 10^4 \text{ L}}{293 \text{ K}} \times \dfrac{K \cdot mol}{0.08206 \text{ L} \cdot atm}$

           $\times \dfrac{6.022 \times 10^{23} \text{ molecules}}{mol} = 2.249 \times 10^{19} = 2 \times 10^{19} \text{ molecule}$

18.71   According to Equation [14.12], $\ln([A]_t / [A]_o) = -kt.$ $[A]_t = 0.10 [A]_o$

       $\ln(0.10 [A]_o / [A]_o) = \ln(0.10) = -(2 \times 10^{-6} \text{ s}^{-1}) t$

       $t = -\ln(0.10) / 2 \times 10^{-6} \text{ s}^{-1} = 1.151 \times 10^6 \text{ s}$

       $1.151 \times 10^6 \text{ s} \times \dfrac{1 \text{ min}}{60 \text{ s}} \times \dfrac{1 \text{ hr}}{60 \text{ min}} \times \dfrac{1 \text{day}}{24 \text{ hr}} = 13.3 \text{ days} (1 \times 10 \text{ days})$

       The value of the rate constant limits the result to 1 sig fig. This implies that there is minimum uncertainty of ± 1 in the tens place of our answer. Realistically, the remediation could take anywhere from 1 to 20 days.

18.73   (a)   Assume the density of water at 20°C is the same as at 25°C.

           $1.00 \text{ gal} \times \dfrac{4 \text{ qt}}{1 \text{ gal}} \times \dfrac{1 L}{1.057 \text{ qt}} \times \dfrac{1000 \text{ mL}}{1 L} \times \dfrac{0.99707 \text{ g } H_2O}{1 \text{ mL}} = 3773$

                                            $= 3.77 \times 10^3 \text{ g } H_2O$

The $H_2O(l)$ must be heated from 20°C to 100°C and then vaporized at 100°C.

$$3.773 \times 10^3 \, g \, H_2O \times \frac{4.184 \, J}{g \, °C} \times 80 \, °C \times \frac{1 \, kJ}{1000 \, J} = 1263 = 1.3 \times 10^3 \, kJ$$

$$3.773 \times 10^3 \, g \, H_2O \times \frac{1 \, mol \, H_2O}{18.02 \, g \, H_2O} \times \frac{40.67 \, kJ}{mol \, H_2O} = 8516 = 8.52 \times 10^3 \, kJ$$

energy = 1263 kJ + 8516 kJ = 9779 = 9.8×10^3 kJ/gal H_2O

(b) According to Solution 5.14, 1 kwh = 3.6×10^6 J.

$$\frac{9779 \, kJ}{gal \, H_2O} \times \frac{1000 \, J}{kJ} \times \frac{1 \, kwh}{3.6 \times 10^6 \, J} \times \frac{\$0.085}{kwh} = \$0.23/gal$$

(c) $\dfrac{\$0.23}{\$1.26} \times 100 = 18\%$ of the total cost is energy

18.75 (a) 17 e^-, 8.5 e^- pairs

$$\ddot{O}\!=\!\dot{N}\!-\!\ddot{O}\!: \longleftrightarrow :\ddot{O}\!-\!\dot{N}\!=\!\ddot{O}$$

Owing to its lower electronegativity, N is more likely to be electron deficient and to accommodate the odd electron.

(b) The fact that NO_2 is an electron deficient molecule indicates that it will be highly reactive. Dimerization results in formation of a N—N single bond which completes the octet of both N atoms. NO_2 and N_2O_4 exist in equilibrium in a closed system. The reaction is exothermic, Equation [22.65]. In an urban environment, NO_2 is produced from hot automobile combustion. At these temperatures, equilibrium favors the monomer because the reaction is exothermic.

(c) $2NO_2(g) + 4CO(g) \rightarrow N_2(g) + 4CO_2(g)$

$NO_2(g) + CO(g) \rightarrow NO(g) + CO_2(g)$

NO_2 is an oxidizing agent and CO is a reducing agent, so we expect products to contain N in a more reduced form, NO or N_2, and C in a more oxidized form, CO_2.

(d) No. Because it is an odd-electron molecule, NO_2 is very reactive. We expect it to undergo chemical reactions or photodissociate before it can migrate to the stratosphere. The expected half-life of an NO_2 molecule is short.

18.77 (a) According to Table 18.1, the mole fraction of CO_2 in air is 0.000375.

$$P_{CO_2} = \chi_{CO_2} \cdot P_{atm} = 0.000375 \, (1.00 \, atm) = 3.75 \times 10^{-4} \, atm$$

$$C_{CO_2} = kP_{CO_2} = 3.1 \times 10^{-2} \, M/atm \times 3.75 \times 10^{-4} \, atm = 1.16 \times 10^{-5} = 1.2 \times 10^{-5} \, M$$

(b) H_2CO_3 is a weak acid, so the $[H^+]$ is regulated by the equilibria:

$H_2CO_3(aq) \rightleftharpoons H^+(aq) + HCO_3^-(aq) \, K_{a1} = 4.3 \times 10^{-7}$

$HCO_3^-(aq) \rightleftharpoons H^+(aq) + CO_3^{2-}(aq) \, K_{a2} = 5.6 \times 10^{-11}$

Since the value of K_{a2} is small compared to K_{a1}, we will assume that most of the $H^+(aq)$ is produced by the first dissociation.

$$K_{a1} = 4.3 \times 10^{-7} = \frac{[H^+][HCO_3^-]}{[H_2CO_3]}; [H^+] = [HCO_3^-] = x, [H_2CO_3] = 1.2 \times 10^{-5} - x$$

Since K_{a1} and $[H_2CO_3]$ have similar values, we cannot assume x is small compared to 1.2×10^{-5}.

$$4.3 \times 10^{-7} = \frac{x^2}{(1.2 \times 10^{-5} - x)}; 5.00 \times 10^{-12} - 4.3 \times 10^{-7} x = x^2$$

$$0 = x^2 + 4.3 \times 10^{-7} - 5.00 \times 10^{-12}$$

$$x = \frac{-4.3 \times 10^{-7} \pm \sqrt{(4.3 \times 10^{-7})^2 - 4(1)(-5.00 \times 10^{-12})}}{2(1)}$$

$$x = \frac{-4.3 \times 10^{-7} \pm \sqrt{1.85 \times 10^{-13} + 2.00 \times 10^{-11}}}{2} = \frac{-4.3 \times 10^{-7} \pm 4.49 \times 10^{-6}}{2}$$

The negative result is meaningless; $x = 2.03 \times 10^{-6} = 2.0 \times 10^{-6} M H^+$; pH = 5.69

Since this $[H^+]$ is quite small, the $[H^+]$ from the autoionization of water might be significant. Calculation shows that for $[H^+] = 2.0 \times 10^{-6} M$ from H_2CO_3, $[H^+]$ from $H_2O = 5.2 \times 10^{-9} M$, which we can ignore.

19 Chemical Thermodynamics

Visualizing Concepts

19.1   (a)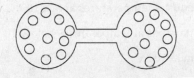

     (b) ΔS is positive, because the disorder of the system increases. Each gas has greater motional freedom as it expands into the second bulb, and there are many more possible arrangements for the mixed gases.

         By definition, ideal gases experience no attractive or repulsive intermolecular interactions, so ΔH for the mixing of ideal gases is zero, assuming heat exchange only between the two bulbs.

     (c)   The process is irreversible. It is inconceivable that the gases would reseparate.

     (d)   The entropy change of the surroundings is related to ΔH for the system. Since we are mixing ideal gases and $\Delta H = 0$, ΔH_{surr} is also zero, assuming heat exchange only between the two bulbs.

19.4   (a)   At 300 K, $\Delta H = T\Delta S$. Since $\Delta G = \Delta H - T\Delta S$, $\Delta G = 0$ at this point. When $\Delta G = 0$, the system is at equilibrium.

     (b)   The reaction is spontaneous when ΔG is negative. This condition is met when $T\Delta S > \Delta H$. From the diagram, $T\Delta S > \Delta H$ when $T > 300$ K. The reaction is spontaneous at temperatures above 300 K.

Spontaneous Processes

19.7   *Analyze/Plan.* Follow the logic in Sample Exercise 19.1.   *Solve.*

     (a)   Nonspontaneous; –5°C is below the melting point of ice, so melting does not happen without continuous intervention.

     (b)   Spontaneous; sugar is soluble in water, and even more soluble in hot coffee.

     (c)   Spontaneous; N_2 molecules are stable relative to isolated N atoms.

     (d)   Spontaneous; the filings organize in a magnetic field without intervention.

     (e)   Nonspontaneous; CO_2 and H_2O are in contact continuously at atmospheric conditions in nature and do not form CH_4 and O_2.

19.9   (a) $NH_4NO_3(s)$ dissolves in water, as in a chemical cold pack. Naphthalene (mothballs) sublimes at room temperature.

        (b)   Melting of a solid is spontaneous above its melting point but nonspontaneous below its melting point.

19.11   *Analyze/Plan.* Define the system and surroundings. Use the appropriate definition to answer the specific questions.   *Solve.*

        (a)   Water is the system. Heat must be added to the system to evaporate the water. The process is endothermic.

        (b)   At 1 atm, the reaction is spontaneous at temperatures above 100°C.

        (c)   At 1 atm, the reaction is nonspontaneous at temperatures below 100°C.

        (d)   The two phases are in equilibrium at 100°C.

19.13   *Analyze/Plan.* Define the system and surroundings. Use the appropriate definition to answer the specific questions.   *Solve.*

        (a)   For a *reversible* process, the forward and reverse changes occur by the same path. In a reversible process, both the system and the surroundings are restored to their original condition by exactly reversing the change. A reversible change produces the maximum amount of work.

        (b)   If a system is returned to its original state via a reversible path, the surroundings are also returned to their original state. That is, there is no net change in the surroundings.

        (c)   The vaporization of water to steam is reversible if it occurs at the boiling temperature of water for a specified external (atmospheric) pressure, and only if the needed heat is added infinitely slowly.

19.15   No. ΔE is a state function. $\Delta E = q + w$; q and w are not state functions. Their values do depend on path, but their sum, ΔE, does not.

19.17   *Analyze/Plan.* Define the system and surroundings. Use the appropriate definition to answer the specific questions.   *Solve.*

        (a)   An ice cube can melt reversibly at the conditions of temperature and pressure where the solid and liquid are in equilibrium. At 1 atm external pressure, the normal melting point of water is 0°C.

        (b)   We know that melting is a process that increases the energy of the system, even though there is no change in temperature. ΔE is not zero for the process.

Entropy and the Second Law of Thermodynamics

19.19   (a)   For a process that occurs at constant temperature, an isothermal process, $\Delta S = q_{rev}/T$. Here q_{rev} is the heat that would be transferred if the process were reversible. Since ΔS is a state function, it is independent of path, so ΔS for the reversible path must equal ΔS for any path.

        (b)   No. ΔS is a state function, so it is independent of path.

19.21   (a) $CH_3OH(l) \rightarrow CH_3OH(g)$, entropy increases, more mol gas in products, greater motional freedom.

   (b) $\Delta S = \dfrac{\Delta H}{T} = \dfrac{71.8\,kJ}{mol\,CH_3OH(l)} \times 1.00\,mol\,CH_3OH(l) \times \dfrac{1}{(273.15+64.7)K} \times \dfrac{1000\,J}{1\,kJ} = 213\,J/K$

19.23   (a)   For a spontaneous process, the entropy of the universe increases; for a reversible process, the entropy of the universe does not change.

   (b)   In a reversible process, $\Delta S_{system} + \Delta S_{surroundings} = 0$. If ΔS_{system} is positive, $\Delta S_{surroundings}$ must be negative.

   (c)   Since $\Delta S_{universe}$ must be positive for a spontaneous process, $\Delta S_{surroundings}$ must be greater than $-42\,J/K$.

19.25   *Analyze.* Calculate ΔS for the isothermal expansion of 0.100 mol He from 2.00 L to 5.00 L at 27°C.

   Plan. Use the relationship $\Delta S_{sys} = nR \ln(V_2/V)$.

   Solve. $\Delta S_{sys} = 0.100\,(8.314\,J/mol{\cdot}K)(\ln\,[5.00\,L/2.00\,L]) = 0.762\,J/K$.

   Check. We expect ΔS to be positive when the motional freedom of a gas increases, and our calculation agrees with this prediction.

The Molecular Interpretation of Entropy

19.27   (a)   The higher the temperature, the broader the distribution of molecular speeds and kinetic energies available to the particles. This wider range of accessible kinetic energies leads to more microstates for the system.

   (b)   An increase in volume generates more possible positions for the particles and leads to more microstates for the system.

   (c)   Going from solid to liquid to gas, particles have greater translational motion, which increases the number of positions available to the particles and the number of microstates for the system.

19.29   *Analyze/Plan.* Consider the conditions that lead to an increase in entropy: more mol gas in products than reactants, increase in volume of sample and, therefore, number of possible arrangements, more motional freedom of molecules, etc.   *Solve.*

   (a)   More gaseous particles means more possible arrangements and greater disorder; ΔS is positive.

   (b) ΔS is positive for Exercise 19.8 (a) and (c). Both processes represent an increase in volume and possible arrangements for the sample. [In (e), even though HCl(aq) is a mixture, there are fewer moles of gas in the product, so ΔS is not positive.]

19.31   *Analyze/Plan.* Consider the conditions that lead to an increase in entropy: more mol gas in products than reactants, increase in volume of sample and, therefore, number of possible arrangements, more motional freedom of molecules, etc.   *Solve.*

   (a)   S increases; translational motion is greater in the liquid than the solid.

(b)   S decreases; volume and translational motion decrease going from the gas to the liquid.

(c)   S increases; volume and translational motion are greater in the gas than the solid.

19.33   (a)   The entropy of a pure crystalline substance at absolute zero is zero.

(b)   In *translational* motion, the entire molecule moves in a single direction; in *rotational* motion, the molecule rotates or spins around a fixed axis. *Vibrational* motion is reciprocating motion. The bonds within a molecule stretch and bend, but the average position of the atoms does not change.

(c)

19.35   *Analyze/Plan.* Consider the physical changes that occur when a substance is heated. How do these changes affect the entropy of the substance?

(a)   Both 1 and 2 represent changes in entropy at constant temperature; these are phase changes. Since 1 happens at lower temperature, it represents melting (fusion), and 2 represents vaporization.

(b)   The substance changes from solid to liquid in 1, from liquid to gas in 2. The larger volume and greater motional freedom of the gas phase causes ΔS for vaporization to (always) be larger than ΔS for fusion.

19.37   *Analyze/Plan.* Consider the factors that lead to higher entropy: more mol gas in products than reactants, increase in volume of sample and, therefore, number of possible arrangements, more motional freedom of molecules, etc.   *Solve.*

(a)   Ar(g) (gases have higher entropy due primarily to much larger volume)

(b)   He(g) at 1.5 atm (larger volume and more motional freedom)

(c)   1 mol of Ne(g) in 15.0 L (larger volume provides more motional freedom)

(d) $CO_2(g)$ (more motional freedom)

19.39   *Analyze/Plan.* Consider the markers of an increase in entropy for a chemical reaction: liquids or solutions formed from solids, gases formed from either solids or liquids, increase in moles gas during reaction.   *Solve.*

(a) ΔS negative (moles of gas decrease)

(b) ΔS positive (gas produced, increased disorder)

(c) ΔS negative (moles of gas decrease)

(d) ΔS positive (moles of gas increase)

Entropy Changes in Chemical Reactions

19.41　*Analyze/Plan.* Given two molecules in the same state, predict which will have the higher molar entropy. In general, for molecules in the same state, the more atoms in the molecule, the more degrees of freedom, the greater the number of microstates and the higher the standard entropy, $S°$.

(a)　$C_2H_6(g)$ has more degrees of freedom and larger $S°$.

(b)　$CO_2(g)$ has more degrees of freedom and larger $S°$.

19.43　*Analyze/Plan.* Consider the conditions that lead to an increase in entropy: more mol gas in products than reactants, increase in volume of sample and, therefore, number of possible arrangements, more motional freedom of molecules, etc. *Solve.*

(a)　$Sc(s)$, 34.6 J/mol•K; $Sc(g)$, 174.7 J/mol•K. In general, the gas phase of a substance has a larger $S°$ than the solid phase because of the greater volume and motional freedom of the molecules.

(b)　$NH_3(g)$, 192.5 J/mol•K; $NH_3(aq)$, 111.3 J/mol•K. Molecules in the gas phase have more motional freedom than molecules in solution.

(c)　1 mol of $P_4(g)$, 280 J/K; 2 mol of $P_2(g)$, 2(218.1) = 436.2 J/K. More particles have greater motional energy (more available microstates).

(d)　$C(diamond)$, 2.43 J/mol•K; $C(graphite)$ 5.69 J/mol•K. Diamond is a network covalent solid with each C atom tetrahedrally bound to four other C atoms. Graphite consists of sheets of fused planar 6-membered rings with each C atom bound in a trigonal planar arrangement to three other C atoms. The internal entropy in graphite is greater because there is translational freedom among the planar sheets of C atoms while there is very little vibrational freedom within the network covalent diamond lattice.

19.45　For elements with similar structures, the heavier the atoms, the lower the vibrational frequencies at a given temperature. This means that more vibrations can be accessed at a particular temperature resulting in a greater absolute entropy for the heavier elements.

19.47　*Analyze/Plan.* Follow the logic in Sample Exercise 19.5.　*Solve.*

(a)　$\Delta S° = S° \, C_2H_6(g) - S° \, C_2H_4(g) - S° \, H_2(g)$

$= 229.5 - 219.4 - 130.58 = -120.5 \, J/K$

$\Delta S°$ is negative because there are fewer moles of gas in the products.

(b)　$\Delta S° = 2S° \, NO_2(g) - \Delta S° \, N_2O_4(g) = 2(240.45) - 304.3 = +176.6 \, J/K$

$\Delta S°$ is positive because there are more moles of gas in the products.

(c)　$\Delta S° = \Delta S° \, BeO(s) + \Delta S° \, H_2O(g) - \Delta S° \, Be(OH)_2(s)$

$= 13.77 + 188.83 - 50.21 = +152.39 \, J/K$

$\Delta S°$ is positive because the product contains more total particles and more moles of gas.

(d) $\Delta S° = 2S° \, CO_2(g) + 4S° \, H_2O(g) - 2S° \, CH_3OH(g) - 3S° \, O_2(g)$

$= 2(213.6) + 4(188.83) - 2(237.6) - 3(205.0) = +92.3 \, J/K$

$\Delta S°$ is positive because the product contains more total particles and more moles of gas.

Gibbs Free Energy

19.49 (a) $\Delta G = \Delta H - T\Delta S$

(b) If ΔG is positive, the process is nonspontaneous, but the reverse process is spontaneous.

(c) There is no relationship between ΔG and rate of reaction. A spontaneous reaction, one with a $-\Delta G$, may occur at a very slow rate. For example: $2H_2(g) + O_2(g) \rightarrow 2H_2O(g)$, $\Delta G = -457 \, kJ$ is very slow if not initiated by a spark.

19.51 *Analyze/Plan.* Consider the definitions of $\Delta H°$, $\Delta S°$ and $\Delta G°$, along with sign conventions. $\Delta G° = \Delta H° - T\Delta S°$.   *Solve.*

(a) $\Delta H°$ is negative; the reaction is exothermic.

(b) $\Delta S°$ is negative; the reaction leads to decrease in disorder (increase in order) of the system.

(c) $\Delta G° = \Delta H° - T\Delta S° = -35.4 \, kJ - 298 \, K \, (-0.0855 \, kJ/K) = -9.921 = -9.9 \, kJ$

(d) At 298 K, $\Delta G°$ is negative. If all reactants and products are present in their standard states, the reaction is spontaneous (in the forward direction) at this temperature.

19.53 *Analyze/Plan.* Follow the logic in Sample Exercise 19.7. Calculate $\Delta H°$ according to Equation [5.31], $\Delta S°$ by Equation [19.8] and $\Delta G°$ by Equation [19.13]. Then use $\Delta H°$ and $\Delta S°$ to calculate $\Delta G°$ using Equation [19.20], $\Delta G° = \Delta H° - T\Delta S°$.   *Solve.*

(a) $\Delta H° = 2(-268.61) - [0 + 0] = -537.22 \, kJ$

$\Delta S° = 2(173.51) - [130.58 + 202.7] = 13.74 = 13.7 \, J/K$

$\Delta G° = 2(-270.70) - [0 + 0] = -541.40 \, kJ$

$\Delta G° = -537.22 \, kJ - 298(0.01374) \, kJ = -541.31 \, kJ$

(b) $\Delta H° = -106.7 - [0 + 2(0)] = -106.7 \, kJ$

$\Delta S° = 309.4 - [5.69 + 2(222.96)] = -142.21 = -142.2 \, J/K$

$\Delta G° = -64.0 - [0 + 2(0)] = -64.0 \, kJ$

$\Delta G° = -106.7 \, kJ - 298(-0.14221) \, kJ = -64.3 \, kJ$

(c) $\Delta H° = 2(-542.2) - [2(-288.07) + 0] = -508.26 = -508.3 \, kJ$

$\Delta S° = 2(325) - [2(311.7) + 205.0] = -178.4 = -178 \, J/K$

$\Delta G° = 2(-502.5) - [2(-269.6) + 0] = -465.8 \, kJ$

$\Delta G° = -508.26 \, kJ - 298(-0.1784) \, kJ = -455.097 = -455.1 \, kJ$

(The discrepancy in $\Delta G°$ values is due to experimental uncertainties in the tabulated thermodynamic data.)

(d) $\Delta H° = -84.68 + 2(-241.82) - [2(-201.2) + 0] = -165.92 = -165.9$ kJ

$\Delta S° = 229.5 + 2(188.83) - [2(237.6) + 130.58] = 1.38 = 1.4$ J/K

$\Delta G° = -32.89 + 2(-228.57) - [2(-161.9) + 0] = -166.23 = -166.2$ kJ

$\Delta G° = -165.92$ kJ $- 298(0.00138)$ kJ $= -166.33 = -166.3$ kJ

19.55   *Analyze/Plan.* Follow the logic in Sample Exercise 19.6.   *Solve.*

(a) $\Delta G° = 2\Delta G° \, SO_3(g) - [2\Delta G° \, SO_2(g) + \Delta G° \, O_2(g)]$

$= 2(-370.4) - [2(-300.4) + 0] = -140.0$ kJ, spontaneous

(b) $\Delta G° = 3\Delta G° \, NO(g) - [\Delta G° \, NO_2(g) + \Delta G° \, N_2O(g)]$

$= 3(86.71) - [51.84 + 103.59] = +104.70$ kJ, nonspontaneous

(c) $\Delta G° = 4\Delta G° \, FeCl_3(s) + 3\Delta G° \, O_2(g) - [6\Delta G° \, Cl_2(g) + 2\Delta G° \, Fe_2O_3(s)]$

$= 4(-334) + 3(0) - [6(0) + 2(-740.98)] = +146$ kJ, nonspontaneous

(d) $\Delta G° = \Delta G° \, S(s) + 2\Delta G° \, H_2O(g) - [\Delta G° \, SO_2(g) + 2\Delta G° \, H_2(g)]$

$= 0 + 2(-228.57) - [(-300.4) + 2(0)] = -156.7$ kJ, spontaneous

19.57   *Analyze/Plan.* Follow the logic in Sample Exercise 19.8(a).   *Solve.*

(a) $C_6H_{12}(l) + 9O_2(g) \rightarrow 6CO_2(g) + 6H_2O(l)$

(b)   Because there are fewer moles of gas in the products, $\Delta S°$ is negative, which makes $-T\Delta S$ positive. $\Delta G°$ is less negative (more positive) than $\Delta H°$.

19.59   *Analyze/Plan.* Based on the signs of ΔH and ΔS for a particular reaction, assign a category from Table 19.4 to each reaction.   *Solve.*

(a) ΔG is negative at low temperatures, positive at high temperatures. That is, the reaction proceeds in the forward direction spontaneously at lower temperatures but spontaneously reverses at higher temperatures.

(b) ΔG is positive at all temperatures. The reaction is nonspontaneous in the forward direction at all temperatures.

(c) ΔG is positive at low temperatures, negative at high temperatures. That is, the reaction will proceed spontaneously in the forward direction at high temperature.

19.61   *Analyze/Plan.* We are told that the reaction is spontaneous and endothermic, and asked to estimate the sign and magnitude of ΔS. If a reaction is spontaneous, $\Delta G < 0$. Use this information with Equation [19.20] to solve the problem.   *Solve.*

At 450 K, $\Delta G < 0$; $\Delta G = \Delta H - T\Delta S < 0$

34.5 kJ $- 450$ K $(\Delta S) < 0$; 34.5 kJ < 450 K (ΔS); $\Delta S > 34.5$ kJ/450 K

$\Delta S > 0.0767$ kJ/K or $\Delta S > 76.7$ J/K

19.63   *Analyze/Plan.* Use Equation [19.20] to calculate T when $\Delta G = 0$. This is similar to calculating the temperature of a phase trasition in Sample Exercise 19.9. Use Table 19.4 to determine whether the reaction is spontaneous or non-spontaneous above this temperature.   *Solve.*

(a) $\Delta G = \Delta H - T\Delta S$; $0 = -32$ kJ $- T(-98$ J/K); 32×10^3 J $= T(98$ J/K)

   $T = 32 \times 10^3$ J$/(98$ J/K$) = 326.5 = 330$ K

(b)   Nonspontaneous. The sign of ΔS is negative, so as T increases, ΔG becomes more positive.

19.65   *Analyze/Plan.* Given a chemical equation and thermodynamic data (values of $\Delta H_f^\circ, \Delta G_f^\circ$ and S°) for reactants and products, predict the variation of ΔG° with temperature and calculate ΔG° at 800 K and 1000 K. Use Equations [5.31] and [19.8] to calculate ΔH° and ΔS°, respectively; use these values to calculate ΔG° at various temperatures, using Equation [19.20]. The signs of ΔH° and ΔS° determine the variation of ΔG° with temperature. *Solve.*

(a)   Calculate ΔH° and ΔS° to determine the sign of $T\Delta S^\circ$.

   $\Delta H^\circ = 3\Delta H^\circ \, NO(g) - \Delta H^\circ \, NO_2(g) - \Delta H^\circ \, N_2O(g)$

      $= 3(90.37) - 33.84 - 81.6 = 155.7$ kJ

   $\Delta S^\circ = 3S^\circ \, NO(g) - S^\circ \, NO_2(g) - S^\circ \, N_2O(g)$

      $= 3(210.62) - 240.45 - 220.0 = 171.4$ J/K

   $\Delta G^\circ = \Delta H^\circ - T\Delta S^\circ$. Since ΔS° is positive, $-T\Delta S^\circ$ becomes more negative as T increases and ΔG° becomes more negative.

(b) $\Delta G^\circ = \Delta H^\circ - T\Delta S^\circ = 155.7$ kJ $- (800$ K$)(0.1714$ kJ/K$)$

   $\Delta G^\circ = 155.7$ kJ $- 137$ kJ $= 19$ kJ

   Since ΔG° is positive at 800 K, the reaction is not spontaneous at this temperature.

(c) $\Delta G^\circ = 155.7$ kJ $- (1000$ K$)(0.1714$ kJ/K$) = 155.7$ kJ $- 171.4$ kJ $= -15.7$ kJ

   ΔG° is negative at 1000 K and the reaction is spontaneous at this temperature.

19.67   *Analyze/Plan.* Follow the logic in Sample Exercise 19.9.   *Solve.*

(a) $\Delta S_{vap}^\circ = \Delta H_{vap}^\circ / T_b$; $T_b = \Delta H_{vap}^\circ / \Delta S_{vap}^\circ$

   $\Delta H_{vap}^\circ = \Delta H^\circ \, C_6H_6(g) - \Delta H^\circ \, C_6H_6(l) = 82.9 - 49.0 = 33.9$ kJ

   $\Delta S_{vap}^\circ = S^\circ \, C_6H_6(g) - S^\circ \, C_6H_6(l) = 269.2 - 172.8 = 96.4$ J/K

   $T_b = 33.9 \times 10^3$ J$/96.4$ J/K $= 351.66 = 352$ K $= 79°C$

(b)   From the *Handbook of Chemistry and Physics*, 74th Edition, $T_b = 80.1°C$. The values are remarkably close; the small difference is due to deviation from ideal behavior by $C_6H_6(g)$ and experimental uncertainty in the boiling point measurement and the thermodynamic data.

19.69   *Analyze/Plan.* We are asked to write a balanced equation for the combustion of acetylene, calculate ΔH° for this reaction and calculate maximum useful work possible by the system. Combustion is combination with O_2 to produce CO_2 and H_2O. Calculate ΔH° using data from Appendix C and Equation 5.31. The maximum obtainable work is ΔG (Equation [19.19]), which can be calculated from data in Appendix C and Equation [19.13].   *Solve.*

(a) $C_2H_2(g) + 5/2 \, O_2(g) \rightarrow 2CO_2(g) + H_2O(l)$

(b) $\Delta H° = 2\Delta H° \, CO_2(g) + \Delta H° \, H_2O(l) - \Delta H° \, C_2H_2(g) - 5/2\Delta H° \, O_2(g)$

       $= 2(-393.5) - 285.83 - 226.7 - 5/2(0) = -1299.5$ kJ produced/mol C_2H_2 burned

(c) $w_{max} = \Delta G° = 2\Delta G° \, CO_2(g) + \Delta G° \, H_2O(l) - \Delta G° \, C_2H_2(g) - 5/2 \, \Delta G° \, O_2(g)$

       $= 2(-394.4) - 237.13 - 209.2 - 5/2(0) = -1235.1$ kJ

The negative sign indicates that the system does work on the surroundings; the system can accomplish a maximum of 1235.1 kJ of work on its surroundings.

Free Energy and Equilibrium

19.71   *Analyze/Plan.* We are given a chemical reaction and asked to predict the effect of the partial pressure of $O_2(g)$ on the value of ΔG for the system. Consider the relationship $\Delta G = \Delta G° + RT \ln Q$ where Q is the reaction quotient.   *Solve.*

(a) $O_2(g)$ appears in the denominator of Q for this reaction. An increase in pressure of O_2 decreases Q and ΔG becomes smaller or more negative. Increasing the concentration of a reactant increases the tendency for a reaction to occur.

(b) $O_2(g)$ appears in the numerator of Q for this reaction. Increasing the pressure of O_2 increases Q and ΔG becomes more positive. Increasing the concentration of a product decreases the tendency for the reaction to occur.

(c) $O_2(g)$ appears in the numerator of Q for this reaction. An increase in pressure of O_2 increases Q and ΔG becomes more positive. Since pressure of O_2 is raised to the third power in Q, an increase in pressure of O_2 will have the largest effect on ΔG for this reaction.

19.73   *Analyze/Plan.* Given a chemical reaction, we are asked to calculate $\Delta G°$ from Appendix C data, and ΔG for a given set of initial conditions. Use Equation [19.13] to calculate $\Delta G°$, and Equation [19.21] to calculate ΔG. Follow the logic in Sample Exercise 19.10 when calculating ΔG.   *Solve.*

(a) $\Delta G° = \Delta G° \, N_2O_4(g) - 2\Delta G° \, NO_2(g) = 98.28 - 2(51.84) = -5.40$ kJ

(b) $\Delta G = \Delta G° + RT \ln P_{N_2O_4} / P_{NO_2}^2$

       $= -5.40 \text{ kJ} + \dfrac{8.314 \times 10^{-3} \text{ kJ}}{\text{K} \cdot \text{mol}} \times 298 \text{ K} \times \ln[1.60/(0.40)^2] = 0.3048 = 0.3 \text{ kJ}$

19.75   *Analyze/Plan.* Given a chemical reaction, we are asked to calculate K using $\Delta G_f°$ data from Appendix C. Follow the logic in Sample Exercise 19.13. $\Delta G° = -RT \ln K$, Equation [19.22]; $\ln K = -\Delta G°/RT$ *Solve.*

(a) $\Delta G° = 2\Delta G° \, HI(g) - \Delta G° \, H_2(g) - \Delta G° \, I_2(g)$

       $= 2(1.30) - 0 - 19.37 = -16.77$ kJ

       $\ln K = \dfrac{-(-16.77 \text{ kJ}) \times 10^3 \text{ J/kJ}}{8.314 \text{ J/K} \times 298 \text{ K}} = 6.76876 = 6.769; \quad K = 870$

(b) $\Delta G° = \Delta G° \, C_2H_4(g) + \Delta G° \, H_2O(g) - \Delta G° \, C_2H_5OH(g)$

       $= 68.11 - 228.57 - (-168.5) = 8.04 = 8.0$ kJ

(c) $\Delta G° = \Delta G° \, C_6H_6(g) - 3\Delta G° \, C_2H_2(g) = 129.7 - 3(209.2) = -497.9 \text{ kJ}$

$$\ln K = \frac{-\Delta G°}{RT} = \frac{-(-497.9 \text{ kJ}) \times 10^3 \text{ J/KJ}}{8.314 \text{ J/K} \times 298 \text{ K}} = 200.963 = 201.0; \, K = 2 \times 10^{87}$$

19.77   *Analyze/Plan.* Given a chemical reaction and thermodynamic data in Appendix C, calculate the equilibrium pressure of $CO_2(g)$ at two temperatures. $K = P_{CO_2}$. Calculate $\Delta G°$ at the two temperatures using $\Delta G° = \Delta H° - T\Delta S°$ and then calculate K and P_{CO_2}. *Solve.*

$\Delta H° = \Delta H° \, BaO(s) + \Delta H° \, CO_2(g) - \Delta H° \, BaCO_3(s)$

     $= -553.5 + -393.5 - (-1216.3) = +269.3 \text{ kJ}$

$\Delta S° = S° \, BaO(s) + S° \, CO_2(g) - S° \, BaCO_3(s)$

     $= 70.42 + 213.6 - 112.1 = 171.92 \text{ J/K} = 0.1719 \text{ kJ/K}$

(a) ΔG at 298 K $= 269.3 \text{ kJ} - 298 \text{ K} (0.17192 \text{ kJ/K}) = 218.07 = 218.1 \text{ kJ}$

$$\ln K = \frac{-\Delta G°}{RT} = \frac{-218.07 \times 10^3 \text{ J}}{8.314 \text{ J/K} \times 298 \text{ K}} = -88.017 = -88.02$$

$K = 6.0 \times 10^{-39}; \quad P_{CO_2} = 6.0 \times 10^{-39} \text{ atm}$

(b) ΔG at 1100 K $= 269.3 \text{ kJ} - 1100 \text{ K} (0.17192 \text{ kJ}) = 80.19 = +80.2 \text{ kJ}$

$$\ln K = \frac{-\Delta G°}{RT} = \frac{-80.19 \times 10^3 \text{ J}}{8.314 \text{ J/K} \times 1100 \text{ K}} = -8.768 = -8.77$$

$K = 1.6 \times 10^{-4}; \quad P_{CO_2} = 1.6 \times 10^{-4} \text{ atm}$

19.79   *Analyze/Plan.* Given an acid dissociation equilibrium and the corresponding K_a value, calculate $\Delta G°$ and ΔG for a given set of concentrations. Use Equation [19.22] to calculate $\Delta G°$ and Equation [19.21] to calculate ΔG.   *Solve.*

(a) $HNO_2(aq) \;\rightleftharpoons\; H^+(aq) + NO_2^-(aq)$

(b) $\Delta G° = -RT \ln K_a = -(8.314 \times 10^{-3})(298) \ln (4.5 \times 10^{-4}) = 19.0928 = 19.1 \text{ kJ}$

(c) $\Delta G = 0$ at equilibrium

(d) $\Delta G = \Delta G° + RT \ln Q$

$$= 19.09 \text{ kJ} + (8.314 \times 10^{-3})(298) \ln \frac{(5.0 \times 10^{-2})(6.0 \times 10^{-4})}{0.20} = -2.72 = -3 \text{ kJ}$$

Additional Exercises

19.82

Process	ΔH	ΔS
(a)	+	+
(b)	−	−
(c)	+	+
(d)	+	+
(e)	−	+

19.86   (a)   Formation reactions are the synthesis of 1 mole of compound from elements in their standard states.

$$1/2\,N_2(g) + 3/2\,H_2(g) \rightarrow NH_3(g)$$

$$C(s) + 2Cl_2(g) \rightarrow CCl_4(l)$$

$$K(s) + 1/2\,N_2(g) + 3/2\,O_2(g) \rightarrow KNO_3(s)$$

In each of these formation reactions, there are fewer moles of gas in the products than the reactants, so we expect $\Delta S°$ to be negative. If $\Delta G_f° = \Delta H_f° - T\Delta S_f°$ and $\Delta S_f°$ is negative, $-T\Delta S_f°$ is positive and $\Delta G_f°$ is more positive than $\Delta H_f°$.

       (b) $C(s) + 1/2\,O_2(g) \rightarrow CO(g)$

In this reaction, there are more moles of gas in products, $\Delta S_f°$ is positive, $-T\Delta S_f°$ is negative and $\Delta G_f°$ is more negative than $\Delta H_f°$.

19.90   (a) $K = \dfrac{\chi_{CH_3COOH}}{\chi_{CH_3OH}P_{CO}}$

$\Delta G° = -RT \ln K; \ln K = -\Delta G/RT$

$\Delta G° = \Delta G°\,CH_3COOH(l) - \Delta G°\,CH_3OH(l) - \Delta G°\,CO(g)$

$\quad = -392.4 - (-166.23) - (-137.2) = -89.0\,kJ$

$\ln K = \dfrac{-(-89.0\,kJ)}{(8.314 \times 10^{-3}\,kJ/K)(298\,K)} = 35.922 = 35.9; \; K = 4 \times 10^{15}$

       (b) $\Delta H° = \Delta H°\,CH_3COOH(l) - \Delta H°\,CH_3OH(l) - \Delta H°\,CO(g)$

$\quad = -487.0 - (-238.6) - (-110.5) = -137.9\,kJ$

The reaction is exothermic, so the value of K will decrease with increasing temperature, and the mole fraction of CH_3COOH will also decrease. Elevated temperatures must be used to increase the speed of the reaction. Thermodynamics cannot predict the rate at which a reaction reaches equilibrium.

       (c) $\Delta G° = -RT \ln K; K = 1, \ln K = 0, \Delta G° = 0$

$\Delta G° = \Delta H° - T\Delta S°;$ when $\Delta G° = 0, \Delta H° = T\Delta S°$

$\Delta S° = S°\,CH_3COOH(l) - S°\,CH_3OH(l) - S°\,CO(g)$

$\quad = 159.8 - 126.8 - 197.9 = -164.9\,J/K = -0.1649\,kJ/K$

$-137.9\,kJ = T(-0.1649\,kJ/K), T = 836.3\,K$

The equilibrium favors products up to 836 K or 563°C, so the elevated temperatures to increase the rate of reaction can be safely employed.

19.94   (a)   The equilibrium of interest here can be written as:

K^+ (plasma) $\rightleftharpoons$ K^+ (muscle)

Since an aqueous solution is involved in both cases, assume that the equilibrium constant for the above process is exactly 1, that is, $\Delta G° = 0$. However, ΔG is not zero because the concentrations are not the same on both sides of the membrane. Use Equation [19.21] to calculate ΔG:

$$\Delta G = \Delta G^\circ + RT \ln \frac{[K^+ \text{ (muscle)}]}{[K^+ \text{ (plasma)}]}$$

$$= 0 + (8.314)(310) \ln \frac{(0.15)}{(5.0 \times 10^{-3})} = 8766 \text{ J} = 8.8 \text{ kJ}$$

(b)   Note that ΔG is positive. This means that work must be done on the system (blood plasma plus muscle cells) to move the K^+ ions "uphill," as it were. The minimum amount of work possible is given by the value for ΔG. This value represents the minimum amount of work required to transfer one mole of K^+ ions from the blood plasma at 5×10^{-3} M to muscle cell fluids at 0.15 M, assuming constancy of concentrations. In practice, a larger than minimum amount of work is required.

Integrative Exercises

19.98   (a)   At the boiling point, vaporization is a reversible process, so $\Delta S^\circ_{vap} = \Delta H^\circ_{vap}/T$.

acetone: $\Delta S^\circ_{vap} = \Delta H^\circ_{vap}/T = (29.1 \text{ kJ/mol}) / 329.25 \text{ K} = 88.4 \text{ J/mol} \bullet \text{K}$

dimethyl ether: $\Delta S^\circ_{vap} = (21.5 \text{ kJ/mol}) / 248.35 \text{ K} = 86.6 \text{ J/mol} \bullet \text{K}$

ethanol: $\Delta S^\circ_{vap} = (38.6 \text{ kJ/mol}) / 351.6 \text{ K} = 110 \text{ J/mol} \bullet \text{K}$

octane: $\Delta S^\circ_{vap} = (34.4 \text{ kJ/mol}) / 398.75 \text{ K} = 86.3 \text{ J/mol} \bullet \text{K}$

pyridine: $\Delta S^\circ_{vap} = (35.1 \text{ kJ/mol}) / 388.45 \text{ K} = 90.4 \text{ J/mol} \bullet \text{K}$

(b)   Ethanol is the only liquid listed that doesn't follow *Trouton's rule* and it is also the only substance that exhibits hydrogen bonding in the pure liquid. Hydrogen bonding leads to more ordering in the liquid state and a greater than usual increase in entropy upon vaporization. The rule appears to hold for liquids with London dispersion forces (octane) and ordinary dipole-dipole forces (acetone, dimethyl ether, pyridine), but not for those with hydrogen bonding.

(c)   Owing to strong hydrogen bonding interactions, water probably does not obey Trouton's rule.

From Appendix B, ΔH°_{vap} at 100°C = 40.67 kJ/mol.

$\Delta S^\circ_{vap} = (40.67 \text{ kJ/mol}) / 373.15 \text{ K} = 109.0 \text{ J/mol} \bullet \text{K}$

(d)   Use $\Delta S^{\circ}_{vap} = 88 \text{ J/mol·K}$, the middle of the range for Trouton's rule, to estimate ΔH°_{vap} for chlorobenzene.

$$\Delta H^{\circ}_{vap} = \Delta S^{\circ}_{vap} \times T = 88 \text{ J/mol·K} \times 404.95 \text{ K} = 36 \text{ kJ/mol}$$

19.101   (a) $O_2(g) \xrightarrow{h\nu} 2O(g)$; S increases because there are more moles of gas in the products.

         (b) $O_2(g) + O(g) \rightarrow O_3(g)$, S decreases because there are fewer moles of gas in the products.

         (c)   S increases as the gas molecules diffuse into the larger volume of the stratosphere; there are more possible positions and therefore more motional freedom.

         (d) $NaCl(aq) \rightarrow NaCl(s) + H_2O(l)$; ΔS decreases as the mixture (seawater, greater disorder) is separated into pure substances (fewer possible arrangements, more order).

19.104   (a) $K = P^2_{NO_2}/P_{N_2O_4}$

            Assume equal amounts means equal number of moles. For gases, $P = n(RT/V)$. In an equilibrium mixture, RT/V is a constant, so moles of gas are directly proportional to partial pressure. Gases with equal partial pressures will have equal moles of gas present. The condition $P_{NO_2} = P_{N_2O_4}$ leads to the expression $K = P_{NO_2}$. The value of K then depends on P_T for the mixture. For any particular value of P_T, the condition of equal moles of the two gases can be achieved at some temperature. For example, $P_{NO_2} = P_{N_2O_4} = 1.0$ atm, $P_T = 2.0$ atm.

$$K = \frac{(1.0)^2}{1.0} = 1.0; \quad \ln K_{eq} = 0; \quad \Delta G^{\circ} = 0 = \Delta H^{\circ} - T\Delta S^{\circ}; \quad T = \Delta H^{\circ}/\Delta S^{\circ}$$

$$\Delta H^{\circ} = 2\Delta H^{\circ} NO_2(g) - \Delta H^{\circ} N_2O_4(g) = 2(33.84) - 9.66 = +58.02 \text{ kJ}$$

$$\Delta S^{\circ} = 2S^{\circ} NO_2(g) - S^{\circ} N_2O_4(g) = 2(240.45) - 304.3 = 0.1766 \text{ kJ/K}$$

$$T = \frac{58.02 \text{ kJ}}{0.1766 \text{ kJ/K}} = 328.5 \text{ K or } 55.5^{\circ}C$$

         (b) $P_T = 1.00$ atm; $P_{N_2O_4} = x$, $P_{NO_2} = 2x$; $x + 2x = 1.00$ atm

$$x = P_{N_2O_4} = 0.3333 = 0.333 \text{ atm}; \quad P_{NO_2} = 0.6667 = 0.667 \text{ atm}$$

$$K = \frac{(0.6667)^2}{0.3333} = 1.334 = 1.33; \quad \Delta G^{\circ} = -RT \ln K = \Delta H^{\circ} - T\Delta S^{\circ}$$

$$-(8.314 \times 10^{-3} \text{ kJ/K})(\ln 1.334) T = 58.02 \text{ kJ} - (0.1766 \text{ kJ/K}) T$$

$$(-0.00239 \text{ kJ/K}) T + (0.1766 \text{ kJ/K}) T = 58.02 \text{ kJ}$$

$$(0.1742 \text{ kJ/K}) T = 58.02 \text{ kJ}; \quad T = 333.0 \text{ K}$$

(c) $P_T = 10.00$ atm; $x + 2x = 10.00$ atm

$x = P_{N_2O_4} = 3.3333 = 3.333$ atm; $P_{NO_2} = 6.6667 = 6.667$ atm

$K = \dfrac{(6.6667)^2}{3.3333} = 13.334 = 13.33$; $-RT \ln K = \Delta H^\circ - T\Delta S^\circ$

$-(8.314 \times 10^{-3} \text{ kJ/K})(\ln 13.334) \text{ T} = 58.02 \text{ kJ} - (0.1766 \text{ kJ/K}) \text{ T}$

$(-0.02154 \text{ kJ/K}) \text{ T} + (0.1766 \text{ kJ/K}) \text{ T} = 58.02 \text{ kJ}$

$(0.15506 \text{ kJ/K}) \text{ T} = 58.02 \text{ kJ}$; $\text{T} = 374.2 \text{ K}$

(d)   The reaction is endothermic, so an increase in the value of K as calculated in parts
(b) and (c) should be accompanied by an increase in T.

20 Electrochemistry

Visualizing Concepts

20.1   In a Brønsted-Lowry acid-base reaction, H^+ is transferred from the acid to the base. In a redox reaction, the substance being oxidized (the reductant) loses electrons and the substance being reduced (the oxidant) gains electrons. Furthermore, the number of electrons gained and lost must be equal. The concept of electron transfer from reductant to oxidant is clearly applicable to redox reactions. (The path of the transfer may or may not be direct, but ultimately electrons are transferred during redox reactions.)

20.3 $Ni^{2+}(aq) + 2e^- \rightarrow Ni(s)$, $E^{\circ}_{red} = -0.28 \, V$, cathode

$Fe^{2+}(aq) + 2e^- \rightarrow Fe(s)$, $E^{\circ}_{red} = -0.44 \, V$, anode

$Ni^{2+}(aq)$ has the larger E°_{red}, so it will be reduced in the redox reaction. Reduction occurs at the cathode, so $Ni^{2+}(aq)$ and $Ni(s)$ will be in the cathode compartment and $Fe^{2+}(aq)$ and $Fe(s)$ will be in the anode compartment. The voltmeter will read:

$E^{\circ}_{cell} = E^{\circ}_{red}(cathode) - E^{\circ}_{red}(anode) = -0.28 \, V - (-044 \, V) = 0.16 \, V$

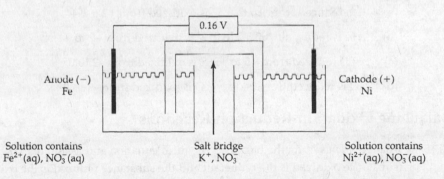

20.5 $A(aq) + B(aq) \rightarrow A^-(aq) + B^+(aq)$

(a)   A gains electrons; it is being reduced. B loses electrons; it is being oxidized.

(b)   Reduction occurs at the cathode; oxidation occurs at the anode.

A(aq) + $1e^- \rightarrow A^-(aq)$ occurs at the cathode.

$B(aq) \rightarrow B^+(aq) + 1e^-$ occurs at the anode.

(c)   In a voltaic cell, the anode is at higher potential energy than the cathode. The anode reaction, $B(aq) \rightarrow B^+(aq) + 1e^-$, is higher in potential energy.

20.7   Zinc, $E_{red}^{\circ} = -0.763\,V$, is more easily oxidized than iron, $E_{red}^{\circ} = -0.440\,V$. If conditions are favorable for oxidation, zinc will be preferentially oxidized, preventing iron from corroding. The protection lasts until all the Zn coating has reacted.

Oxidation States

20.9   (a)   *Oxidation* is the loss of electrons.

(b)   The electrons appear on the products side (right side) of an oxidation half-reaction.

(c)   The *oxidant* is the reactant that is reduced; it gains the electrons that are lost by the substance being oxidized.

(d)   An *oxidizing agent* is the substance that promotes oxidation. That is, it gains electrons that are lost by the substance being oxidized. It is the same as the oxidant.

20.11   (a)   True.

(b)   False. Fe^{3+} is reduced to Fe^{2+}, so it is the oxidizing agent, and Co^{2+} is the reducing agent.

(c)   True.

20.13   *Analyze/Plan.* Given a chemical equation, we are asked to indicate which elements undergo a change in oxidation number and the magnitude of the change. Assign oxidation numbers according to the rules given in Section 4.4. Note the changes and report the magnitudes.   *Solve.*

(a)   I is reduced from +5 to 0; C is oxidized from +2 to +4.

(b)   Hg is reduced from +2 to 0; N is oxidized from –2 to 0.

(c)   N is reduced from +5 to +2; S is oxidized from –2 to 0.

(d)   Cl is reduced from +4 to +3; O is oxidized from –1 to 0.

Balancing Oxidation-Reduction Reactions

20.15   *Analyze/Plan.* Write the balanced chemical equation and assign oxidation numbers. The substance oxidized is the reductant and the substance reduced is the oxidant.   *Solve.*

(a) $TiCl_4(g) + 2Mg(l) \rightarrow Ti(s) + 2MgCl_2(l)$

(b) $Mg(l)$ is oxidized; $TiCl_4(g)$ is reduced.

(c) $Mg(l)$ is the reductant; $TiCl_4(g)$ is the oxidant.

20.17   *Analyze/Plan.* Follow the logic in Sample Exercises 20.2 and 20.3. If the half-reaction occurs in basic solution, balance as in acid, then add OH^- to each side.   *Solve.*

(a) $Sn^{2+}(aq) \rightarrow Sn^{4+}(aq) + 2e^-$, oxidation

(b) $TiO_2(s) + 4H^+(aq) + 2e^- \rightarrow Ti^{2+}(aq) + 2H_2O(l)$, reduction

(c) $ClO_3^-(aq) + 6H^+(aq) + 6e^- \rightarrow Cl^-(aq) + 3H_2O(l)$, reduction

(d) $4OH^-(aq) \rightarrow O_2(g) + 2H_2O(l) + 4e^-$, oxidation

(e) $SO_3^{2-}(aq) + 2OH^-(aq) \rightarrow SO_4^{2-}(aq) + H_2O(l) + 2e^-$, oxidation

(f) $N_2(g) + 8H^+(aq) + 6e^- \rightarrow 2NH_4^+(aq)$, reduction

(g) $N_2(g) + 6H_2O + 2e^- \rightarrow 2NH_3(g) + OH^-(aq)$, reduction

20.19   *Analyze/Plan.* Follow the logic in Sample Exercises 20.2 and 20.3 to balance the given equations. Use the method in Sample Exercise 20.1 to identify oxidizing and reducing agents.   *Solve.*

(a) $Cr_2O_7^{2-}(aq) + I^-(aq) + 8H^+ \rightarrow 2Cr^{3+}(aq) + IO_3^-(aq) + 4H_2O(l)$

   oxidizing agent, $Cr_2O_7^{2-}$; reducing agent, I^-

(b)   The half-reactions are:

   $4[MnO_4^-(aq) + 8H^+(aq) + 5e^- \rightarrow Mn^{2+}(aq) + 4H_2O(l)]$

   $5[CH_3OH(aq) + H_2O(l) \rightarrow HCO_2H(aq) + 4H^+(aq) + 4e^-]$
   _____

   $4MnO_4^-(aq) + 5CH_3OH(aq) + 12H^+(aq) \rightarrow 4Mn^{2+}(aq) + 5HCO_2H(aq) + 11H_2O(l)$

   oxidizing agent, MnO_4^-; reducing agent, CH_3OH

(c) $I_2(s) + 6H_2O(l) \rightarrow 2IO_3^-(aq) + 12H^+(aq) + 10e^-$

   $5[OCl^-(aq) + 2H^+(aq) + 2e^- \rightarrow Cl^-(aq) + H_2O(l)$
   _____

   $I_2(s) + 5OCl^-(aq) + H_2O(l) \rightarrow 2IO_3^-(aq) + 5Cl^-(aq) + 2H^+(aq)]$

   oxidizing agent, OCl^-; reducing agent, I_2

(d) $As_2O_3(s) + 5H_2O(l) \rightarrow 2H_3AsO_4(aq) + 4H^+(aq) + 4e^-$

   $2NO_3^-(aq) + 6H^+(aq) + 4e \rightarrow N_2O_3(aq) + 3H_2O(l)$
   _____

   $As_2O_3(s) + 2NO_3^-(aq) + 2H_2O(l) + 2H^+(aq) \rightarrow 2H_3AsO_4(aq) + N_2O_3(aq)$

   oxidizing agent, NO_3^-; reducing agent, As_2O_3

(e) $2[MnO_4^-(aq) + 2H_2O(l) + 3e^- \rightarrow MnO_2(s) + 4OH^-]$

   $Br^-(aq) + 6OH^-(aq) \rightarrow BrO_3^-(aq) + 3H_2O(l) + 6e^-$
   _____

   $2MnO_4^-(aq) + Br^-(aq) + H_2O(l) \rightarrow 2MnO_2(s) + BrO_3^-(aq) + 2OH^-(aq)$

   oxidizing agent, MnO_4^-; reducing agent, Br^-

(f) $Pb(OH)_4^{2-}(aq) + ClO^-(aq) \rightarrow PbO_2(s) + Cl^-(aq) + 2OH^-(aq) + H_2O(l)$

   oxidizing agent, ClO^-; reducing agent, $Pb(OH)_4^{2-}$

Voltaic Cells

20.21   (a)   The reaction $Cu^{2+}(aq) + Zn(s) \rightarrow Cu(s) + Zn^{2+}(aq)$ is occurring in both figures. In Figure 20.3, the reactants are in contact, and the concentrations of the ions in solution aren't specified. In Figure 20.4 the oxidation half-reaction and reduction half-reaction are occurring in separate compartments, joined by a porous connector. The concentrations of the two solutions are initially 1.0 M. In Figure 20.4, electrical current is isolated and flows through the voltmeter. In Figure 20.3, the flow of electrons cannot be isolated or utilized.

(b)   In the cathode compartment of the voltaic cell in Figure 20.5, Cu^{2+} cations are reduced to Cu atoms, decreasing the number of positively charged particles in the compartment. Na^+ cations are drawn into the compartment to maintain charge balance as Cu^{2+} ions are removed.

20.23   *Analyze/Plan.* Follow the logic in Sample Exercise 20.4.   *Solve.*

(a)   Fe(s) is oxidized, $Ag^+(aq)$ is reduced.

(b) $Ag^+(aq) + 1e^- \rightarrow Ag(s)$; $Fe(s) \rightarrow Fe^{2+}(aq) + 2e^-$

(c)   Fe(s) is the anode, Ag(s) is the cathode.

(d)   Fe(s) is negative; Ag(s) is positive.

(e)   Electrons flow from the Fe(–) electrode toward the Ag(+) electrode.

(f)   Cations migrate toward the Ag(s) cathode; anions migrate toward the Fe(s) anode.

Cell EMF under Standard Conditions

20.25   (a)   *Electromotive force*, emf, is the driving force that causes electrons to flow through the external circuit of a voltaic cell. It is the potential energy difference between an electron at the anode and an electron at the cathode.

(b)   One *volt* is the potential energy difference required to impart 1 J of energy to a charge of 1 coulomb. 1 V = 1 J/C.

(c)   *Cell potential*, E_{cell}, is the emf of an electrochemical cell.

20.27   (a) $2H^+(aq) + 2e^- \rightarrow H_2(g)$

(b)   A *standard* hydrogen electrode is a hydrogen electrode where the components are at standard conditions, $1\ M\ H^+(aq)$ and $H_2(g)$ at 1 atm.

(c)   The platinum foil in an SHE serves as an inert electron carrier and a solid reaction surface.

20.29   (a)   A *standard reduction potential* is the relative potential of a reduction half-reaction measured at standard conditions, $1\ M$ aqueous solution and 1 atm gas pressure.

(b) $E_{red}^{\circ} = 0\ V$ for a standard hydrogen electrode.

(c)   The reduction of $Ag^+(aq)$ to Ag(s) is much more energetically favorable, because it has a substantially more positive E_{red}° (0.799 V) than the reduction of $Sn^{2+}(aq)$ to Sn(s) (–0.136 V).

20.31   *Analyze/Plan.* Follow the logic in Sample Exercise 20.5.   *Solve.*

(a)   The two half-reactions are:

$$TI^{3+}(aq) + 2e^- \rightarrow TI^+(aq) \qquad\qquad \text{cathode } E_{red}^{\circ} = ?$$
$$2[Cr^{2+}(aq) \rightarrow Cr^{3+}(aq) + e^-] \qquad\qquad \text{anode } E_{red}^{\circ} = -0.41\ V$$

(b) $\quad E_{cell}^{\circ} = E_{red}^{\circ} \text{ (cathode)} - E_{red}^{\circ} \text{ (anode)}; 1.19 \text{ V} = E_{red}^{\circ} - (-0.41 \text{ V});$

$\quad E_{red}^{\circ} = 1.19 \text{ V} - 0.41 \text{ V} = 0.78 \text{ V}$

(c)

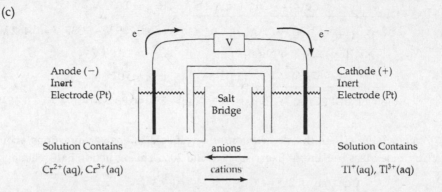

Solution Contains   Solution Contains

$Cr^{2+}(aq), Cr^{3+}(aq)$ $Tl^{+}(aq), Tl^{3+}(aq)$

Note that because $Cr^{2+}(aq)$ is readily oxidized, it would be necessary to keep oxygen out of the left-hand cell compartment.

20.33 *Analyze/Plan.* Follow the logic in Sample Exercise 20.6.   *Solve.*

(a) $\quad Cl_2(g) \rightarrow 2Cl^-(aq) + 2e^- \qquad\qquad E_{red}^{\circ} = 1.359 \text{ V}$

$\quad I_2(s) + 2e^- \rightarrow 2I^-(aq) \qquad\qquad E_{red}^{\circ} = 0.536 \text{ V}$

$\quad E^{\circ} = 1.359 \text{ V} - 0.536 \text{ V} = 0.823 \text{ V}$

(b) $\quad\quad Ni(s) \rightarrow Ni^{2+}(aq) + 2e^- \qquad\qquad E_{red}^{\circ} = -0.28 \text{ V}$

$\quad 2[Ce^{4+}(aq) + 1e^- \rightarrow Ce^{3+}(aq)] \qquad\quad E_{red}^{\circ} = 1.61 \text{ V}$

$\quad E^{\circ} = 1.61 \text{ V} - (-0.28 \text{ V}) = 1.89 \text{ V}$

(c) $\quad\quad Fe(s) \rightarrow Fe^{2+}(aq) + 2e^- \qquad\qquad E_{red}^{\circ} = -0.440 \text{ V}$

$\quad 2[Fe^{3+}(aq) + 1e^- \rightarrow Fe^{2+}(aq)] \qquad\quad E_{red}^{\circ} = 0.771 \text{ V}$

$\quad E^{\circ} = 0.771 \text{ V} - (-0.440 \text{ V}) = 1.211 \text{ V}$

(d) $\quad\quad 3[Ca(s) \rightarrow Ca^{2+}(aq) + 2e^-] \qquad\quad E_{red}^{\circ} = -2.87 \text{ V}$

$\quad 2[Al^{3+}(aq) + 3e^- \rightarrow Al(s)] \qquad\qquad E_{red}^{\circ} = -1.66 \text{ V}$

$\quad E^{\circ} = -1.66 \text{ V} - (-2.87 \text{ V})] = 1.21 \text{ V}$

20.35 *Analyze/Plan.* Given four half-reactions, find E_{red}° from Appendix E and combine them to obtain a desired E_{cell}. (a) The largest E_{cell} will combine the half-reaction with the most positive E_{red}° as the cathode reaction and the one with the most negative E_{red}° as the anode reaction. (b) The smallest positive E_{cell}° will combine two half-reactions whose E_{red}° values are closest in magnitude and sign.   *Solve.*

(a) $\quad 3[Ag^+(aq) + 1e^- \rightarrow Ag(s)] \qquad\qquad E_{red}^{\circ} = 0.799$

$\quad\quad\quad Cr(s) \rightarrow Cr^{3+}(aq) + 3e^- \qquad\qquad E_{red}^{\circ} = -0.74$

$\quad\overline{3Ag^+(aq) + Cr(s) \rightarrow 3Ag(s) + Cr^{3+}(aq)} \quad E^{\circ} = 0.799 - (-0.74) = 1.54 \text{ V}$

(b)   Two of the combinations have essentially equal E° values.

$$2[Ag^+(aq) + 1e^- \rightarrow Ag(s)] \qquad E^{\circ}_{red} = 0.799 \text{ V}$$

$$\underline{Cu(s) \rightarrow Cu^{2+}(aq) + 2e^- \qquad E^{\circ}_{red} = 0.337 \text{ V}}$$

$$2Ag^+(aq) + Cu(s) \rightarrow 2Ag(s) + Cu^{2+}(aq) \quad E^{\circ} = 0.799 \text{ V} - 0.337 \text{ V} = 0.462 \text{ V}$$

$$3[Ni^{2+}(aq) + 2e^- \rightarrow Ni(s)] \qquad E^{\circ}_{red} = -0.28 \text{ V}$$

$$\underline{2[Cr(s) \rightarrow Cr^{3+}(aq) + 3e^-] \qquad E^{\circ}_{red} = -0.74 \text{ V}}$$

$$3Ni^{2+}(aq) + 2Cr(s) \rightarrow 3Ni(s) + 2Cr^{3+}(aq) \quad E^{\circ} = -0.28 \text{ V} - (-0.74 \text{ V}) = 0.46 \text{ V}$$

20.37   *Analyze/Plan.* Given the description of a voltaic cell, answer questions about this cell. Combine ideas in Sample Exercises 20.4 and 20.7. The reduction half-reactions are:

$$Cu^{2+}(aq) + 2e^- \rightarrow Cu(s) \qquad E^{\circ} = 0.337 \text{ V}$$

$$Sn^{2+}(aq) + 2e^- \rightarrow Sn(s) \qquad E^{\circ} = -0.136 \text{ V}$$

Solve.

(a)   It is evident that Cu^{2+} is more readily reduced. Therefore, Cu serves as the cathode, Sn as the anode.

(b)   The copper electrode gains mass as Cu is plated out, the Sn electrode loses mass as Sn is oxidized.

(c)   The overall cell reaction is $Cu^{2+}(aq) + Sn(s) \rightarrow Cu(s) + Sn^{2+}(aq)$

(d) $E^{\circ} = 0.337 \text{ V} - (-0.136 \text{ V}) = 0.473 \text{ V}$

Strengths of Oxidizing and Reducing Agents

20.39   *Analyze/Plan.* Follow the logic in Sample Exercise 20.8. In each case, choose the half-reaction with the more positive reduction potential and with the given substance on the left. *Solve.*

(a) $Cl_2(g)$ (1.359 V vs. 1.065 V)

(b) $Ni^{2+}(aq)$ (–0.28V vs. –0.403 V)

(c) $BrO_3^-(aq)$ (1.52 V vs. 1.195 V)

(d) $O_3(g)$ (2.07 V vs. 1.776 V)

20.41   *Analyze/Plan.* If the substance is on the left of a reduction half-reaction, it will be an oxidant; if it is on the right, it will be a reductant. The sign and magnitude of the E°_{red} determines whether it is strong or weak.   *Solve.*

(a) $Cl_2(aq)$: strong oxidant (on the left, $E^{\circ}_{red} = 1.359 \text{ V}$)

(b) $MnO_4^-(aq, acidic)$: strong oxidant (on the left, $E^{\circ}_{red} = 1.51 \text{ V}$)

(c)   Ba(s): strong reductant (on the right, $E^{\circ}_{red} = -2.90 \text{ V}$)

(d)   Zn(s): reductant (on the right, $E^{\circ}_{red} = -0.763 \text{ V}$)

20.43   *Analyze/Plan.* Follow the logic in Sample Exercise 20.8.   *Solve.*

(a)   Arranged in order of increasing strength as oxidizing agents (and increasing reduction potential):

$$Cu^{2+}(aq) < O_2(g) < Cr_2O_7^{2-}(aq) < Cl_2(g) < H_2O_2(aq)$$

(b)   Arranged in order of increasing strength as reducing agents (and decreasing reduction potential):

$$H_2O_2(aq) < I^-(aq) < Sn^{2+}(aq) < Zn(s) < Al(s)$$

20.45   *Analyze/Plan.* In order to reduce Eu^{3+} to Eu^{2+}, we need an oxidizing agent, one of the reduced species from Table 20.1 or Appendix E. It must have a greater tendency to be oxidized than Eu^{3+} has to be reduced. That is, E°_{red} must be more negative than –0.43 V. *Solve.*

Any of the **reduced** species in Table 20.1 or Appendix E from a half-reaction with a reduction potential more negative than –0.43 V will reduce Eu^{3+} to Eu^{2+}. From the list of possible reductants in the exercise, Al and $H_2C_2O_4$ will reduce Eu^{3+} to Eu^{2+}.

Free Energy and Redox Reactions

20.47   *Analyze/Plan.* In each reaction, $Fe^{2+} \rightarrow Fe^{3+}$ will be the oxidation half-reaction and one of the other given half-reactions will be the reduction half-reaction. Follow the logic in Sample Exercise 20.10 to calculate E° and ΔG° for each reaction.   *Solve.*

(a) $2Fe^{2+}(aq) + S_2O_6^{2-}(aq) + 4H^+(aq) \rightarrow 2Fe^{3+}(aq) + 2H_2SO_3(aq)$

$E^{\circ} = 0.60\ V - 0.77\ V = -0.17\ V$

$2Fe^{2+}(aq) + N_2O(aq) + 2H^+(aq) \rightarrow 2Fe^{3+}(aq) + N_2(g) + H_2O(l)$

$E^{\circ} = -1.77\ V - 0.77\ V = -2.54\ V$

$Fe^{2+}(aq) + VO_2^+(aq) + 2H^+(aq) \rightarrow Fe^{3+}(aq) + VO^{2+}(aq) + H_2O(l)$

$E^{\circ} = 1.00\ V - 0.77\ V = +0.23\ V$

(b) $\Delta G^{\circ} = -nFE^{\circ}$ For the first reaction,

$$\Delta G^{\circ} = -2\ mol \times \frac{96{,}500\ J}{1\ V \cdot mol} \times (-0.17\ V) = 3.281 \times 10^4 = 3.3 \times 10^4\ J\ or\ 33\ kJ$$

For the second reaction, $\Delta G^{\circ} = -2(96{,}500)(-2.54) = 4.902 \times 10^5 = 4.90 \times 10^2\ kJ$

For the third reaction, $\Delta G^{\circ} = -1(96{,}500)(0.23) = -2.22 \times 10^4\ J = -22\ kJ$

(c) $\Delta G^{\circ} = -RT\ lnK;\ lnK = -\Delta G^{\circ}/RT;\ K = e^{-\Delta G^{\circ}/RT}$

For the first reaction,

$$lnK = \frac{-3.281 \times 10^4\ J}{(8.314\ J/mol \cdot K)(298\ K)} = -13.243 = -13;\ K = e^{-13.2428} = 1.78 \times 10^{-6} = 2 \times 10^{-6}$$

[Convert ln to log; the number of decimal places in the log is the number of sig figs in the result.]

For the second reaction,

$$\ln K = \frac{-4.902 \times 10^5 \text{ J}}{8.314 \text{ J/mol} \cdot \text{K} \times 298 \text{ K}} = -197.86 = -198; \ K = e^{-198} = 1.23 \times 10^{-86} = 1 \times 10^{-86}$$

For the third reaction,

$$\ln K = \frac{-(-2.22 \times 10^4 \text{ J})}{8.314 \text{ J/mol} \cdot \text{K} \ \times \ 298 \text{ K}} = 8.958 = 9.0; \ K = e^{9.0} = 7.77 \times 10^3 = 8 \times 10^3$$

Check. The equilibrium constants calculated here are indicators of equilibrium position, but are not particularly precise numerical values.

20.49   *Analyze/Plan.* Given K, calculate $\Delta G°$ and E°. Reverse the logic in Sample Exercise 20.10. According to Equation [19.22], $\Delta G° = -RT \ln K$. According to Equation [20.12], $\Delta G° = -nFE°$, $E° = -\Delta G°/nF$.   *Solve.*

$K = 1.5 \times 10^{-4}$

$\Delta G° = -RT \ln K = -(8.314 \text{ J/mol} \cdot \text{K})(298) \ln (1.5 \times 10^{-4}) = 2.181 \times 10^4 \text{ J} = 21.8 \text{ kJ}$

$E° = -\Delta G°/nF; \ n = 2; \ F = 96.5 \text{ kJ/mol e}^-$

$$E° = \frac{-21.81 \text{ kJ}}{2 \text{ mol e}^- \ \times \ 96.5 \text{ kJ/V} \cdot \text{mol e}^-} = -0.113 \text{ V}$$

Check. The unit of $\Delta G°$ is actually kJ/mol, which means kJ per 'mole of reaction', or for the reaction as written. Since we don't have a specific reaction, we interpret the unit as referring to the overall reaction.

20.51   *Analyze.* Given $E_{red}^°$ values for half reactions, calculate the value of K for a given redox reaction.

Plan. Combine the relationships involving E°, $\Delta G°$ and K to get a direct relationship between E° and K. For each reaction, calculate E° from $E_{red}^°$, then apply the relationship to calculate K.

Solve. $\Delta G° = -nFE°$, $\Delta G° = -RT \ln K; \ \ln K = 2.303 \log K$

$$-nFE° = -RT \ln K, \ E° = \frac{RT}{nF} \ln K = \frac{2.303 \, RT}{nF} \log K$$

From Equation [20.15] and [20.16], 2.303 RT/F = 0.0592.

$$E° = \frac{0.0592}{n} \log K; \ \log K = \frac{nE°}{0.0592}; \ K = 10^{\log K}$$

(a) $E° = -0.28 - (-0.440) = 0.16 \text{ V}, \ n = 2 \ (Ni^{2+} + 2e^- \rightarrow Ni)$

$$\log K = \frac{2(0.16)}{0.0592} = 5.4054 = 5.4; \ K = 2.54 \times 10^5 = 3 \times 10^5$$

(b) $E° = 0 - (-0.277) = 0.277 \text{ V}; \ n = 2 \ (2H^+ + 2e^- \rightarrow H_2)$

$$\log K = \frac{2(0.277)}{0.0592} = 9.358 = 9.36; \ K = 2.3 \times 10^9$$

(c) $E° = 1.51 - 1.065 = 0.445 = 0.45 \text{ V}; \ n = 10 \ (2MnO_4^- + 10e^- \rightarrow 2Mn^{+2})$

$$\log K = \frac{10(0.445)}{0.0592} = 75.169 \approx 75; \ K = 1.5 \times 10^{75} = 10^{75}$$

Check. Note that small differences in E° values lead to large changes in the magnitude of K. Sig fig rules limit precision of K values; using log instead of ln leads to more sig figs in the K value. This result is strictly numerical and does not indicate any greater precision in the data.

20.53   *Analyze/Plan.* $E^\circ = \dfrac{0.0592\text{ V}}{n}\log K$. See Solution 20.51 for a more complete development.

$\log K = \dfrac{nE^\circ}{0.0592\text{ V}}$.   *Solve.*

(a) $\log K = \dfrac{1(0.177\text{ V})}{0.0592\text{ V}} = 2.9899 = 2.99;\ \ K = 9.8 \times 10^2$

(b) $\log K = \dfrac{2(0.177\text{ V})}{0.0592\text{ V}} = 5.9797 = 5.98;\ \ K = 9.5 \times 10^5$

(c) $\log K = \dfrac{3(0.177\text{ V})}{0.0592\text{ V}} = 8.9696 = 8.97;\ \ K = 9.32 \times 10^8 = 9.3 \times 10^8$

Cell EMF under Nonstandard Conditions

20.55   (a)   The *Nernst equation* is applicable when the components of an electrochemical cell are at nonstandard conditions.

(b)   Q = 1 if all reactants and products are at standard conditions.

(c)   If concentration of reactants increases, Q decreases, and E increases.

20.57   *Analyze/Plan.* Given a circumstance, determine its effect on cell emf. Each circumstance changes the value of Q. An increase in Q reduces emf; a decrease in Q increases emf. *Solve.*

$$Zn(s) + 2H^+(aq) \rightarrow Zn^{2+}(aq) + H_2(g);\ E = E^\circ - \frac{0.0592}{n}\log Q;\ \ Q = \frac{[Zn^{2+}]P_{H_2}}{[H^+]^2}$$

(a) P_{H_2} increases, Q increases, E decreases

(b) $[Zn^{2+}]$ increases, Q increases, E decreases

(c) $[H^+]$ decreases, Q increases, E decreases

(d)   No effect; does not appear in the Nernst equation

20.59   *Analyze/Plan.* Follow the logic in Sample Exercise 20.11.   *Solve.*

(a)
$$Ni^{2+}(aq) + 2e^- \rightarrow Ni(s) \qquad\qquad E^\circ_{red} = -0.28\text{ V}$$
$$\underline{Zn(s) \rightarrow Zn^{2+}(aq) + 2e^- \qquad\qquad E^\circ_{red} = -0.763\text{ V}}$$
$$Ni^{2+}(aq) + Zn(s) \rightarrow Ni(s) + Zn^{2+}(aq) \qquad E^\circ = -0.28 - (-0.763) = 0.483 = 0.48\text{ V}$$

(b)
$$E = E^\circ - \frac{0.0592}{n}\log\frac{[Zn^{2+}]}{[Ni^{2+}]};\ n = 2$$

$$E = 0.483 - \frac{0.0592}{2}\log\frac{(0.100)}{(3.00)} = 0.483 - \frac{0.0592}{2}\log(0.0333)$$

$$E = 0.483 - \frac{0.0592(-1.477)}{2} = 0.483 + 0.0437 = 0.527 = 0.53\text{ V}$$

261

(c) $E = 0.483 - \dfrac{0.0592}{2} \log \dfrac{(0.900)}{(0.200)} = 0.483 - 0.0193 = 0.464 = 0.46$ V

20.61   *Analyze/Plan.* Follow the logic in Sample Exercise 20.11.   *Solve.*

(a) $4[Fe^{2+}(aq) \rightarrow Fe^{3+}(aq) + 1e^-]$ $E^{\circ}_{red} = 0.771$ V

$$\dfrac{O_2(g) + 4H^+(aq) + 4e^- \rightarrow 2H_2O(l) \qquad\qquad E^{\circ}_{red} = 1.23 \text{ V}}{4Fe^{2+}(aq) + O_2(g) + 4H^+(aq) \rightarrow 4Fe^{3+}(aq) + 2H_2O(l) \quad E^{\circ} = 1.23 - 0.771 = 0.459 = 0.46}$$

(b) $E = E^{\circ} - \dfrac{0.0592}{n} \log \dfrac{[Fe^{3+}]^4}{[Fe^{2+}]^4 [H^+]^4 P_{O_2}}$; $n = 4$, $[H^+] = 10^{-3.50} = 3.2 \times 10^{-4}$ M

      $E = 0.459 \text{ V} - \dfrac{0.0592}{4} \log \dfrac{(0.010)^4}{(1.3)^4 (3.2 \times 10^{-4})^4 (0.50)} = 0.459 - \dfrac{0.0592}{4} \log (7.0 \times 10^5)$

      $E = 0.459 - \dfrac{0.0592}{4} (5.845) = 0.459 - 0.0865 = 0.3725 = 0.37$ V

20.63   *Analyze/Plan.* We are given a concentration cell with Zn electrodes. Use the definition of a concentration cell in Section 20.6 to answer the stated questions. Use Equation [20.16] to calculate the cell emf. For a concentration cell, Q = [dilute]/[concentrated].   *Solve.*

(a)   The compartment with the more dilute solution will be the anode. That is, the compartment with $[Zn^{2+}] = 1.00 \times 10^{-2}$ M is the anode.

(b)   Since the oxidation half-reaction is the opposite of the reduction half-reaction, E° is zero.

(c) $E = E^{\circ} - \dfrac{0.0592}{n} \log Q$; $Q = [Zn^{2+}, \text{dilute}]/[Zn^{2+}, \text{conc.}]$

      $E = 0 - \dfrac{0.0592}{2} \log \dfrac{(1.00 \times 10^{-2})}{(1.8)} = 0.0668$ V

(d)   In the anode compartment, $Zn(s) \rightarrow Zn^{2+}(aq)$, so $[Zn^{2+}]$ increases from 1.00×10^{-2} M. In the cathode compartment, $Zn^{2+}(aq) \rightarrow Zn(s)$, so $[Zn^{2+}]$ decreases from 1.8 M.

20.65   *Analyze/Plan.* Follow the logic in Sample Exercise 20.12.   *Solve.*

$E = E^{\circ} - \dfrac{0.0592}{2} \log \dfrac{[P_{H_2}][Zn^{2+}]}{[H^+]^2}$; $E^{\circ} = 0.0 \text{ V} - (-0.763 \text{ V}) = 0.763$ V

$0.684 = 0.763 - \dfrac{0.0592}{2} \times (\log [P_{H_2}][Zn^{2+}] - 2 \log [H^+])$

     $= 0.763 - \dfrac{0.0592}{2} \times (-0.5686 - 2 \log [H^+])$

$0.684 = 0.763 + 0.0168 + 0.0592 \log [H^+]$; $\log [H^+] = \dfrac{0.684 - 0.0168 - 0.763}{0.0592}$

$\log [H^+] = -1.6188 = -1.6$; $[H^+] = 0.0241 = 0.02$ M; pH = 1.6

Batteries and Fuel Cells

20.67   (a)   The emf of a battery decreases as it is used. This happens because the concentrations of products increase and the concentrations of reactants decrease. According to the Nernst equation, these changes increase Q and decrease E_{cell}.

       (b)   The major difference between AA- and D-size batteries is the amount of reactants present. The additional reactants in a D-size battery enable it to provide power for a longer time.

20.69   *Analyze/Plan.* Given mass of a reactant (Pb), calculate mass of product (PbO_2). This is a stoichiometry problem; we need the balanced equation for the chemical reaction that occurs in the lead-acid battery. Then, g Pb → mol Pb → mol PbO_2 → g PbO_2.   *Solve.*

The overall cell reaction (Equation [20.19]) is:

$$Pb(s) + PbO_2(s) + 2H^+ (aq) + 2HSO_4^- (aq) \rightarrow 2PbSO_4(s) + 2H_2O(l)$$

$$402 \text{ g Pb} \times \frac{1 \text{ mol Pb}}{207.2 \text{ g Pb}} \times \frac{1 \text{ mol PbO}_2}{1 \text{ mol Pb}} \times \frac{239.2 \text{ g PbO}_2}{1 \text{ mol PbO}_2} = 464 \text{ g PbO}_2$$

20.71   *Analyze/Plan.* We are given a redox reaction and asked to write half-reactions, calculate E°, and indicate whether Li(s) is the anode or cathode. Determine which reactant is oxidized and which is reduced. Separate into half-reactions, find E°_{red} for the half-reactions from Appendix E and calculate E°.   *Solve.*

       (a)   Li(s) is oxidized at the anode.

       (b)

$$Ag_2CrO_4(s) + 2e^- \rightarrow 2Ag(s) + CrO_4^{2-}(aq) \qquad E^{\circ}_{red} = 0.446 \text{ V}$$
$$\underline{2[Li(s) \rightarrow Li^+(aq) + 1e^-] \qquad\qquad\qquad E^{\circ}_{red} = -3.05 \text{ V}}$$
$$Ag_2CrO_4(s) + 2Li(s) \rightarrow 2Ag(s) + CrO_4^-(aq) + 2Li^+(aq)$$

         E° = 0.446 V – (–3.05 V) = 3.496 = 3.50 V

       (c)   The emf of the battery, 3.5 V, is exactly the standard cell potential calculated in part (b).

       (d)   For this battery at ambient conditions, E ≈ E°, so log Q ≈ 0. This makes sense because all reactants and products in the battery are solids and thus present in their standard states. Assuming that E° is relatively constant with temperature, the value of the second term in the Nernst equation is ≈ 0 at 37°C, and E ≈ 3.5 V.

20.73   *Analyze/Plan.* (a) Consider the function of Zn in an alkaline battery. What effect would it have on the redox reaction and cell emf if Cd replaces Zn? (b) Both batteries contain Ni. What is the difference in environmental impact between Cd and the metal hydride? *Solve.*

       (a) E°_{red} for Cd (–0.40 V) is less negative than E°_{red} for Zn (–0.76 V), so E_{cell} will have a smaller (less positive) value.

       (b)   NiMH batteries use an alloy such as $ZrNi_2$ as the anode material. This eliminates the use and concomitant disposal problems associated with Cd, a toxic heavy metal.

20.75   The main advantage of a H_2-O_2 fuel cell over an alkaline battery is that the fuel cell is not a closed system. Fuel, H_2, and oxidant, O_2 are continuously supplied to the fuel cell, so that it can produce electrical current for a time limited only by the amount of available fuel. An alkaline battery contains a finite amount of reactant and produces current only until the reactants are spent, or reach equilibrium.

Alkaline batteries are much more convenient, because they are self-contained. Fuel cells require a means to acquire and store volatile and explosive $H_2(g)$. Disposal of spent alkaline batteries, which contain zinc and manganese solids, is much more problematic. H_2-O_2 fuel cells produce only $H_2O(l)$, which is not a disposal problem.

Corrosion

20.77   *Analyze/Plan.* (a) Decide which reactant is oxidized and which is reduced. Write the balanced half-reactions and assign the appropriate one as anode and cathode. (b) Write the balanced half-reaction for $Fe^{2+}(aq) \rightarrow Fe_2O_3 \cdot 3H_2O$. Use the reduction half-reaction from part (a) to obtain the overall reaction.   *Solve.*

   (a)   anode: $Fe(s) \rightarrow Fe^{2+}(aq) + 2e^-$

          cathode: $O_2(g) + 4H^+(aq) + 4e^- \rightarrow 2H_2O(l)$

   (b) $2Fe^{2+}(aq) + 6H_2O(l) \rightarrow Fe_2O_3 \cdot 3H_2O(s) + 6H^+(aq) + 2e^-$

          $O_2(g) + 4H^+(aq) + 4e^- \rightarrow 2H_2O(l)$

   (Multiply the oxidation half-reaction by two to balance electrons and obtain the overall balanced reaction.)

20.79   (a)   A "sacrificial anode" is a metal that is oxidized in preference to another when the two metals are coupled in an electrochemical cell; the sacrificial anode has a more negative E_{red}° than the other metal. In this case, Mg acts as a sacrificial anode because it is oxidized in preference to the pipe metal; it is sacrificed to preserve the pipe.

     (b) E_{red}° for Mg^{2+} is –2.37 V, more negative than most metals present in pipes, including $Fe (E_{red}^{\circ} = -0.44$ V$)$ and $Zn (E_{red}^{\circ} = -0.763$ V$)$.

20.81   *Analyze/Plan.* Given the materials brass, composed of Zn and Cu, and galvanized steel, determine the possible spontaneous redox reactions that could occur when the materials come in contact. Calculate E° values for these reactions.

Solve. The main metallic component of steel is Fe. Galvanized steel is steel plated with Zn. The three metals in question are Fe, Zn, and Cu; their E_{red}° values are shown below.

E_{red}° $Fe^{2+}(aq) = -0.440$ V

E_{red}° $Zn^{2+}(aq) = -0.763$ V

E_{red}° $Cu^{2+}(aq) = 0.337$ V

Zn, with the most negative E_{red}° value, can act as a sacrificial anode for either Fe or Cu. That is, Zn(s) will be preferentially oxidized when in contact with Fe(s) or Cu(s). For environmental corrosion, the oxidizing agent is usually $O_2(g)$ in acidic solution, $E_{red}^{\circ} = 1.23$ V. The pertinent reactions and their E° values are:

$$2Zn(s) + O_2(g) + 4H^+(aq) \rightarrow 2Zn^{2+}(aq) + 2H_2O(l)$$

$$E° = 1.23 \text{ V} - (-0.763 \text{ V}) = 1.99 \text{ V}$$

$$2Fe(s) + O_2(g) + 4H^+(aq) \rightarrow 2Fe^{2+}(aq) + 2H_2O(l)$$

$$E° = 1.23 \text{ V} - (-0.440 \text{ V}) = 1.67 \text{ V}$$

$$2Cu(s) + O_2(g) + 4H^+(aq) \rightarrow 2Cu^{2+}(aq) + 2H_2O(l)$$

$$E° = 1.23 \text{ V} - (0.337 \text{ V}) = 0.893 \text{ V}$$

Note, however, that Fe has a more negative $E°_{red}$ than Cu so when the two are in contact Fe acts as the sacrificial anode, and corrosion (of Fe) occurs preferentially. This is verified by the larger E° value for the corrosion of Fe, 1.67 V, relative to the corrosion of Cu, 0.893 V. When the three metals Zn, Fe, and Cu are in contact, oxidation of Zn will happen first, followed by oxidation of Fe, and finally Cu.

Electrolysis; Electrical Work

20.83   (a)   *Electrolysis* is an electrochemical process driven by an outside energy source.

(b)   Electrolysis reactions are, by definition, nonspontaneous.

(c) $2Cl^-(l) \rightarrow Cl_2(g) + 2e^-$

20.85   *Analyze/Plan*. Follow the logic in Sample Exercise 20.14, paying close attention to units. Coulombs = **amps** · s; since this is a $3e^-$ reduction, each mole of Cr(s) requires 3 Faradays.

Solve.

(a) $7.60 \text{ A} \times 2.00 \text{ d} \times \dfrac{24 \text{ hr}}{1 \text{ d}} \times \dfrac{60 \text{ min}}{1 \text{ hr}} \times \dfrac{60 \text{ s}}{1 \text{ min}} \times \dfrac{1 \text{ C}}{1 \text{ amp} \cdot \text{s}} \times \dfrac{1 \text{ F}}{96,500 \text{ C}}$

$$\times \dfrac{1 \text{ mol Cr}}{3 \text{ F}} \times \dfrac{52.00 \text{ g Cr}}{1 \text{ mol Cr}} = 236 \text{ g Cr(s)}$$

(b) $0.250 \text{ mol Cr} \times \dfrac{3 \text{ F}}{1 \text{ mol Cr}} \times \dfrac{96,500 \text{ C}}{\text{F}} \times \dfrac{1 \text{ amp} \cdot \text{s}}{1 \text{ C}} \times \dfrac{1}{8.00 \text{ hr}} \times \dfrac{1 \text{ hr}}{60 \text{ min}} \times \dfrac{1 \text{ min}}{60 \text{ s}}$

20.87   *Analyze/Plan*. Given a spontaneous chemical reaction, calculate the maximum possible work for a given amount of reactant at standard conditions. Separate the equation into half-reactions and calculate cell emf. Use Equation [20.19], $w_{max} = -nFE$, to calculate maximum work. At standard conditions, $E = E°$.   *Solve.*

$$I_2(s) + 2e^- \rightarrow 2I^-(aq) \qquad\qquad E°_{red} = 0.536 \text{ V}$$

$$\underline{Sn(s) \rightarrow Sn^{2+}(aq) + 2e^- \qquad\qquad E°_{red} = -0.136 \text{ V}}$$

$$I_2(s) + Sn(s) \rightarrow 2I^-(aq) + Sn^{2+}(aq) \qquad E° = 0.536 - (-0.136) = 0.672 \text{ V}$$

$$w_{max} = -2(96.5)(0.672) = -129.7 = -130 \text{ kJ/mol Sn}$$

$$\dfrac{-129.7 \text{ kJ}}{\text{mol Sn(s)}} \times \dfrac{1 \text{ mol Sn}}{118.71 \text{ g Sn}} \times 75.0 \text{ g Sn} \times \dfrac{1000 \text{ J}}{\text{kJ}} = -8.19 \times 10^4 \text{ J}$$

Check. The (–) sign indicates that work is done by the cell.

20.89   *Analyze/Plan.* Follow the logic in Sample Exercise 20.15, paying close attention to units.

     Solve.

(a) $7.5 \times 10^4 \text{ A} \times 24 \text{ hr} \times \dfrac{3600 \text{ s}}{1 \text{ hr}} \times \dfrac{1 \text{ C}}{1 \text{ amp} \cdot \text{s}} \times \dfrac{1 \text{ F}}{96,500 \text{ C}} \times \dfrac{1 \text{ mol Li}}{1 \text{ F}}$

$$\times \dfrac{6.94 \text{ g Li}}{1 \text{ mol Li}} \times 0.85 = 3.961 \times 10^5 = 4.0 \times 10^5 \text{ g Li}$$

(b)   If the cell is 85% efficient, $\dfrac{96,500 \text{ C}}{\text{F}} \times \dfrac{1 \text{ F}}{0.85 \text{ mol}} = 1.135 \times 10^5$

$$= 1.1 \times 10^5 \text{ C/mol Li required}$$

$$\text{Energy} = 7.5 \text{ V} \times \dfrac{1.135 \times 10^5 \text{ C}}{\text{mol Li}} \times \dfrac{1 \text{ J}}{1 \text{ C} \cdot \text{V}} \times \dfrac{1 \text{ kWh}}{3.6 \times 10^6 \text{ J}} = 0.24 \text{ kWh/mol Li}$$

Additional Exercises

20.91   (a) $Ni^+(aq) + 1e^- \rightarrow Ni(s)$

$$\dfrac{Ni^+(aq) \qquad\qquad \rightarrow Ni^{2+}(aq) + 1e^-}{2Ni^+(aq) \qquad\qquad \rightarrow Ni(s) + Ni^{2+}(aq)}$$

(b) $MnO_4{}^{2-}(aq) + 4H^+(aq) + 2e^- \rightarrow MnO_2(s) + 2H_2O(l)$

$$\dfrac{2[MnO_4{}^{2-}(aq) \rightarrow MnO_4{}^-(aq) + 1e^-]}{3MnO_4{}^{2-}(aq) + 4H^+(aq) \rightarrow 2MnO_4{}^-(aq) + MnO_2(s) + 2H_2O(l)}$$

(c) $H_2SO_3(aq) + 4H^+(aq) + 4e^- \rightarrow S(s) + 3H_2O(l)$

$$\dfrac{2[H_2SO_3(aq) + H_2O(l) \rightarrow HSO_4{}^-(aq) + 3H^+(aq) + 2e^-]}{3H_2SO_3(aq) \rightarrow S(s) + 2HSO_4{}^-(aq) + 2H^+(aq) + H_2O(l)}$$

(d) $Cl_2(aq) + 2H_2O(l) \rightarrow 2ClO^-(aq) + 4H^+(aq) + 2e^-$

$$\dfrac{4OH^-(aq) \qquad + 4OH^-(aq)}{Cl_2(aq) + 4OH^-(aq) \rightarrow 2ClO^-(aq) + 2H_2O(l) + 2e^-}$$

$$\dfrac{Cl_2(aq) + 2e^- \rightarrow 2Cl^-(aq)}{1/2[2Cl_2(aq) + 4OH^-(aq) \rightarrow 2Cl^-(aq) + 2ClO^-(aq) + 2H_2O(l)]}$$

$$Cl_2(aq) + 2OH^-(aq) \rightarrow Cl^-(aq) + ClO^-(aq) + H_2O(l)$$

20.93   (a)

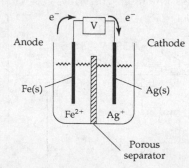

$Fe(s) \rightarrow Fe^{2+}(aq) + 2e^-$

$$\dfrac{2Ag^+(aq) + 2e^- \rightarrow 2Ag(s)}{Fe(s) + 2Ag^+(aq) \rightarrow Fe^{2+}(aq) + 2Ag(s)}$$

(b) $$Zn(s) \rightarrow Zn^{2+}(s) + 2e^-$$
$$\frac{2H^+(aq) + 2e^- \rightarrow H_2(g)}{Zn(s) + 2H^+(aq) \rightarrow Zn^{2+}(aq) + H_2(g)}$$

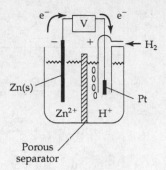

(c) $Cu| Cu^{2+}|| ClO_3^-, Cl^-| Pt$ Here, both the oxidized and reduced forms of the cathode solution are in the same phase, so we separate them by a comma, and then indicate an inert electrode.

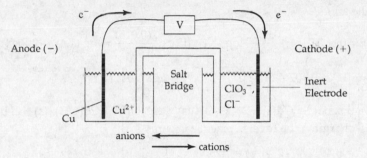

20.95   (a)   The reduction potential for $O_2(g)$ in the presence of acid is 1.23 V. $O_2(g)$ cannot oxidize Au(s) to $Au^+(aq)$ or $Au^{3+}(aq)$, even in the presence of acid.

(b)   The possible oxidizing agents need a reduction potential greater than 1.50 V. These include $Co^{3+}(aq)$, $F_2(g)$, $H_2O_2(aq)$, and $O_3(g)$. Marginal oxidizing agents (those with reduction potential near 1.50 V) from Appendix E are $BrO_3^-(aq)$, $Ce^{4+}(aq)$, $HClO(aq)$, $MnO_4^-(aq)$, and $PbO_2(s)$.

(c) $$4Au(s) + 8NaCN(aq) + 2H_2O(l) + O_2(g) \rightarrow 4Na[Au(CN)_2](aq) + 4NaOH(aq)$$

$$Au(s) + 2CN^-(aq) \rightarrow [Au(CN)_2]^- + 1e^-$$

$$O_2(g) + 2H_2O(l) + 4e^- \rightarrow 4OH^-(aq)$$

Au(s) is being oxidized and $O_2(g)$ is being reduced.

(d) $$2[Na[Au(CN)_2](aq) + 1e^- \rightarrow Au(s) + 2CN^-(aq) + Na^+(aq)]$$
$$\frac{Zn(s) \rightarrow Zn^{2+}(aq) + 2e^-}{2Na[Au(CN)_2](aq) + Zn(s) \rightarrow 2Au(s) + Zn^{2+}(aq) + 2Na^+(aq) + 4CN^-(aq)}$$

Zn(s) is being oxidized and $[Au(CN)_2]^-(aq)$ is being reduced. While $OH^-(aq)$ is not included in the redox reaction above, its presence in the reaction mixture probably causes $Zn(OH)_2(s)$ to form as the product. This increases the driving force   (and   E°)   for   the   overall   reaction. a very precise or useful result.)

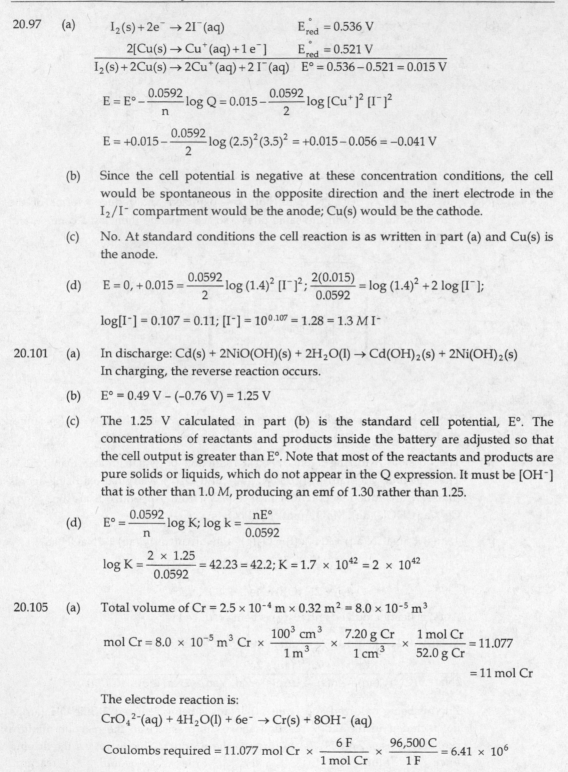

20.97   (a) $I_2(s) + 2e^- \rightarrow 2I^-(aq)$ $E^\circ_{red} = 0.536$ V

$$\frac{2[Cu(s) \rightarrow Cu^+(aq) + 1e^-] \qquad E^\circ_{red} = 0.521 \text{ V}}{I_2(s) + 2Cu(s) \rightarrow 2Cu^+(aq) + 2I^-(aq) \quad E^\circ = 0.536 - 0.521 = 0.015 \text{ V}}$$

$$E = E^\circ - \frac{0.0592}{n} \log Q = 0.015 - \frac{0.0592}{2} \log [Cu^+]^2 [I^-]^2$$

$$E = +0.015 - \frac{0.0592}{2} \log (2.5)^2 (3.5)^2 = +0.015 - 0.056 = -0.041 \text{ V}$$

(b)   Since the cell potential is negative at these concentration conditions, the cell would be spontaneous in the opposite direction and the inert electrode in the I_2/I^- compartment would be the anode; Cu(s) would be the cathode.

(c)   No. At standard conditions the cell reaction is as written in part (a) and Cu(s) is the anode.

(d) $E = 0, +0.015 = \dfrac{0.0592}{2} \log (1.4)^2 [I^-]^2; \dfrac{2(0.015)}{0.0592} = \log (1.4)^2 + 2 \log [I^-];$

$\log[I^-] = 0.107 = 0.11; [I^-] = 10^{0.107} = 1.28 = 1.3 \, M \, I^-$

20.101   (a)   In discharge: $Cd(s) + 2NiO(OH)(s) + 2H_2O(l) \rightarrow Cd(OH)_2(s) + 2Ni(OH)_2(s)$
In charging, the reverse reaction occurs.

(b) $E^\circ = 0.49$ V – (–0.76 V) = 1.25 V

(c)   The 1.25 V calculated in part (b) is the standard cell potential, E°. The concentrations of reactants and products inside the battery are adjusted so that the cell output is greater than E°. Note that most of the reactants and products are pure solids or liquids, which do not appear in the Q expression. It must be $[OH^-]$ that is other than 1.0 M, producing an emf of 1.30 rather than 1.25.

(d) $E^\circ = \dfrac{0.0592}{n} \log K; \log k = \dfrac{nE^\circ}{0.0592}$

$\log K = \dfrac{2 \times 1.25}{0.0592} = 42.23 = 42.2; K = 1.7 \times 10^{42} = 2 \times 10^{42}$

20.105   (a)   Total volume of Cr = 2.5×10^{-4} m $\times$ 0.32 m^2 = 8.0×10^{-5} m^3

$$\text{mol Cr} = 8.0 \times 10^{-5} \text{ m}^3 \text{ Cr} \times \frac{100^3 \text{ cm}^3}{1 \text{ m}^3} \times \frac{7.20 \text{ g Cr}}{1 \text{ cm}^3} \times \frac{1 \text{ mol Cr}}{52.0 \text{ g Cr}} = 11.077$$

$$= 11 \text{ mol Cr}$$

The electrode reaction is:

$CrO_4{}^{2-}(aq) + 4H_2O(l) + 6e^- \rightarrow Cr(s) + 8OH^-(aq)$

$$\text{Coulombs required} = 11.077 \text{ mol Cr} \times \frac{6 \text{ F}}{1 \text{ mol Cr}} \times \frac{96,500 \text{ C}}{1 \text{ F}} = 6.41 \times 10^6$$

$$= 6.4 \times 10^6 \text{ C}$$

(b) $6.41 \times 10^6 \, C \times \dfrac{1 \, amp \cdot s}{1 \, C} \times \dfrac{1}{10.0 \, s} = 6.4 \times 10^5 \, amp$

(c)   If the cell is 65 efficient, $(6.41 \times 10^6 / 0.65) = 9.867 \times 10^6 = 9.9 \times 10^6$ C are required to plate the bumper.

$$6.0 \, V \times 9.867 \times 10^6 \, C \times \dfrac{1 \, J}{1 \, C \cdot V} \times \dfrac{1 \, kWh}{3.6 \times 10^6 \, J} = 16.445 = 16 \, kWh$$

Integrative Exercises

20.109 $N_2(g) + 3H_2(g) \rightarrow 2NH_3(g)$

(a)   The oxidation number of $H_2(g)$ and $N_2(g)$ is 0. The oxidation number of N in NH_3 is –3, H in NH_3 is +1. H_2 is being oxidized and N_2 is being reduced.

(b)   Calculate $\Delta G°$ from $\Delta G_f^°$ values in Appendix C. Use $\Delta G° = -RT \ln K$ to calculate K.

$$\Delta G° = 2\Delta G_f^° \, NH_3(g) - \Delta G_f^° \, N_2(g) - 3\Delta G_f^° \, H_2(g)$$

$$\Delta G° = 2(-16.66 \, kJ) - 0 - 3(0) = -33.32 \, kJ$$

$$\Delta G° = -RT \ln K, \ \ln K = \dfrac{-\Delta G°}{RT} = \dfrac{-(-33.32 \times 10^3 \, J)}{(8.314 \, J/mol \cdot K)(298 \, K)} = 13.4487 = 13.45$$

$$K = e^{13.4487} = 6.9 \times 10^5$$

(c) $\Delta G° = -nfE°, \ E° = \dfrac{-\Delta G°}{nF}$

n = ? 2 N atoms change from 0 to –3, or 6 H atoms change from 0 to +1. Either way, n = 6.

$$E° = \dfrac{-(-33.32 \, kJ)}{6 \times 96.5 \, kJ/V} = 0.05755 \, V$$

20.113   (a)

$$Ag^+(aq) + e^- \rightarrow Ag(s) \qquad\qquad E_{red}^° = 0.799 \, V$$
$$\underline{Fe^{2+}(aq) \rightarrow Fe^{3+}(aq) + 1e^- \qquad E_{red}^° = 0.771 \, V}$$
$$Ag^+(aq) + Fe^{2+}(aq) \rightarrow Ag(s) + Fe^{3+}(aq) \qquad E° = 0.799 \, V - 0.771 \, V = 0.028 \, V$$

(b) $Ag^+(aq)$ is reduced at the cathode and $Fe^{2+}(aq)$ is oxidized at the anode.

(c) $\Delta G° = -nFE° = -(1)(96.5)(0.028) = -2.7 \, kJ$

$$\Delta S° = S° \, Ag(s) + S° \, Fe^{3+}(aq) - S° Ag^+(aq) - S° \, Fe^{2+}(aq)$$

$$= 42.55 \, J + 293.3 \, J - 73.93 \, J - 113.4 \, J = 148.5 \, J$$

$\Delta G° = \Delta H° - T\Delta S°$ Since $\Delta S°$ is positive, $\Delta G°$ will become more negative and $E°$ will become more positive as temperature is increased.

20.117   The reaction can be written as a sum of the steps:

$$Pb^{2+}(aq) + 2e^- \rightarrow Pb(s) \qquad\qquad E^\circ_{red} = -0.126 \text{ V}$$

$$\underline{\qquad PbS(s) \rightarrow Pb^{2+}(aq) + S^{2-}(aq) \qquad \text{``}E^\circ\text{''} = ?}$$

$$PbS(s) + 2e^- \rightarrow Pb(s) + S^{2-}(aq) \qquad E^\circ_{red} = ?$$

"E°" for the second step can be calculated from K_{sp}.

$$E^\circ = \frac{0.0592}{n} \log K_{sp} = \frac{0.0592}{2} \log (8.0 \times 10^{-28}) = \frac{0.0592}{2} (-27.10) = -0.802 \text{ V}$$

E° for the half-reaction = -0.126 V + $(-0.802$ V$)$ = -0.928 V

Calculating an imaginary E° for a nonredox process like step 2 may be a disturbing idea. Alternatively, one could calculate K for step 1 (5.4×10^{-5}), K for the reaction in question (K = $K_1 \times K_{sp}$ = 4.4×10^{-32}) and then E° for the half-reaction. The result is the same.

21 Nuclear Chemistry

Nuclear Chemistry

21.1   *Analyze.* Given the name and mass number of a nuclide, decide if it lies within the belt of stability. If not, suggest a process that moves it toward the belt.

Plan. Calculate the number of protons and neutrons in each nuclide. Locate this point on Figure 21.2. If the point is above the belt, β-decay increases protons and decreases neutrons, decreasing the neutron-to-proton ratio. If the point is below the belt, either positron emission or neutron capture decreases protons and increases neutrons, increasing the neutron-to-proton ratio.   *Solve.*

(a) ^{24}Ne: 10 p, 14 n, just above the belt of stability. Reduce the neutron-to proton ratio via β-decay.

(b) ^{32}Cl: 17 p, 15 n, just below the belt of stability. Increase the neutron-to-proton ratio via positron emission or orbital electron capture.

(c) ^{108}Sn: 50 p, 58 n, just below the belt of stability. Increase the neutron-to-proton ratio via positron emission or orbital electron capture.

(d) ^{216}Po: 84 p, 132 n, just beyond the belt of stability. Nuclei with atomic numbers ≥ 84 tend to decay via alpha emission, which decreases both protons and neutrons.

21.4   *Analyze/Plan.* Write the balanced equation for the decay. Nuclear decay is a first-order process; use appropriate relationships for first-order processes to determine $t_{1/2}$, k and remaining ^{88}Mo after 12 minutes.   *Solve.*

(a) $t_{1/2}$ is the time required for half of the original nuclide to decay. Relative to the graph, this is the time when the amount of ^{88}Mo is reduced from 1.0 to 0.5. This time is 7 minutes.

(b)   For a first-order process, $t_{1/2} = 0.693/k$ or $k = 0.693/t_{1/2}$.

       $k = 0.693/7$ min $= 0.0990 = 0.1$ min^{-1}

(c)   From the graph, the fraction of ^{88}Mo remaining after 12 min is 0.3/1.0 = 0.3

       Check. $\ln(N_t/N_o) = -kt = -(0.099)(12) = -1.188$; $N_t/N_o = e^{-1.188} = 0.30$.

(d) $^{88}_{42}$Mo $\rightarrow$ $^{88}_{41}$Nb $+ ^{0}_{1}$e

Radioactivity

21.7   *Analyze/Plan.* Given various nuclide descriptions, determine the number of protons and neutrons in each nuclide. The left superscript is the mass number, protons plus neutrons. If there is a left subscript, it is the atomic number, the number of protons. Protons can always be determined from chemical symbol; all isotopes of the same element have the same number of protons. A number following the element name, as in part (c) is the mass number.   *Solve.*

p = protons, n = neutrons, e = electrons; number of protons = atomic number; number of neutrons = mass number – atomic number

(a) $^{55}_{25}\text{Mn}$: 25p, 30n   (b) ^{201}Hg: 80p, 121n   (c) ^{39}K: 19p, 20n

21.9   *Analyze/Plan.* See definitions in Section 21.1. In each case, the left superscript is mass number, the left subscript is related to atomic number.   *Solve.*

(a) $^{1}_{1}\text{p}$ or $^{1}_{1}\text{H}$ (b) $^{0}_{1}\text{e}$ (c) $^{0}_{-1}\beta$ or $^{0}_{-1}\text{e}$

21.11   *Analyze/Plan.* Follow the logic in Sample Exercises 21.1 and 21.2. Pay attention to definitions of decay particles and conservation of mass and charge.   *Solve.*

(a) $^{214}_{83}\text{Bi} \rightarrow ^{214}_{84}\text{Po} + ^{0}_{-1}\text{e}$ (b) $^{195}_{79}\text{Au} + ^{0}_{-1}\text{e}\,\text{(orbital electron)} \rightarrow ^{195}_{78}\text{Pt}$

(c) $^{38}_{19}\text{K} \rightarrow ^{38}_{18}\text{Ar} + ^{0}_{1}\text{e}$ (d) $^{242}_{94}\text{Pu} \rightarrow ^{238}_{92}\text{U} + ^{4}_{2}\text{He}$

21.13   *Analyze/Plan.* Using definitions of the decay processes and conservation of mass number and atomic number, work backwards to the reactants in the nuclear reactions.   *Solve.*

(a) $^{211}_{82}\text{Pb} \rightarrow ^{211}_{83}\text{Bi} + ^{0}_{-1}\beta$ (b) $^{50}_{25}\text{Mn} \rightarrow ^{50}_{24}\text{Cr} + ^{0}_{1}\text{e}$

(c) $^{179}_{74}\text{W} + ^{0}_{-1}\text{e} \rightarrow ^{179}_{73}\text{Ta}$ (d) $^{230}_{90}\text{Th} \rightarrow ^{226}_{88}\text{Ra} + ^{4}_{2}\text{He}$

21.15   *Analyze/Plan.* Given the starting and ending nuclides in a nuclear decay sequence, we are asked to determine the number of alpha and beta emissions. Use the total change in A and Z, along with definitions of alpha and beta decay, to answer the question.   *Solve.*

The total mass number change is (235–207) = 28. Since each α particle emission decreases the mass number by four, whereas emission of a β particle does not correspond to a mass change, there are 7 α particle emissions. The change in atomic number in the series is 10. Each α particle results in an atomic number lower by two. The 7 α particle emissions alone would cause a decrease of 14 in atomic number. Each β particle emission raises the atomic number by one. To obtain the observed lowering of 10 in the series, there must be 4 β emissions.

Nuclear Stability

21.17   *Analyze/Plan.* Follow the logic in sample Exercise 21.3, paying attention to the guidelines for neutron-to-proton ratio.   *Solve.*

(a) $^{8}_{5}\text{B}$ - low neutron/proton ratio, positron emission (for low atomic numbers, positron emission is more common than orbital electron capture)

(b) $^{68}_{29}$Cu - high neutron/proton ratio, beta emission

(c) $^{241}_{93}$Np - high neutron/proton ratio, beta emission

(Even though ^{241}Np has an atomic number ≥ 84, the most common decay pathway for nuclides with neutron/proton ratios higher than the isotope listed on the periodic chart is beta decay.)

(d) $^{39}_{17}$Cl - high neutron/proton ratio, beta emission

21.19   *Analyze/Plan.* Use the criteria listed in Table 21.3.   *Solve.*

(a)   Stable: $^{39}_{19}$K   odd proton, even neutron more abundant than odd proton, odd neutron; 20 neutrons is a magic number.

(b)   Stable: $^{209}_{83}$Bi   odd proton, even neutron more abundant than odd proton, odd neutron; 126 neutrons is a magic number.

(c)   Stable: $^{25}_{12}$Mg   even though $^{24}_{10}$Ne is an even proton, even neutron nuclide, it has a very high neutron/proton ratio and lies outside the band of stability.

21.21   *Analyze/Plan.* For each nuclide, determine the number of protons and neutrons and decide if they are magic numbers.   *Solve.*

(a) $^{4}_{2}$He   (b) $^{40}_{20}$Ca   (c) $^{208}_{82}$Pb

(d) $^{58}_{28}$Ni has a magic number of protons, but not neutrons.

21.23   *Analyze/Plan.* For each nuclide, determine the number of protons and neutrons and find the location on Figure 21.2. Rationalize the location based on magic numbers, neutron-to-proton ratio and Z value. Predict radioactivity (nonstability of nucleus).   *Solve.*

Radioactive. $^{14}_{8}$O — low neutron/proton ratio, $^{115}_{52}$Te — low neutron/proton ratio;

$$^{208}_{84}\text{Po} \ -\text{atomic number} \geq 84$$

Stable: $^{32}_{16}$S, $^{78}_{34}$Se — even proton, even neutron, stable neutron/proton ratio

Nuclear Transmutations

21.25   Protons and alpha particles are positively charged and must be moving very fast to overcome electrostatic forces which would repel them from the target nucleus. Neutrons are electrically neutral and not repelled by the nucleus.

21.27   *Analyze/Plan.* Determine A and Z for the missing particle by conservation principles. Find the appropriate symbol for the particle.   *Solve.*

(a) $^{32}_{16}$S + $^{1}_{0}$n → $^{1}_{1}$p + $^{32}_{15}$P   (b) $^{7}_{4}$Be + $^{0}_{-1}$e (orbital electron) → $^{7}_{3}$Li

(c) $^{187}_{75}$Re → $^{187}_{76}$Os + $^{0}_{-1}$e   (d) $^{98}_{42}$Mo + $^{2}_{1}$H → $^{1}_{0}$n + $^{99}_{43}$Tc

(e) $^{235}_{92}$U + $^{1}_{0}$n → $^{135}_{54}$Xe + $^{99}_{38}$Sr + 2 $^{1}_{0}$n

21.29   *Analyze/Plan.* Follow the logic in Sample Exercise 21.5, paying attention to conservation of A and Z.   *Solve.*

(a) $^{238}_{92}U + ^{1}_{0}n \rightarrow ^{239}_{92}U + ^{0}_{0}\gamma$ (b) $^{14}_{7}N + ^{1}_{1}H \rightarrow ^{11}_{6}C + ^{4}_{2}He$

(c) $^{18}_{8}O + ^{1}_{0}n \rightarrow ^{19}_{9}F + ^{0}_{-1}e$

Rates of Radioactive Decay

21.31   Chemical reactions do not affect the character of atomic nuclei. The energy changes involved in chemical reactions are much too small to allow us to alter nuclear properties via chemical processes. Therefore, the nuclei that are formed in a nuclear reaction will continue to be radioactive regardless of any chemical changes we bring to bear. However, we can hope to use chemical means to separate radioactive substances, or remove them from foods or a portion of the environment.

21.33   *Analyze/Plan.* Follow the logic in Sample Exercise 21.6.   *Solve.*

After 12.3 yr, one half-life, there are $(1/2)48.0 = 24.0$ mg. 49.2 yr is exactly four half-lives. There are then $(48.0)(1/2)^4 = 3.0$ mg tritium remaining.

21.35   *Analyze/Plan.* Follow the logic in Sample Exercise 21.7. In this case, we are given initial sample mass as well as mass at time t, so we can proceed directly to calculate k (Equation [21.20]) and then t (Equation [21.19]).   *Solve.*

$k = 0.693 / t_{1/2} = 0.693/27.8 \text{ d} = 0.02493 = 0.0249 \text{ d}^{-1}$

$t = \dfrac{-1}{k} \ln \dfrac{N_t}{N_o} = \dfrac{-1}{0.02493 \text{ d}^{-1}} \ln \dfrac{1.50}{5.75} = 53.9 \text{ d}$

21.37   (a)   *Analyze/Plan.* $^{226}_{88}Ra \rightarrow ^{222}_{86}Rn + ^{4}_{2}He$

1 α particle is produced for each ^{226}Ra that decays. Calculate the mass of ^{226}Ra remaining after 1.0 min, calculate by subtraction the mass that has decayed, and use Avogadro's number to get the number of $^{4}_{2}$He particles.   *Solve.*

Calculate k in min^{-1}. $1600 \text{ yr} \times \dfrac{365 \text{ d}}{1 \text{ yr}} \times \dfrac{24 \text{ hr}}{1 \text{ d}} \times \dfrac{60 \text{ min}}{1 \text{ hr}} = 8.410 \times 10^8 \text{ min}$

$k = \dfrac{0.693}{t_{1/2}} = \dfrac{0.693}{8.410 \times 10^8 \text{ min}} = 8.241 \times 10^{-10} \text{ min}^{-1}$

$\ln \dfrac{N_t}{N_o} = -kt = (-8.241 \times 10^{-10} \text{ min}^{-1})(1.0 \text{ min}) = -8.241 \times 10^{-10} = -8.2 \times 10^{-10}$

$\dfrac{N_t}{N_o} = e^{-8.2 \times 10^{-10}} = (1.000 - 8.2 \times 10^{-10});$ (don't round here!)

$[e^{-8.2 \times 10^{-10}}$ is a number very close to 1. In this calculation, it is conveneint to express the number as $(1 - 8.2 \times 10^{-10})]$.

$N_t = 5.0 \times 10^{-3} \text{ g } (1.00 - 8.2 \times 10^{-10})$ The amount that decays is $N_o - N_t$:

5.0×10^{-3} g $- [5.0 \times 10^{-3} (1.00 - 8.2 \times 10^{-10})] = 5.0 \times 10^{-3}$ g (8.2×10^{-10})

$$= \sim 4.1 \times 10^{-12} \text{ g Ra}$$

(In terms of sig figs, $[N_o - N_t]$ is a very small number, found by subtracting two numbers known only to two sig figs and one decimal place. At best, we can express the result as an order of magnitude.)

$$[N_o - N_t] = 4.1 \times 10^{-12} \text{ g Ra} \times \frac{1 \text{ mol Ra}}{226 \text{ g Ra}} \times \frac{6.022 \times 10^{23} \text{ Ra atoms}}{1 \text{ mol Ra}} \times \frac{1 \, ^4_2\text{He}}{1 \text{ Ra atom}}$$

$$= 1.1 \times 10^{10} = \sim 10^{10} \, \alpha \text{ particles emitted in 1 min}$$

(b)　*Plan.* The result from (a) is disintegrations/min. Change this to dis/s and apply the definition 1 Ci $= 2.7 \times 10^{10}$ dis/s.

$$1.1 \times 10^{10} \frac{\text{dis}}{\text{min}} \times \frac{1 \text{ min}}{60 \text{ s}} \times \frac{1 \text{ Ci}}{3.7 \times 10^{10} \text{ dis/s}} \times \frac{1000 \text{ mCi}}{\text{Ci}} = 4.945 = \sim 5 \text{ mCi}$$

Realistically, the activity is best expressed as between 0.01 and 0.001 Ci.

21.39　*Analyze/Plan.* Follow the logic in Sample Exercise 21.7. *Solve.*

$$t = \frac{-1}{k} \ln \frac{N_t}{N_o}; k = 0.693/5715 \text{ yr} = 1.213 \times 10^{-4} \text{ yr}^{-1}$$

$$t = \frac{-1}{1.213 \times 10^{-4} \text{ yr}^{-1}} \ln \frac{24.9}{32.5} = 2.20 \times 10^3 \text{ yr}$$

21.41　*Analyze/Plan.* Follow the procedure outlined in Sample Exercise 21.7. The original quantity of ^{238}U is 50.0 mg plus the amount that gave rise to 14.0 mg of ^{206}Pb. This amount is 14.0(238/206) = 16.2 mg. *Solve.*

$$k = 0.693/4.5 \times 10^9 \text{ yr} = 1.54 \times 10^{-10} = 1.5 \times 10^{-10} \text{ yr}^{-1}$$

$$t = \frac{-1}{k} \ln \frac{N_t}{N_o} = \frac{-1}{1.54 \times 10^{-10} \text{ yr}^{-1}} \ln \frac{50.0}{66.2} = 1.8 \times 10^9 \text{ yr}$$

Energy Changes

21.43　*Analyze/Plan.* Given an energy change, find the corresponding change in mass. Use Equation [21.22], $E = mc^2$. *Solve.*

$\Delta E = c^2 \Delta m; \Delta m = \Delta E/c^2; 1 \text{ J} = \text{kg} \cdot \text{m}^2/\text{s}^2$

$$\Delta m = \frac{393.5 \times 10^3 \text{ kg} \cdot \text{m}^2/\text{s}^2}{(2.9979 \times 10^8 \text{ m/s})^2} \times \frac{1000 \text{ g}}{1 \text{ kg}} = 4.378 \times 10^{-9} \text{ g}$$

21.45　*Analyze/Plan.* Given the mass of a ^{23}Na nucleus, find the energy required to separate the nucleus into protons and neutrons. This corresponds to the binding energy of the nucleus. Calculate the total mass of the separate particles and subtract the mass of the nucleus. Convert the difference to energy using Equation [21.22]. Use Avogadro's number to calculate energy per mole of nuclei. *Solve.*

Δm = mass of individual protons and neutrons – mass of nucleus

Δm = 11(1.0072765 amu) + 12(1.0086649 amu) – 22.983733 amu = 0.2002873

$$= 0.200287 \text{ amu}$$

$$\Delta E = (2.9979246 \times 10^8 \text{ m/s})^2 \times 0.2002873 \text{ amu} \times \frac{1 \text{ g}}{6.0221421 \times 10^{23} \text{ amu}} \times \frac{1 \text{ kg}}{1 \times 10^3 \text{ g}}$$

$$= 2.989123 \times 10^{-11} = 2.98912 \times 10^{-11} \text{ J/} ^{23}\text{Na nucleus required}$$

$$2.989123 \times 10^{-11} \frac{\text{J}}{\text{nucleus}} \times \frac{6.0221421 \times 10^{23} \text{ atoms}}{\text{mol}} = 1.80009 \times 10^{13} \text{ J/mol } ^{23}\text{Na}$$

21.47   *Analyze/Plan.* In each case, calculate the mass defect (Δm), total nuclear binding energy and then binding energy per nucleon.   *Solve.*

(a) Δm = 6(1.0072765) + 6(1.0086649) – 11.996708 = 0.0989404 = 0.098940 amu

$$\Delta E = 0.0989404 \text{ amu} \times \frac{1 \text{ g}}{6.0221421 \times 10^{23} \text{ amu}} \times \frac{1 \text{ kg}}{1000 \text{ g}} \times \frac{8.987551 \times 10^{16} \text{ m}^2}{\text{s}^2}$$

$$= 1.476604 \times 10^{-11} = 1.4766 \times 10^{-11} \text{ J}$$

binding energy/nucleon = 1.476604×10^{-11} J/12 = 1.2305×10^{-12} J/nucleon

(b) Δm = 17(1.0072765) + 20(1.0086649) – 36.956576 = 0.3404225 = 0.340423 amu

$$\Delta E = 0.3404225 \text{ amu} \times \frac{1 \text{ g}}{6.0221421 \times 10^{23} \text{ amu}} \times \frac{1 \text{ kg}}{1000 \text{ g}} \times \frac{8.987551 \times 10^{16} \text{ m}^2}{\text{s}^2}$$

$$= 5.080525 \times 10^{-11} = 5.08053 \times 10^{-11} \text{ J}$$

binding energy/ nucleon = 5.080525×10^{-11} J / 37 = 1.37312×10^{-12} J/nucleon

(c)   Calculate the nuclear mass by subtracting the electron mass from the atomic mass.   136.905812 amu – 56(5.485799×10^{-4} amu) = 136.875092 amu

Δm = 56(1.0072765) + 81(1.0086649) – 136.875092 = 1.2342489 = 1.234249 amu

$$\Delta E = 1.2342489 \text{ amu} \times \frac{1 \text{ g}}{6.0221421 \times 10^{23} \text{ amu}} \times \frac{1 \text{ kg}}{1000 \text{ g}} \times \frac{8.987551 \times 10^{16} \text{ m}^2}{\text{s}^2}$$

$$= 1.842014 \times 10^{-10} \text{ J}$$

binding energy/nucleon = 1.842014×10^{-10} J / 137 = 1.344536×10^{-12} J/nucleon

21.49   *Analyze/Plan.* Use Equation [21.22] to calculate the mass equivalence of the solar radiation.   *Solve.*

(a) $\dfrac{1.07 \times 10^{16}\,\text{kJ}}{1\,\text{min}} \times \dfrac{60\,\text{min}}{1\,\text{hr}} \times \dfrac{24\,\text{hr}}{1\,\text{day}} = 1.541 \times 10^{19}\,\dfrac{\text{kJ}}{\text{day}} = 1.54 \times 10^{22}\,\text{J/day}$

$\Delta m = \dfrac{1.541 \times 10^{22}\,\text{kg}\cdot\text{m}^2/\text{s}^2/\text{d}}{(2.998 \times 10^8\,\text{m/s})^2} = 1.714 \times 10^5 = 1.71 \times 10^5\,\text{kg/d}$

(b)   *Analyze/Plan.* Calculate the mass change in the given nuclear reaction, then a conversion factor for g ^{235}U to mass equivalent.   *Solve.*

$\Delta m = 140.8833 + 91.9021 + 2(1.0086649) - 234.9935 = -0.19077 = -0.1908\,\text{amu}$

Converting from atoms to moles and amu to grams, it requires 1.000 mol or 235.0 g ^{235}U to produce energy equivalent to a change in mass of 0.1908 g.

0.10% of 1.714×10^5 kg is 1.714×10^2 kg $= 1.714 \times 10^5$ g

$1.714 \times 10^5\,\text{g} \times \dfrac{235.0\,\text{g}\ ^{235}\text{U}}{0.1908\,\text{g}} = 2.111 \times 10^8 = 2.1 \times 10^8\,\text{g}\ ^{235}\text{U}$

(This is about 230 tons of ^{235}U **per day**.)

21.51   We can use Figure 21.13 to see that the binding energy per nucleon (which gives rise to the mass defect) is greatest for nuclei of mass numbers around 50. Thus (a) $^{59}_{27}$Co should possess the greatest mass defect per nucleon.

Effects and Uses of Radioisotopes

21.53   The ^{59}Fe would be incorporated into the diet component, which in turn is fed to the rabbits. After a time blood samples could be removed from the animals, the red blood cells separated, and the radioactivity of the sample measured. If the iron in the dietary compound has been incorporated into blood hemoglobin, the blood cell sample should show beta emission. Samples could be taken at various times to determine the rate of iron uptake, rate of loss of the iron from the blood, and so forth.

21.55   (a)   Control rods control neutron flux so that there are enough neutrons to sustain the chain reaction but not so many that the core overheats.

         (b)   A moderator slows neutrons so that they are more easily captured by fissioning nuclei.

21.57   *Analyze/Plan.* Use conservation of A and Z to complete the equations, keeping in mind the symbols and definitions of various decay products.   *Solve.*

         (a) $^{235}_{92}\text{U} + ^{1}_{0}\text{n} \rightarrow\ ^{160}_{62}\text{Sm} + ^{72}_{30}\text{Zn} + 4\ ^{1}_{0}\text{n}$

         (b) $^{239}_{94}\text{Pu} + ^{1}_{0}\text{n} \rightarrow\ ^{144}_{58}\text{Ce} + ^{94}_{36}\text{Kr} + 2\ ^{1}_{0}\text{n}$

21.59   The extremely high temperature is required to overcome the electrostatic charge repulsions between the nuclei so that they come together to react.

21.61   •OH is a free radical; it contains an unpaired (free) electron, which makes it an extremely reactive species. (As an odd electron molecule, it violates the octet rule.) It can react with almost any particle (atom, molecule, ion) to acquire an electron and become OH⁻. This often starts a disruptive chain of reactions, each producing a different free radical.

Hydroxide ion, OH⁻, on the other hand, will be readily neutralized in the buffered cell environment. Its most common reaction is ubiquitous and innocuous:
$H^+ + OH^- \rightarrow H_2O$. The acid-base reactions of OH⁻ are usually much less disruptive to the organism than the chain of redox reactions initiated by •OH radical.

21.63   *Analyze/Plan.* Use definitions of the various radiation units and conversion factors to calculate the specified quantities. Pay particular attention to units.   *Solve.*

(a) $1 \text{ Ci} = 3.7 \times 10^{10}$ disintegrations(dis)/s; 1 Bq = 1 dis/s

$$8.7 \text{ mCi} \times \frac{1 \text{ Ci}}{1000 \text{ mCi}} \times \frac{3.7 \times 10^{10} \text{ dis/s}}{\text{Ci}} = 3.22 \times 10^8 = 3.2 \times 10^8 \text{ dis/s} = 3.2 \times 10^8 \text{ Bq}$$

(b) $1 \text{ rad} = 1 \times 10^{-2}$ J/kg; 1 Gy = 1 J/kg = 100 rad. From part (a), the activity of the source is 3.2×10^8 dis/s.

$$3.22 \times 10^8 \text{ dis/s} \times 2.0 \text{ s} \times 0.65 \times \frac{9.12 \times 10^{-13} \text{ J}}{\text{dis}} \times \frac{1}{0.250 \text{ kg}} = 1.53 \times 10^{-3} = 1.5 \times 10^{-3} \text{ J/kg}$$

$$1.5 \times 10^{-3} \text{ J/kg} \times \frac{1 \text{ rad}}{1 \times 10^{-2} \text{ J/kg}} \times \frac{1000 \text{ mrad}}{\text{rad}} = 1.5 \times 10^2 \text{ mrad}$$

$$1.5 \times 10^{-3} \text{ J/kg} \times \frac{1 \text{ Gy}}{1 \text{ J/kg}} = 1.5 \times 10^{-3} \text{ Gy}$$

(c)   rem = rad (RBE); Sv = Gy (RBE) , where 1 Sv = 100 rem

mrem = 1.53×10^2 mrad (9.5) = $1.45 \times 10^3 = 1.5 \times 10^3$ mrem (or 1.5 rem)

Sv = 1.53×10^{-3} Gy (9.5) = $1.45 \times 10^{-2} = 1.5 \times 10^{-2}$ Sv

Additional Exercises

21.65 $^{222}_{86}\text{Rn} \rightarrow X + 3\,^4_2\text{He} + 2\,^0_{-1}\beta$

This corresponds to a reduction in mass number of (3 × 4 =) 12 and a reduction in atomic number of (3 × 2 – 2) = 4. The stable nucleus is $^{210}_{82}\text{Pb}$. (This is part of the sequence in Figure [21.4].)

21.67   The most massive radionuclides will have the highest neutron/proton ratios. Thus, they are most likely to decay by a process that lowers this ratio, beta emission. The least massive nuclides, on the other hand, will decay by a process that increases the neutron/proton ratio, positron emission, or orbital electron capture.

Time (hr)	N_t (dis/min)	ln N_t
0	180	5.193
2.5	130	4.868
5.0	104	4.644
7.5	77	4.34
10.0	59	4.08
12.5	46	3.83
17.5	24	3.18

21.70

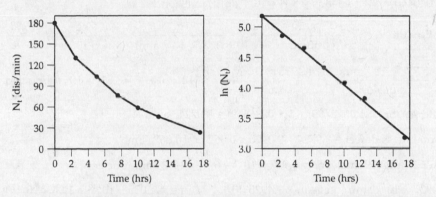

The plot on the left is a graph of activity (disintegrations per minute) vs. time. Choose $t_{1/2}$ at the time where $N_t = 1/2\, N_o = 90$ dis/min. $t_{1/2} \approx 6.0$ hr.

Rearrange Equation [21.19] to obtain the linear relationship shown on the right.

$\ln(N_t / N_o) = -kt$; $\ln N_t - \ln N_o = -kt$; $\ln N_t = -kt + \ln N_o$

The slope of this line $= -k = -0.11$; $t_{1/2} = 0.693/0.11 = 6.3$ hr.

21.71 1×10^{-6} curie $\times \dfrac{3.7 \times 10^{10} \text{ dis/s}}{\text{curie}} = 3.7 \times 10^4$ dis/s

rate $= 3.7 \times 10^4$ nuclei/s $= kN$

$k = \dfrac{0.693}{t_{1/2}} = \dfrac{0.693}{28.8 \text{ yr}} \times \dfrac{1 \text{ yr}}{365 \times 24 \times 3600 \text{ sec}} = 7.630 \times 10^{-10} = 7.63 \times 10^{-10}$ s^{-1}

3.7×10^4 nuclei/s $= (7.63 \times 10^{-10}/\text{s})$ N; N $= 4.849 \times 10^{13} = 4.8 \times 10^{13}$ nuclei

mass ^{90}Sr $= 4.849 \times 10^{13}$ nuclei $\times \dfrac{90 \text{ g Sr}}{6.022 \times 10^{23} \text{ nuclei}} = 7.2 \times 10^{-9}$ g Sr

21.74   Assume that no depletion of iodide from the water due to plant uptake has occurred. Then the activity after 30 days would be:

$k = 0.693/t_{1/2} = 0.693/8.04$ d $= 0.0862$ d^{-1}

$$\ln \frac{N_t}{N_o} = -(0.0862 \text{ d}^{-1})(30 \text{ d}) = -2.586 = -2.6; \frac{N_t}{N_o} = 0.0753 = 0.08$$

We thus expect $N_t = 0.0753(184) = 13.9 = 1 \times 10^1$ counts/min. Although the theoretical counts for no iodide absorption is slightly higher than the observed 13.5 counts/min, the uncertainty of the theoretical value is greater than the difference in the two values. We conclude that very little, if any, iodide is absorbed by the plant.

21.76   Because of the relationship $\Delta E = \Delta mc^2$, the mass defect (Δm) is directly related to the binding energy (ΔE) of the nucleus.

^{7}Be: 4p, 3n; $4(1.0072765) + 3(1.0086649) = 7.05510$ amu

Total mass defect = $7.0551 - 7.0147 = 0.0404$ amu

0.0404 amu/7 nucleons = 5.77×10^{-3} amu/nucleon

$$\Delta E = \Delta m \times c^2 = \frac{5.77 \times 10^{-3} \text{ amu}}{\text{nucleon}} \times \frac{1 \text{g}}{6.022 \times 10^{23} \text{ amu}} \times \frac{1 \text{ kg}}{1 \times 10^3 \text{ g}} \times \frac{8.988 \times 10^{16} \text{ m}^2}{\text{sec}^2}$$

$$= \frac{5.77 \times 10^{-3} \text{ amu}}{\text{nucleon}} \times \frac{1.4925 \times 10^{-10} \text{ J}}{1 \text{amu}} = 8.612 \times 10^{-13} = 8.61 \times 10^{-13} \text{ J/nucleon}$$

^{9}Be: 4p, 5n; $4(1.0072765) + 5(1.0086649) = 9.07243$ amu

Total mass defect = $9.0724 - 9.0100 = 0.06243 = 0.0624$ amu

0.0624 amu/9 nucleons = $6.937 \times 10^{-3} = 6.94 \times 10^{-3}$ amu/nucleon

6.937×10^{-3} amu/nucleon $\times 1.4925 \times 10^{-10}$ J/amu = $1.035 \times 10^{-12} = 1.04 \times 10^{-12}$ J/nucleon

^{10}Be: 4p, 6n; $4(1.0072765) + 6(1.0086649) = 10.0811$ amu

Total mass defect = $10.0811 - 10.0113 = 0.0698$ amu

0.0698 amu/10 nucleons = 6.98×10^{-3} amu/nucleon

6.98×10^{-3} amu/nucleon $\times 1.4925 \times 10^{-10}$ J/amu = $1.042 \times 10^{-12} = 1.04 \times 10^{-12}$ J/nucleon

The binding energies/nucleon for ^{9}Be and ^{10}Be are very similar; that for ^{10}Be is slightly higher.

Integrative Exercises

21.81   Calculate the amount of energy produced by the nuclear fusion reaction, the enthalpy of combustion, $\Delta H°$, of C_3H_8, and then the mass of C_3H_8 required.

Δm for the reaction $\quad 4\,^1_1\text{H} \rightarrow\,^4_2\text{He} + 2\,^0_1\text{e} \quad$ is:

$4(1.00782) - 4.00260$ amu $- 2(5.4858 \times 10^{-4}$ amu$)\ = 0.027583 = 0.02758$ amu

$$\Delta E = \Delta mc^2 = 0.027583 \text{ amu} \times \frac{1 \text{g}}{6.02214 \times 10^{23} \text{ amu}} \times \frac{1 \text{ kg}}{1000 \text{ g}} \times (2.9979246 \times 10^8 \text{ m/s})^2$$

$$= 4.11654 \times 10^{-12} = 4.117 \times 10^{-12} \text{ J/4 }^1\text{H nuclei}$$

$$1.0 \text{ g }^1\text{H} \times \frac{1 \text{ }^1\text{H nucleus}}{1.00782 \text{ amu}} \times \frac{6.02214 \times 10^{23} \text{ amu}}{\text{g}} \times \frac{4.11654 \times 10^{-12} \text{ J}}{4 \text{ }^1\text{H nuclei}}$$

$$= 6.1495 \times 10^{11} \text{ J} = 6.1 \times 10^8 \text{ kJ produced by the fusion of } 1.0 \text{ g }^1\text{H}.$$

$$\text{C}_3\text{H}_8(g) + 5\text{O}_2(g) \rightarrow 3\text{CO}_2(g) + 4\text{H}_2\text{O}(g)$$

$$\Delta H° = 3(-393.5 \text{ kJ}) + 4(-241.82 \text{ kJ}) - (-103.85) - (0) = -2043.9 \text{ kJ}$$

$$6.1495 \times 10^8 \text{ kJ} \times \frac{1 \text{ mol C}_3\text{H}_8(g)}{2043.9 \text{ kJ}} \times \frac{44.094 \text{ g C}_3\text{H}_8}{\text{mol C}_3\text{H}_8} = 1.327 \times 10^7 \text{ g} = 1.3 \times 10^4 \text{ kg C}_3\text{H}_8$$

13,000 kg $\text{C}_3\text{H}_8(g)$ would have to be burned to produce the same amount of energy as fusion of 1.0 g ^1H.

21.84   Determine the wavelengths of the photons by first calculating the energy equivalent of the mass of an electron or positron. (Since **two** photons are formed by annihilation of **two** particles of equal mass, we need to calculate the energy equivalent of just one particle.) The mass of an electron is 9.109×10^{-31} kg.

$$\Delta E = (9.109 \times 10^{-31} \text{ kg}) \times (2.998 \times 10^8 \text{ m/s})^2 = 8.187 \times 10^{-14} \text{ J}$$

Also, $\Delta E = h\nu$; $\Delta E = hc/\lambda$; $\lambda = hc/\Delta E$

$$\lambda = \frac{(6.626 \times 10^{-34} \text{ J} \cdot \text{s}) (2.998 \times 10^8 \text{ m/s})}{8.187 \times 10^{-14} \text{ J}} = 2.426 \times 10^{-12} \text{ m} = 2.426 \times 10^{-3} \text{ nm}$$

This is a very short wavelength indeed; it lies at the short wavelength end of the range of observed gamma ray wavelengths (see Figure 6.4).

22 Chemistry of the Nonmetals

Chemistry of Nonmetals

22.2   (a)   Acid-base (Brønsted)

      (b)   Charges on species from left to right in the reaction: 0, 0, 1+, 1–

      (c) $NH_3(aq) + H_2O(l) \rightleftharpoons NH_4^+(aq) + OH^-(aq)$

22.5   *Analyze.* Given: space-filling models of molecules containing nitrogen and oxygen atoms.

       Find: molecular formulas and Lewis structures.

       Plan. Nitrogen atoms are blue, and oxygen atoms are red. Count the number of spheres of each color to determine the molecular formula. From each molecular formula, count the valence electrons (N = 5, O = 6) and draw a correct Lewis structure. Resonance structures are likely.

       Solve.

      (a) N_2O_5 40 valence electrons, 20 e^- pairs

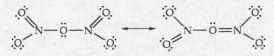

       Many other resonance structures are possible. Those with double bonds to the central oxygen (like the right-hand structure above) do not minimize formal charge and are less significant in the net bonding model.

      (b) N_2O_4 34 e^-, 17 e^- pairs

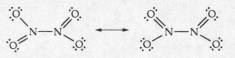

       Other equivalent resonance structures with different arrangement of the double bonds are possible.

      (c) NO_2 17 e^-, 8.5 e^- pairs

       We place the odd electron on N because of electronegativity arguments.

(d) N_2O_3 28 e⁻, 14 e⁻ pairs

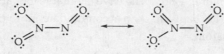

(e)   NO   11 e⁻, 5.5 e⁻ pairs

$\ddot{N}=\ddot{O}$

We place the odd electron on N because of electronegativity arguments.

(f) N_2O 16 e⁻, 8 e⁻ pairs

:N≡N—Ö: ⟷ :N̈=N=Ö: ⟷ :N̈—N≡O:

The right-most structure above does not minimize formal charge and makes smaller contribution to the net bonding model.

22.8   *Analyze/Plan.* Evaluate the graph, describe the trend in data, recall the general trend for each of the properties listed, and use details of the data to discriminate between possibilities.

Solve. The general trend is an increase in value moving from left to right across the period, with a small discontinuity at S. Considering just this overall feature, both (a) first ionization energy and (c) electronegativity increase moving from left to right, so these are possibilities. (b) Atomic radius decreases, and can be eliminated. Since Si is a solid and Cl and Ar are gases at room temperature, melting points must decrease across the row; (d) melting point can be eliminated. According to data in Tables 22.2, 22.5, 22.7, and 22.8, (e) X-X single bond enthalpies show no consistent trend. Furthermore, there is no known Ar-Ar single bond, so no value for this property can be known; (e) can be eliminated.

Now let's examine trends in (a) first ionization energy and (c) electronegativity more closely. From electronegativity values in Chapter 8, we see a continuous increase with no discontinuity at S, and no value for Ar. Values in (a) first ionization energy from Chapter 7 do match the pattern in the figure. The slightly lower value of I_1 for S is due to a decrease in repulsion by removing an electron from a fully occupied orbital. In summary, only (a) first ionization energy fits the property depicted in the graph.

Periodic Trends and Chemical Reactions

22.11   *Analyze/Plan.* Use the color-coded periodic chart on the front-inside cover of the text to classify the given elements.   *Solve.*

Metals: (b) Sr, (c) Mn, (e) Rh;   nonmetals: (a) P, (d) Se, (f) Kr;   metalloids: none

22.13   *Analyze/Plan.* Follow the logic in Sample Exercise 22.1.   *Solve.*

(a)   O   (b)   Br   (c)   Ba

(d)   O   (e)   Co

22.15   *Analyze/Plan.* Use the position of the specified elements on the periodic chart, periodic trends, and the arguments in Sample Exercise 22.1 to explain the observations.   *Solve.*

(a)   Nitrogen is too small to accommodate five fluorine atoms about it. The P and As atoms are larger. Furthermore, P and As have available 3d and 4d orbitals, respectively, to form hybrid orbitals that can accommodate more than an octet of electrons about the central atom.

(b)   Si does not readily form π bonds, which would be necessary to satisfy the octet rule for both atoms in SiO.

(c)   A reducing agent is a substance that readily loses electrons. As has a lower electronegativity than N; that is, it more readily gives up electrons to an acceptor and is more easily oxidized.

22.17   *Analyze/Plan.* Follow the logic in Sample Exercise 22.2.   *Solve.*

(a) $Mg_3N_2(s) + 6H_2O(l) \rightarrow 2NH_3(g) + 3Mg(OH)_2(s)$

Because $H_2O(l)$ is a reactant, the state of NH_3 in the products could be expressed as $NH_3(aq)$.

(b) $2C_3H_7OH(l) + 9O_2(g) \rightarrow 6CO_2(g) + 8H_2O(l)$

(c) $MnO_2(s) + C(s) \overset{\Delta}{\rightarrow} CO(g) + MnO(s)$ or

$MnO_2(s) + 2C(s) \overset{\Delta}{\rightarrow} 2CO(g) + Mn(s)$ or

$MnO_2(s) + C(s) \overset{\Delta}{\rightarrow} CO_2(g) + Mn(s)$

(d) $AlP(s) + 3H_2O(l) \rightarrow PH_3(g) + Al(OH)_3(s)$

(e) $Na_2S(s) + 2HCl(aq) \rightarrow H_2S(g) + 2NaCl(aq)$

Hydrogen, the Noble Gases, and the Halogens

22.19   *Analyze/Plan.* Use information on the isotopes of hydrogen in Section 22.2 to list their symbols, names, and relative abundances.   *Solve.*

a) $_1^1H$ - protium; $_1^2H$ - deuterium; $_1^3H$ - tritium

b)   The order of abundance is proteum > deuterium > tritium.

22.21   *Analyze/Plan.* Consider the electron configuration of hydrogen and the Group 1A elements.   *Solve.*

Like other elements in group 1A, hydrogen has only one valence electron and its most common oxidation number is +1.

22.23   *Analyze/Plan.* Use information on the descriptive chemistry of hydrogen in Section 22.2 to formulate the required equations. Steam is $H_2O(g)$.   *Solve.*

(a) $Mg(s) + 2H^+(aq) \rightarrow Mg^{2+}(aq) + H_2(g)$

(b) $C(s) + H_2O(g) \xrightarrow{1000°C} CO(g) + H_2(g)$

(c) $CH_4(g) + H_2O(g) \xrightarrow{1100°C} CO(g) + 3H_2(g)$

22.25   *Analyze/Plan.* Use information on the descriptive chemistry of hydrogen given in Section 22.2 to complete and balance the equations.   *Solve.*

(a) $NaH(s) + H_2O(l) \rightarrow NaOH(aq) + H_2(g)$

(b) $Fe(s) + H_2SO_4(aq) \rightarrow Fe^{2+}(aq) + H_2(g) + SO_4^{2-}(aq)$

(c) $H_2(g) + Br_2(g) \rightarrow 2HBr(g)$

(d) $2Na(l) + H_2(g) \rightarrow 2NaH(s)$

(e) $PbO(s) + H_2(g) \xrightarrow{\Delta} Pb(s) + H_2O(g)$

22.27   *Analyze/Plan.* If the element bound to H is a nonmetal, the hydride is molecular. If H is bound to a metal with integer stoichiometry, the hydride is ionic; with noninteger stoichiometry, the hydride is metallic.   *Solve.*

(a)   ionic (metal hydride)

(b)   molecular (nonmetal hydride)

(c)   metallic (nonstoichiometric transition metal hydride)

22.29   *Analyze/Plan.* Consider the periodic properties of Xe and Ar.   *Solve.*

Xenon is larger, and can more readily accommodate an expanded octet. More important is the lower ionization energy of xenon; because the valence electrons are a greater average distance from the nucleus, they are more readily promoted to a state in which the Xe atom can form bonds with fluorine.

22.31   *Analyze/Plan.* Follow the rules for assigning oxidation numbers in Section 4.4 and the logic in Sample Exercise 4.8.   *Solve.*

(a) ClO_3^-, +5   (b)   HI, –1   (c) ICl_3; I, +3; Cl, –1

(d)   NaOCl, +1   (e) $HClO_4$, +7   (f) XeF_4, +4; F, –1

22.33   *Analyze/Plan.* Review the nomenclature rules and ion names in Section 2.8.   *Solve.*

(a)   iron(III) chlorate   (b)   chlorous acid

(c)   xenon hexafluoride   (d)   bromine pentafluoride

(e)   xenon oxide tetrafluoride   (f)   iodic acid

22.35   *Analyze/Plan.* Consider intermolecular forces and periodic properties, including oxidizing power, of the listed substances.   *Solve.*

(a)   Van der Waals intermolecular attractive forces increase with increasing numbers of electrons in the atoms.

(b) F_2 reacts with water: $F_2(g) + H_2O(l) \rightarrow 2HF(aq) + 1/2\ O_2(g)$. That is, fluorine is too strong an oxidizing agent to exist in water.

(c)   HF has extensive hydrogen bonding.

(d)   Oxidizing power is related to electronegativity. Electronegativity decreases in the order given.

22.37   Vehicle fuels produce energy via combustion reactions. The reaction $H_2(g) + 1/2\ O_2(g) \rightarrow H_2O(g)$ is very exothermic, producing 242 kJ per mole of H_2 burned. The only product of combustion is H_2O, a nonpollutant (but like CO_2, a greenhouse gas).

22.39   If the lifetime of perchlorate anion, ClO_4^-, in soils and water is decades, the ion must be extremely stable (unreactive) in aqueous solutions and aerobic environments; it is not easily oxidized by O_2. Although chlorine is in a very high oxidation state in ClO_4^-, it is not readily reduced, because the ion has a stable, symmetric structure that protects it against reactions. The anion is a symmetrical tetrahedron with several plausible Lewis structures when expanded octets about Cl are considered.

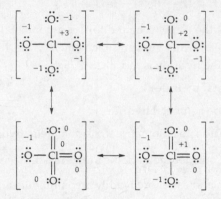

(minimum formal charges)

More structures with alternate locations of the single and double bonds can be drawn. Resonance stabilization could certainly contribute to the ion's high stability.

Oxygen and the Group 6A Elements

22.41   *Analyze/Plan.* Consider the industrial uses of oxygen and ozone given in Section 22.5. *Solve.*

(a)   As an oxidizing agent in steel-making; to bleach pulp and paper; in oxyacetylene torches; in medicine to assist in breathing

(b)   Synthesis of pharmaceuticals, lubricants, and other organic compounds where $C=C$ bonds are cleaved; in water treatment

22.43   *Analyze/Plan.* Use information on the descriptive chemistry of oxygen given in Section 22.5 to complete and balance the equations.   *Solve.*

(a) $2HgO(s) \xrightarrow{\Delta} 2Hg(l) + O_2(g)$

(b) $2Cu(NO_3)_2(s) \xrightarrow{\Delta} 2CuO(s) + 4NO_2(g) + O_2(g)$

(c) $PbS(s) + 4O_3(g) \rightarrow PbSO_4(s) + 4O_2(g)$

(d) $2ZnS(s) + 3O_2(g) \xrightarrow{\Delta} 2ZnO(s) + 2SO_2(g)$

(e) $2K_2O_2(s) + 2CO_2(g) \rightarrow 2K_2CO_3(s) + O_2(g)$

22.45   *Analyze/Plan.* Oxides of metals are bases, oxides of nonmetals are acids, oxides that act as both acids and bases are amphoteric and oxides that act as neither acids nor bases are neutral.   *Solve.*

(a)   acidic (oxide of a nonmetal)

(b)   acidic (oxide of a nonmetal)

(c)   basic (oxide of a metal)

(d)   amphoteric

22.47   *Analyze/Plan.* Follow the rules for assigning oxidation numbers in Section 4.4 and the logic in Sample Exercise 4.8.   *Solve.*

(a) H_2SeO_3, +4   (b) $KHSO_3$, +4   (c) H_2Te, –2

(d) CS_2, –2   (e) $CaSO_4$, +6

Oxygen (a group 6A element) is in the –2 oxidation state in compounds (a), (b), and (e).

22.49   *Analyze/Plan.* The half-reaction for oxidation in all these cases is:

$H_2S(aq) \rightarrow S(s) + 2H^+ + 2e^-$ (The product could be written as $S_8(s)$, but this is not necessary. In fact it is not necessarily the case that S_8 would be formed, rather than some other allotropic form of the element.) Combine this half-reaction with the given reductions to write complete equations. The reduction in (c) happens only in acid solution. The reactants in (d) are acids, so the medium is acidic.   *Solve.*

(a) $2Fe^{3+}(aq) + H_2S(aq) \rightarrow 2Fe^{2+}(aq) + S(s) + 2H^+(aq)$

(b) $Br_2(l) + H_2S(aq) \rightarrow 2Br^-(aq) + S(s) + 2H^+(aq)$

(c) $2MnO_4^-(aq) + 6H^+(aq) + 5H_2S(aq) \rightarrow 2Mn^{2+}(aq) + 5S(s) + 8H_2O(l)$

(d) $2NO_3^-(aq) + H_2S(aq) + 2H^+(aq) \rightarrow 2NO_2(aq) + S(s) + 2H_2O(l)$

22.51   *Analyze/Plan.* For each substance, count valence electrons, draw the correct Lewis structure, and apply the rules of VSEPR to decide electron domain geometry and geometric structure.   *Solve.*

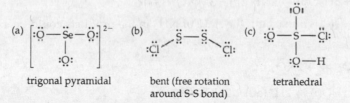

trigonal pyramidal   bent (free rotation   tetrahedral
                                      around S-S bond)

22.53   *Analyze/Plan.* Use information on the descriptive chemistry of sulfur given in Section 22.6 to complete and balance the equations.   *Solve.*

(a) $SO_2(s) + H_2O(l) \rightarrow H_2SO_3(aq) \rightleftharpoons H^+(aq) + HSO_3^-(aq)$

(b) $ZnS(s) + 2HCl(aq) \rightarrow ZnCl_2(aq) + H_2S(g)$

(c) $8SO_3^{2-}(aq) + S_8(s) \rightarrow 8S_2O_3^{2-}(aq)$

(d) $SO_3(aq) + H_2SO_4(l) \rightarrow H_2S_2O_7(l)$

Nitrogen and the Group 5A Elements

22.55   *Analyze/Plan.* Follow the rules for assigning oxidation numbers in Section 4.4 and the logic in Sample Exercise 4.8.   *Solve.*

(a) $NaNO_2$, +3   (b) NH_3, –3   (c) N_2O, +1

(d)   NaCN, –3   (e) HNO_3, +5   (f) NO_2, +4

22.57   *Analyze/Plan.* For each substance, count valence electrons, draw the correct Lewis structure, and apply the rules of VSEPR to decide electron domain geometry and geometric structure.   *Solve.*

(a) $:\ddot{O}\!\!=\!\!\ddot{N}\!-\!\ddot{O}\!-\!H \longleftrightarrow :\ddot{O}\!-\!\ddot{N}\!\!=\!\!\ddot{O}\!-\!H$

The molecule is bent around the central oxygen and nitrogen atoms; the four atoms need not lie in a plane. The right-most form does not minimize formal charges and is less important in the actual bonding model.

(b) $\left[:\ddot{N}\!\!=\!\!N\!\!=\!\!\ddot{N}:\right]^{-} \longleftrightarrow \left[:N\!\!\equiv\!\!N\!-\!\ddot{\ddot{N}}:\right]^{-} \longleftrightarrow \left[:\ddot{\ddot{N}}\!-\!N\!\!\equiv\!\!N:\right]^{-}$

The molecule is linear.

(c)
$$\left[\begin{array}{c} \text{H} \quad \text{H} \\ | \quad\; | \\ \text{H}\!-\!\text{N}\!-\!\text{N}: \\ | \quad\; | \\ \text{H} \quad \text{H} \end{array}\right]^{+}$$

(d)
$$\left[\begin{array}{c} :\ddot{O}: \\ | \\ :\ddot{O}\!-\!N\!\!=\!\!\ddot{O} \end{array}\right]^{-}$$

The geometry is tetrahedral around the left nitrogen, trigonal pyramidal around the right.   (three equivalent resonance forms) The ion is trigonal planar.

22.59   *Analyze/Plan.* Use information on the descriptive chemistry of nitrogen given in Section 22.7 to complete and balance the equations.   *Solve.*

(a) $Mg_3N_2(s) + 6H_2O(l) \rightarrow 2NH_3(g) + 3Mg(OH)_2(s)$

Because $H_2O(l)$ is a reactant, the state of NH_3 in the products could be expressed as $NH_3(aq)$.

(b) $2NO(g) + O_2(g) \rightarrow 2NO_2(g)$

(c) $N_2O_5(g) + H_2O(l) \rightarrow 2H^+(aq) + 2NO_3^-(aq)$

(d) $NH_3(aq) + H^+(aq) \rightarrow NH_4^+(aq)$

(e) $N_2H_4(l) + O_2(g) \rightarrow N_2(g) + 2H_2O(g)$

22.61   *Analyze/Plan.* Follow the method for writing balanced half-reactions given in Section 20.2 and Sample Exercises 20.2 and 20.3. Find standard reduction potentials in Figure 22.28.

Solve. For nonadjacent half-reactions on Figure 22.28, use the relationship $\Delta G° = -nFE°$. Because $\Delta G°$ is a state function, $\Delta G°$ for the overall process is the sum of the $\Delta G°$ values for the steps. Then, $\Delta G° = \Delta G_1° + \Delta G_2°$; $-nFE° = -n_1FE_1° - n_2FE_2°$; $n = n_1 + n_2$;

$$E° = \frac{n_1E_1° + N_2E_2°}{n_1 + n_2}$$

(a) $HNO_2(aq) + H_2O(l) \rightarrow NO_3^-(aq) + 3H^+(aq) + 2e^-$, $E_{red}^\circ = 0.96$ V

The steps are: (1) $NO_3^- \rightarrow NO_2$, $1e^-$, 0.79 V;

(2) $NO_2 \rightarrow HNO_2$, $1e^-$, 1.12 V

$E_{red}^\circ = [(1)\ 0.79\ V + (1)\ 1.12\ V]/(1+1) = 0.955 = 0.96$ V

(b) $N_2(g) + H_2O(l) \rightarrow N_2O(g) + 2H^+(aq) + 2e^-$, $E_{red}^\circ = 1.77$ V

E_{red}° can be read directly from Figure 22.28.

22.63   *Analyze/Plan.* Follow the rules for assigning oxidation numbers in Section 4.4 and the logic in Sample Exercise 4.8.   *Solve.*

(a) H_3PO_3, +3   (b) $H_4P_2O_7$, +5   (c) $SbCl_3$, +3

(d) Mg_3As_2, −3   (e) P_2O_5, +5

22.65   *Analyze/Plan.* Consider the structures of the compounds of interest when explaining the observations.   *Solve.*

(a)   Phosphorus is a larger atom and can more easily accommodate five surrounding atoms and an expanded octet of electrons than nitrogen can. Also, P has energetically "available" 3d orbitals which participate in the bonding, but nitrogen does not.

(b)   Only one of the three hydrogens in H_3PO_2 is bonded to oxygen. The other two are bonded directly to phosphorus and are not easily ionized because the P—H bond is not very polar.

(c) PH_3 is a weaker base than H_2O (PH_4^+ is a stronger acid than H_3O^+). Any attempt to add H^+ to PH_3 in the presence of H_2O merely causes protonation of H_2O.

(d)   White phosphorus consists of P_4 molecules, with P—P—P bond angles of 60°. Each P atom has four VSEPR pairs of electrons, so the predicted electron pair geometry is tetrahedral and the preferred bond angle is 109°. Because of the severely strained bond angles in P_4 molecules, white phosphorus is highly reactive.

22.67   *Analyze/Plan.* Use information on the descriptive chemistry of phosphorus given in Section 22.8 to complete and balance the equations.   *Solve.*

(a) $2Ca_3(PO_4)_2(s) + 6SiO_2(s) + 10C(s) \overset{\Delta}{\rightarrow} P_4(g) + 6CaSiO_3(l) + 10CO(g)$

(b) $PBr_3(l) + 3H_2O(l) \rightarrow H_3PO_3(aq) + 3HBr(aq)$

(c) $4PBr_3(g) + 6H_2(g) \rightarrow P_4(g) + 12HBr(g)$

Carbon, the Other Group 4A Elements, and Boron

22.69   *Analyze/Plan.* Review the nomenclature rules and ion names in Section 2.8.   *Solve.*
(a)   HCN   (b) $Ni(CO)_4$ (c) $Ba(HCO_3)_2$ (d) CaC_2

22.71   *Analyze/Plan.* Use information on the descriptive chemistry of carbon given in Section 22.9 to complete and balance the equations.   *Solve.*

(a) $ZnCO_3(s) \xrightarrow{\Delta} ZnO(s) + CO_2(g)$

(b) $BaC_2(s) + 2H_2O(l) \rightarrow Ba^{2+}(aq) + 2OH^-(aq) + C_2H_2(g)$

(c) $2C_2H_2(g) + 5O_2(g) \rightarrow 4CO_2(g) + 2H_2O(g)$

(d) $CS_2(g) + 3O_2(g) \rightarrow CO_2(g) + 2SO_2(g)$

(e) $Ca(CN)_2(s) + 2HBr(aq) \rightarrow CaBr_2(aq) + 2HCN(aq)$

22.73   *Analyze/Plan.* Use information on the descriptive chemistry of carbon given in Section 22.9 to complete and balance the equations.   *Solve.*

(a) $2CH_4(g) + 2NH_3(g) + 3O_2(g) \xrightarrow[\text{cat}]{800°C} 2HCN(g) + 6H_2O(g)$

(b) $NaHCO_3(s) + H^+(aq) \rightarrow CO_2(g) + H_2O(l) + Na^+(aq)$

(c) $2BaCO_3(s) + O_2(g) + 2SO_2(g) \rightarrow 2BaSO_4(s) + 2CO_2(g)$

22.75   *Analyze/Plan.* Follow the rules for assigning oxidation numbers in Section 4.4 and the logic in Sample Exercise 4.8.   *Solve.*

(a) H_3BO_3, +3   (b) $SiBr_4$, +4   (c) $PbCl_2$, +2

(d) $Na_2B_4O_7 \cdot 10H_2O$, +3   (e) B_2O_3, +3

22.77   *Analyze/Plan.* Consider periodic trends within a family, particularly metallic character, as well as descriptive chemistry in Sections 22.9 and 22.10.   *Solve.*

(a)   Lead; the most metallic element has the lowest first ionization energy.

(b)   Carbon, Si, and Ge; these are the nonmetal and metalloids in group 4A. They form compounds ranging from XH_4 (–4) to XO_2 (+4). The metals Sn and Pb are not found in negative oxidation states.

(c)   Silicon; silicates are the main component of sand.

22.79   *Analyze/Plan.* Consider the structural chemistry of silicates discussed in Section 22.10 and shown in Figures 22.47-22.48.   *Solve.*

(a)   Tetrahedral

(b)   Metasilicic acid will probably adopt the single-strand silicate chain structure shown in Figure 22.48(a). The empirical formula shows 3 O and 2 H atoms per Si atom. The chain has the same Si to O ratio as metasilicic acid. Furthermore, in the chain structure, there are two terminal (not bridging) O atoms on each Si. These can accommodate the 2 H atoms associated with each Si atom of the acid. The sheet structure does not fulfill these requirements.

22.81   (a)   Diborane (Figure 22.50 and below) has bridging H atoms linking the two B atoms. The structure of ethane shown below has the C atoms bound directly, with no bridging atoms.

(b) B_2H_6 is an electron deficient molecule. It has 12 valence electrons, while C_2H_6 has 14 valence electrons. The 6 valence electron pairs in B_2H_6 are all involved in B—H sigma bonding, so the only way to satisfy the octet rule at B is to have the bridging H atoms shown in Figure 22.50.

(c)   A hydride ion, H^-, has two electrons while an H atom has one. The term *hydridic* indicates that the H atoms in B_2H_6 have more than the usual amount of electron density for a covalently bound H atom.

Additional Exercises

22.84   (a) $10.0 \text{ lb FeTi} \times \dfrac{453.6 \text{ g}}{1 \text{ lb}} \times \dfrac{1 \text{ mol FeTi}}{103.7 \text{ g FeTi}} \times \dfrac{1 \text{ mol } H_2}{1 \text{ mol FeTi}} \times \dfrac{2.016 \text{ g H}}{1 \text{ mol } H_2} = 88.18 = 88.2 \text{ g } H_2$

(b) $V = \dfrac{88.18 \text{ g } H_2}{2.016 \text{ g/mol } H_2} \times \dfrac{0.08206 \text{ L} \cdot \text{atm}}{\text{mol} \cdot \text{K}} \times \dfrac{273 \text{ K}}{1 \text{ atm}} = 979.9 = 980 \text{ L}$

22.87   Substances that will burn in O_2: SiH_4, CO, Mg.

The others, SiO_2, CO_2, and CaO, have Si, C, and Ca in maximum oxidation states, so O_2 cannot act as an oxidizing agent.

22.89   (a) $H_2SO_4 - H_2O \rightarrow SO_3$

(b) $2HClO_3 - H_2O \rightarrow Cl_2O_5$

(c) $2HNO_2 - H_2O \rightarrow N_2O_3$

(d) $H_2CO_3 - H_2O \rightarrow CO_2$

(e) $2H_3PO_4 - 3H_2O \rightarrow P_2O_5$

22.95 $GeO_2(s) + C(s) \xrightarrow{\Delta} Ge(l) + CO_2(g)$

$Ge(l) + 2Cl_2(g) \rightarrow GeCl_4(l)$

$GeCl_4(l) + 2H_2O(l) \rightarrow GeO_2(s) + 4HCl(g)$

$GeO_2(s) + 2H_2(g) \rightarrow Ge(s) + 2H_2O(l)$

22.97   Assume that the reactions occur in acidic solution. The half-reaction for reduction of H_2O_2 is in all cases $H_2O_2(aq) + 2H^+(aq) + 2e^- \rightarrow 2H_2O(aq)$.

(a) $N_2H_4(aq) + 2H_2O_2(aq) \rightarrow N_2(g) + 4H_2O(l)$

(b) $SO_2(g) + H_2O_2(aq) \rightarrow SO_4^{2-}(aq) + 2H^+(aq)$

(c) $NO_2^-(aq) + H_2O_2(aq) \rightarrow NO_3^-(aq) + H_2O(l)$

(d) $H_2S(g) + H_2O_2(aq) \rightarrow S(s) + 2H_2O(l)$

(e)

$2H^+(aq) + H_2O_2(aq) + 2e^- \rightarrow 2H_2O(l)$

$\underline{\qquad 2[Fe^{2+}(aq) \rightarrow Fe^{3+}(aq) + e^-] \qquad}$

$2Fe^{2+}(aq) + H_2O_2(aq) + 2H^+(aq) \rightarrow 2Fe^{3+}(aq) + 2H_2O(l)$

Integrative Exercises

22.99   From Appendix C, we need only ΔH_f° for F(g), so that we can estimate ΔH for the process:

$$F_2(g) \rightarrow F(g) + F(g); \qquad \Delta H^\circ = 160 \text{ kJ}$$
$$\underline{XeF_2(g) \rightarrow Xe(g) + F_2(g) \qquad -\Delta H_f^\circ = 109 \text{ kJ}}$$
$$XeF_2(g) \rightarrow Xe(g) + 2F(g) \qquad \Delta H^\circ = 269 \text{ kJ}$$

The average Xe—F bond enthalpy is thus 269/2 = 134 kJ. Similarly,

$$XeF_4(g) \rightarrow Xe(g) + 2F_2(g) \qquad -\Delta H_f^\circ = 218 \text{ kJ}$$
$$\underline{2F_2(g) \rightarrow 4F(g) \qquad \Delta H^\circ = 320 \text{ kJ}}$$
$$XeF_4(g) \rightarrow Xe(g) + 4F(g) \qquad \Delta H^\circ = 538 \text{ kJ}$$

Average Xe—F bond energy = 538/4 = 134 kJ

$$XeF_6(g) \rightarrow Xe(g) + 3F_2(g) \qquad -\Delta H_f^\circ = 298 \text{ kJ}$$
$$\underline{3F_2(g) \rightarrow 6F(g) \qquad \Delta H^\circ = 480 \text{ kJ}}$$
$$XeF_6(g) \rightarrow Xe(g) + 6F(g) \qquad \Delta H^\circ = 778 \text{ kJ}$$

Average Xe—F bond energy = 778/6 = 130 kJ

The average bond enthalpies are: XeF_2, 134 kJ; XeF_4, 134 kJ; XeF_6, 130 kJ. They are remarkably constant in the series.

22.101 First calculate the molar solubility of Cl_2 in water.

$$n = \frac{1 \text{ atm } (0.310 \text{ L})}{\dfrac{0.08206 \text{ L} \cdot \text{atm}}{1 \text{ mol} \cdot \text{K}} \times 273 \text{ K}} = 0.01384 = 0.0138 \text{ mol } Cl_2$$

$$M = \frac{0.01384 \text{ mol}}{0.100 \text{ L}} = 0.1384 = 0.138 M$$

$[Cl^-] = [HOCl] = [H^+]$ Let this quantity = x. Then, $\dfrac{x^3}{(0.1384 - x)} = 4.7 \times 10^{-4}$

Assuming that x is small compared with 0.1384:

$x^3 = (0.1384)(4.7 \times 10^{-4}) = 6.504 \times 10^{-5}$; x = 0.0402 = 0.040 M

We can correct the denominator using this value, to get a better estimate of x:

$$\frac{x^3}{0.1384 - 0.0402} = 4.7 \times 10^{-4}; x = 0.0359 = 0.036 M$$

One more round of approximation gives x = 0.0364 = 0.036 M. This is the equilibrium concentration of HClO.

22.104 (a) $SO_2(g) + 2H_2S(aq) \rightarrow 3S(s) + 2H_2O(g)$ or, if we assume S_8 is the product,

$8SO_2(g) + 16H_2S(aq) \rightarrow 3S_8(s) + 16H_2O(g)$.

(b)   Assume that all S in the coal becomes SO_2 upon combustion, so that

1 mol S (coal) = 1 mol SO_2.

$$2000 \text{ lb coal} \times \frac{0.035 \text{ lb S}}{1 \text{ lb coal}} \times \frac{453.6 \text{ g S}}{1 \text{ lb S}} \times \frac{1 \text{ mol S (coal)}}{32.07 \text{ g S}} \times \frac{1 \text{ mol SO}_2}{1 \text{ mol S (coal)}} \times \frac{2 \text{ mol H}_2\text{S}}{1 \text{ mol SO}_2}$$

$$= 1.98 \times 10^3 = 2.0 \times 10^3 \text{ mol H}_2\text{S}$$

$$V = \frac{1.98 \times 10^3 \text{ mol } (0.08206 \text{ L} \cdot \text{atm/mol} \cdot \text{K})(300 \text{ K})}{(740/760) \text{ atm}} = 5.01 \times 10^4 = 5.0 \times 10^4 \text{ L}$$

(c) $1.98 \times 10^3 \text{ mol H}_2\text{S} \times \dfrac{3 \text{ mol S}}{2 \text{ mol H}_2\text{S}} \times \dfrac{32.07 \text{ g S}}{1 \text{ mol S}} = 9.5 \times 10^4 \text{ g S}$

This is about 210 lb S per ton of coal combusted. (However, two-thirds of this comes from the H_2S, which presumably at some point was also obtained from coal.)

22.106   The reactions can be written as follows:

$$H_2(g) + X(\text{std state}) \rightarrow H_2X(g) \qquad \Delta H_f^\circ$$

$$2H(g) \rightarrow H_2(g) \qquad \Delta H_f^\circ(H-H)$$

$$\underline{X(g) \rightarrow X(\text{std state}) \qquad \Delta H_3}$$

$$\text{Add}: 2H(g) + X(g) \rightarrow H_2X(g) \qquad \Delta H = \Delta H_f^\circ + \Delta H_f^\circ(H-H) + \Delta H_3$$

These are all the necessary ΔH values. Thus,

Compound	ΔH	D H$-$X
H_2O	$\Delta H = 242 \text{ kJ} - 436 \text{ kJ} - 248 \text{ kJ} = -926 \text{ kJ}$	463 kJ
H_2S	$\Delta H = -20 \text{ kJ} - 436 \text{ kJ} - 277 \text{ kJ} = -733 \text{ kJ}$	367 kJ
H_2Se	$\Delta H = +30 \text{ kJ} - 436 \text{ kJ} - 227 \text{ kJ} = -633 \text{ kJ}$	316 kJ
H_2Te	$\Delta H = +100 \text{ kJ} - 436 \text{ kJ} - 197 \text{ kJ} = -533 \text{ kJ}$	266 kJ

The average H$-$X bond energy in each case is just half of ΔH. The H$-$X bond energy decreases steadily in the series. The origin of this effect is probably the increasing size of the orbital from X with which the hydrogen 1s orbital must overlap.

22.109 $(CH_3)_2N_2H_2(g) + 2N_2O_4(g) \rightarrow 2CO_2(g) + 3N_2(g) + 4H_2O(g)$

$$4.0 \text{ tons } (CH_3)_2N_2H_2 \times \frac{2000 \text{ lb}}{1 \text{ ton}} \times \frac{453.6 \text{ g}}{1 \text{ lb}} \times \frac{1 \text{ mol } (CH_3)_2N_2H_2}{60.10 \text{ g } (CH_3)_2N_2H_2}$$

$$\times \frac{2 \text{ mol } N_2O_4}{1 \text{ mol } (CH_3)_2N_2H_2} \times \frac{92.02 \text{ g } N_2O_4}{1 \text{ mol } N_2O_4} \times \frac{1 \text{ lb}}{453.6 \text{ g}} \times \frac{1 \text{ ton}}{2000 \text{ lb}} = 12 \text{ tons } N_2O_4$$

23 Metals and Metallurgy

Visualizing Concepts

23.1   *Analyze.* Given the formulas of substances, decide which are likely to be found together in nature.

Plan. Consider the physical and chemical properties of the two substances, whether they are likely to have formed and to exist under the same environmental conditions.

Solve. Most likely to be found together: (c) Al_2O_3 and Fe_2O_3.

The compounds have the same stoichiometry, both metals are in high oxidation states, so they could have been formed in the same environment. Both compounds are inert solids that would survive environmental degradation.

Reasons to eliminate other choices:

(a) Al^{3+} in highly oxidized state, Cu^+ in reduced state, unlikely to be produced by similar environmental conditions

(b)   Second most likely pair, ionic radii similar. CaS could be water soluble, which would render this an unlikely pair to exist together.

(c) Na^+ has much larger ionic radius than Ag^+, unlikely to share same crystal lattice. NaCl is water soluble, AgCl is not. Unlikely to survive same environmental conditions.

(d) KO_2 is potassium superoxide, extremely reactive and unlikely to exist in nature.

23.5   The close-packed forms, both cubic and hexagonal, maximize the number of metal-metal contacts at 12. This maximizes the orbital interactions among metal atoms and increases delocalization. Substances with delocalized bonding have a special stability (Section 9.6) relative to those with localized bonding. The most common solid-state structure for metals is one of the close-packed forms because they result in maximum delocalization in metallic bonding.

Metallurgy

23.9   *Analyze/Plan.* Use Table 23.1 and other information in Section 23.1 to find important natural sources of Al and Fe. Use the rules for assigning oxidation numbers in Section 4.4 to determine the oxidation state of the metal in each natural source.   *Solve.*

The important sources of iron are **hematite** (Fe_2O_3) and **magnetite** (Fe_3O_4). The major source of aluminum is **bauxite** ($Al_2O_3 \bullet xH_2O$). In ores, iron is present as the +3 ion, or in both the +2 and +3 states, as in magnetite. Aluminum is always present in the +3 oxidation state.

23.11 An ore consists of a little bit of the stuff we want, (chalcopyrite, $CuFeS_2$) and lots of other junk (gangue).

23.13 *Analyze/Plan.* Use principles of writing and balancing chemical equations from Chapter 3 to complete and balance the given reactions. The Δ above each arrow indicates that the reactions take place at elevated temperature. Information in Section 23.2 on *pyrometallurgy* will probably be useful. *Solve.*

    (a) $Cr_2O_3(s) + 6Na(l) \rightarrow 2Cr(s) + 3Na_2O(s)$

    (b) $PbCO_3(s) \overset{\Delta}{\rightarrow} PbO(s) + CO_2(g)$

    (c) $2CdS(s) + 3O_2(g) \overset{\Delta}{\rightarrow} 2CdO(s) + 2SO_2(g)$

    (d) $ZnO(s) + CO(g) \overset{\Delta}{\rightarrow} Zn(l) + CO_2(g)$

23.15 *Analyze/Plan.* Use information on *pyrometallurgy* in Section 23.2, along with principles of writing and balancing equations to provide the requested information. *Solve.*

    (a) $SO_3(g)$

    (b) $CO(g)$ provides a reducing environment for the transformation of Pb^{2+} to Pb.

    (c) $PbSO_4(s) \rightarrow PbO(s) + SO_3(g)$

         $PbO(s) + CO(g) \rightarrow Pb(s) + CO_2(g)$

23.17 *Analyze/Plan.* Use information on *pyrometallurgy* in Section 23.2, along with principles of writing and balancing equations to provide the requested information. *Solve.*

The major reducing agent is CO, formed by partial oxidation of the coke (C) with which the furnace is charged.

$Fe_2O_3(s) + 3CO(g) \rightarrow 2Fe(l) + 3CO_2(g)$

$Fe_3O_4(s) + 4CO(g) \rightarrow 3Fe(l) + 4CO_2(g)$

23.19 *Analyze/Plan.* Use information on *pyrometallurgy* in Section 23.2, along with principles of writing and balancing equations to provide the requested information. *Solve.*

    (a)   *Air* serves primarily to oxidize coke (C) to CO, the main reducing agent in the blast furnace. This exothermic reaction also provides heat for the furnace.

         $2C(s) + O_2(g) \rightarrow 2CO(g) \; \Delta H = -221 \text{ kJ}$

    (b)   *Limestone*, $CaCO_3$, is the source of basic oxide for slag formation.

         $CaCO_3(s) \overset{\Delta}{\rightarrow} CaO(s) + CO_2(g); \; CaO(l) + SiO_2(l) \rightarrow CaSiO_3(l)$

    (c)   *Coke* is the fuel for the blast furnace, and the source of CO, the major reducing agent in the furnace.

         $2C(s) + O_2(g) \rightarrow 2CO(g); \; 4CO(g) + Fe_3O_4(s) \rightarrow 4CO_2(g) + 3Fe(l)$

    (d)   *Water* acts as a source of hydrogen, and as a means of controlling temperature. (See Equation [23.8].) $C(s) + H_2O(g) \rightarrow CO(g) + H_2(g) \; \Delta H = +131 \text{ kJ}$

23.21 *Analyze/Plan.* Use information on the *electrometallurgy* of Cu as a model for describing how electrometallurgy can be employed to purify pure Co. Compare the ease of oxidation and reduction of cobalt with that of water. *Solve.*

Cobalt could be purified by constructing an electrolysis cell in which the crude metal was the anode and a thin sheet of pure cobalt was the cathode. The electrolysis solution is aqueous with a soluble cobalt salt such as $CoSO_4 \cdot 7H_2O$ serving as the electrolyte. (Other soluble salts with anions that do not participate in the cell reactions could be used.) Anode reaction: $Co(s) \rightarrow Co^{2+}(aq) + 2e^-$; cathode reaction: $Co^{2+}(aq) + 2e^- \rightarrow Co(s)$. Although $E°$ for reduction of $Co^{2+}(aq)$ is slightly negative (–0.277 V), it is less than the standard reduction potential for $H_2O(l)$, –0.83 V.

Metals and Alloys

23.23   *Analyze/Plan*. Compare the bonding characteristics of metallic sodium and ionic sodium chloride and use them to explain the difference in malleability. *Solve*.

Sodium is metallic; each atom is bonded to many nearest neighbor atoms by metallic bonding involving just one electron per atom, and delocalized over the entire three-dimensional structure. When sodium metal is distorted, each atom continues to have bonding interactions with many nearest neighbors. In NaCl the ionic forces are strong, and the arrangement of ions in the solid is very regular. When subjected to physical stress, the three-dimensional lattice tends to cleave along the very regular lattice planes, rather than undergo the large distortions characteristic of metals.

23.25   *Analyze/Plan*. Apply the description of the electron-sea model of metallic bonding given in Section 23.5 to the conductivity of metals. *Solve*.

In the electron-sea model for metallic bonding, valence electrons move about the three-dimensional metallic lattice, while the metal atoms maintain regular lattice positions.

Under the influence of an applied potential the electrons can move throughout the structure, giving rise to high electrical conductivity. The mobility of the electrons facilitates the transfer of kinetic energy and leads to high thermal conductivity.

23.27   *Analyze/Plan*. Consider trends in atomic mass and volume of the elements listed to explain the variation in density. *Solve*.

Moving left to right in the period, atomic mass and Z_{eff} increase. The increase in Z_{eff} leads to smaller bonding atomic radii and thus atomic volume. Mass increases, volume decreases, and density increases in the series.

The variation in densities reflects shorter metal-metal bond distances. These shorter distances suggest that the extent of metal-metal bonding increases in the series. This is consistent with greater occupancy of the bonding band as the number of valence electrons increases up to 6.

23.29   *Analyze/Plan*. Consider the definition of ductility, as well as the discussion of band theory in Chapter 12. *Solve*.

Ductility is the property related to the ease with which a solid can be drawn into a wire. Basically, the softer the solid the more ductile it is. The more rigid the solid, the less ductile it is. For metals, ductility decreases as the number of bonding electrons per atom increases, producing a stiffer lattice less susceptible to distortion.

(a)   K is more ductile. Cr, with 6 valence electrons, has a filled bonding band, strong metal-metal interactions, and a rigid lattice. This predicts high hardness and low ductility.

(b)   Zn is more ductile. Si is a covalent-network solid with all valence electrons localized in bonds between Si atoms. Covalent-network substances are high-melting, hard, and not particularly ductile.

23.31   *Analyze/Plan.* Recall the diamond and closest-packed structures described in Section 11.7. Use these structures to draw conclusions about Sn–Sn distance and electrical conductivity in the two allotropes.   *Solve.*

White tin, with a characteristic metallic structure, is expected to be more metallic in character. The electrical conductivity of the white allotropic form is higher because the valence electrons are shared with 12 nearest neighbors rather than being localized in four bonds to nearest neighbors as in gray tin. The Sn–Sn distance should be longer in white tin; there are only four valence electrons from each atom, and 12 nearest neighbors. The **average** tin–tin bond order can, therefore, be only about 1/3, whereas in gray tin the bond order is one. Gray tin, with the higher bond order, has a shorter Sn–Sn distance, 2.81 Å. The bond length in white tin, with the lower bond order, is 3.02 Å.

23.33   *Analyze/Plan.* Use information in Section 23.6 to define *alloy*, and compare the various types of alloys.   *Solve.*

An *alloy* contains atoms of more than one element and has the properties of a metal. *Solution alloys* are homogeneous mixtures with different kinds of atoms dispersed randomly and uniformly. In *heterogeneous alloys* the components (elements or compounds) are not evenly dispersed and their properties depend not only on composition but methods of preparation. In an *intermetallic compound* the component elements have interacted to form a compound substance, for example, Cu_3As. As with more familiar compounds, these are homogeneous and have definite composition and properties.

Transition Metals

23.35   *Analyze/Plan.* Consider the definitions of the properties listed (Chapter 7 and Chapter 23) and whether they refer to single, isolated atoms or bulk material.   *Solve.*

Of the properties listed, (b) the first ionization energy and (f) electron affinity are characteristic of isolated atoms. Electrical conductivity (a), atomic radius (c), melting point (d), and heat of vaporization (e) are properties of the bulk metal. Although it seems that atomic radius would be a property of isolated atoms, it can only be measured in bulk samples.

23.37   *Analyze/Plan.* Examine the electron configurations, Z and Z_{eff}, of the two elements to account for their similar atomic radii.

Solve. Zr: $[Kr]5s^2 4d^2$, Z = 40; Hf: $[Xe]6s^2 4f^{14} 5d^2$, Z = 72

Moving down a family of the periodic chart, atomic size increases because the valence electrons are in a higher principle quantum level (and thus further from the nucleus) and are more effectively shielded from the nuclear charge by a larger core electron cloud. However, the build-up in Z that accompanies the filling of the 4f orbitals causes the valence electrons in Hf to experience a much greater relative nuclear charge than those in La, its neighbor to the left. This increase in Z offsets the usual effect of the increase in *n* value of the valence electrons and the radii of Zr and Hf atoms are similar.

23.39 *Analyze/Plan.* Use Figure 23.20 to determine the highest oxidation state of each metal. Write formulas of the metal fluorides, given that fluoride ion is F^-. *Solve.*

(a) ScF_3 (b) CoF (c) ZnF_2

(d) MoF_6 (The oxidation states of Mo are similar to those of Cr.)

23.41 *Analyze/Plan.* Consider the electron configurations of Cr and Al to rationize observed oxidation states. *Solve.*

Chromium, $[Ar]4s^1 3d^5$, has six valence-shell electrons, some or all of which can be involved in bonding, leading to multiple stable oxidation states. By contrast, aluminum, $[Ne]3s^2 3p^1$, has only three valence electrons which are all lost or shared during bonding, producing the +3 state exclusively.

23.43 *Analyze/Plan.* Write electron configurations for the neutral elements and their positive ions recalling that valence electrons are last in order of descending *n*-value. *Solve.*

(a) Cr^{3+}: $[Ar]3d^3$ (b) Au^{3+}: $[Xe]4f^{14}5d^8$ (c) Ru^{2+}: $[Kr]4d^6$

(d) Cu^+: $[Ar]3d^{10}$ (e) Mn^{4+}: $[Ar]3d^3$ (f) Ir^+: $[Xe]4f^{14}5d^8$

23.45 *Analyze/Plan.* Oxidation is loss of electrons. Which periodic trend determines how tightly a valence electron is held in a particular atom or ion? *Solve.*

Ease of oxidation decreases from left to right across a period (owing to increasing effective nuclear charge); Ti^{2+} should be more easily oxidized than Ni^{2+}.

23.47 *Analyze/Plan.* Consider Equation [23.26] regarding the oxidation states of iron. *Solve.*
Fe^{2+} is a reducing agent that is readily oxidized to Fe^{3+} in the presence of O_2 from air.

23.49 *Analyze/Plan.* Consider information on the descriptive chemistry of iron in Section 23.8. *Solve.*

(a) $Fe(s) + 2HCl(aq) \rightarrow FeCl_2(aq) + H_2(g)$

(b) $Fe(s) + 4HNO_3(aq) \rightarrow Fe(NO_3)_3(aq) + NO(g) + 2H_2O(l)$

(See net ionic equation, Equation [23.28].) In concentrated nitric acid, the reaction can produce $NO_2(g)$ according to the reaction:

$Fe(s) + 6HNO_3(aq) \rightarrow Fe(NO_3)_3(aq) + 3NO_2(g) + 3H_2O(l)$

23.51 *Analyze/Plan.* Consider the definitions of paramagnetic and diamagnetic. *Solve.*

The unpaired electrons in a *paramagnetic* material cause it to be weakly attracted into a magnetic field. A *diamagnetic* material, where all electrons are paired, is very weakly repelled by a magnetic field.

Additional Exercises

23.53 $PbS(s) + O_2(g) \rightarrow Pb(l) + SO_2(g)$

Regardless of the metal of interest, $SO_2(g)$ is a product of roasting sulfide ores. In an oxygen rich environment, $SO_2(g)$ is oxidized to $SO_3(g)$, which dissolves in $H_2O(l)$ to form sulfuric acid, $H_2SO_4(aq)$. Because of its corrosive nature, $SO_2(g)$ is a dangerous environmental pollutant (Section 18.4) and cannot be freely released into the atmosphere. A sulfuric acid plant near a roasting plant would provide a means for disposing of $SO_2(g)$ that would also generate a profit.

23.55 $CO(g)$: $Pb(s)$; $H_2(g)$: $Fe(s)$; $Zn(s)$: $Au(s)$

23.58   Because selenium and tellurium are both nonmetals, we expect them to be difficult to oxidize. Thus, both Se and Te are likely to accumulate as the free elements in the so-called anode slime, along with noble metals that are not oxidized.

23.61   (a)   The very low boiling point (130°C) of OsO_4 indicates that it is a covalent compound. An ionic compound would have a much higher boiling point. For a covalent molecule, we can apply VSEPR to predict that OsO_4 will be tetrahedral. Any molecule with four attached groups will be tetrahedral as long as there are no lone pairs on the central atom. Metal atoms like Os are unlikely to have lone pairs, so a tetrahedral structure for OsO_4 is likely.

(b)   The oxidation state of Os in OsO_4 is +8. According to Figure 23.20, +8 is not a common oxidation state for Fe. Also, in a reversal of the typical trend for representative elements, the electronegativity of Fe is less than that of Os (see Figure 8.6). The electronegativity difference between Fe and O is greater than that between Os and O, so the iron oxides are more likely to be ionic.

23.64   The equilibrium of interest is $[ZnL_4] \rightleftharpoons Zn^{2+}(aq) + 4L$ $K = 1/K_f$

Since $Zn(H_2O)_4{}^{2+}$ is $Zn^{2+}(aq)$, its reduction potential is –0.763 V. As the stability (K_f) of the complexes increases, K decreases. Since E° is directly proportional to log K (Equation [20.16]), E° values for the complexes will become more negative as K_f increases.

23.67   (a) $Mn(s) + 2HNO_3(aq) \rightarrow Mn(NO_3)_2(aq) + H_2(g)$

(b) $Mn(NO_3)_2(s) \xrightarrow{\Delta} MnO_2(s) + 2NO_2(g)$

(c) $3MnO_2(s) \xrightarrow{\Delta} Mn_3O_4(s) + O_2(g)$

(d) $2MnCl_2(s) + 9F_2(g) \rightarrow 2MnF_3(s) + 4ClF_3(g)$

23.70   (a)   insulator   (b)   semiconductor   (c)   metallic conductor

(d)   metallic conductor   (e)   insulator   (f)   metallic conductor

Integrative Exercises

23.71   (a)   Calculate mass Cu_2S and FeS, then mass SO_2 from each.

3.3×10^6 kg sample $\times 0.27 = 8.91 \times 10^5 = 8.9 \times 10^5$ kg $= 8.9 \times 10^8$ g Cu_2S

3.3×10^6 kg sample $\times 0.13 = 4.29 \times 10^5 = 4.3 \times 10^5$ kg $= 4.3 \times 10^8$ g FeS

8.91×10^8 g $Cu_2S \times \dfrac{1 \text{ mol } Cu_2S}{159.1 \text{ g } Cu_2S} \times \dfrac{1 \text{mol } SO_2}{1 \text{ mol } Cu_2S} \times \dfrac{64.07 \text{ g } SO_2}{1 \text{ mol } SO_2} = 3.588 \times 10^8$

$= 3.6 \times 10^8$ g SO_2

$$4.29 \times 10^8 \text{ g FeS} \times \frac{1 \text{ mol FeS}}{87.9 \text{ g FeS}} \times \frac{1 \text{ mol SO}_2}{1 \text{ mol FeS}} \times \frac{64.07 \text{ g SO}_2}{1 \text{ mol SO}_2} = 3.127 \times 10^8 \text{ g}$$

$$= 3.1 \times 10^8 \text{ g SO}_2$$

$$\text{g SO}_2 = 3.588 \times 10^8 + 3.127 \times 10^8 = 6.715 \times 10^8 = 6.7 \times 10^8 \text{ g SO}_2$$

(b)   Calculate mol Cu, mol Fe and mole ratio Cu: Fe.

$$8.91 \times 10^8 \text{ g Cu}_2\text{S} \times \frac{1 \text{ mol Cu}_2\text{S}}{159.1 \text{ g Cu}_2\text{S}} \times \frac{2 \text{ mol Cu}}{1 \text{ mol Cu}_2\text{S}} = 1.12 \times 10^7 = 1.1 \times 10^7 \text{ mol Cu}$$

$$4.29 \times 10^8 \text{ g FeS} \times \frac{1 \text{ mol FeS}}{87.9 \text{ g FeS}} \times \frac{1 \text{ mol Fe}}{1 \text{ mol FeS}} = 4.88 \times 10^6 = 4.9 \times 10^6 \text{ mol Fe}$$

$$1.12 \times 10^7 \text{ mol Cu} / 4.88 \times 10^6 \text{ mol Fe} = 2.3 \text{ mol Cu/mol Fe}$$

(c)   The oxidizing environment of the converter is likely to produce CuO and Fe_2O_3.

(d) $Cu_2S(s) + 2O_2(g) \rightarrow 2CuO(s) + SO_2(g)$

$4FeS(s) + 7O_2(g) \rightarrow 2Fe_2O_3(s) + 4SO_2(g)$

23.73   The first equation indicates that one mole Ni^{2+} is formed from passage of two moles of electrons, and the second equation indicates the same thing. Thus, the simple ratio (1 mol Ni^{2+}/2F).

$$67 \text{ A} \times 11.0 \text{ hr} \times \frac{3600 \text{ s}}{1 \text{ hr}} \times \frac{1 \text{ C}}{1 \text{ A} \bullet \text{s}} \times \frac{1 \text{ F}}{96,500 \text{ C}} \times \frac{1 \text{ mol Ni}^{2+}}{2 \text{ F}} \times \frac{58.7 \text{ g Ni}^{2+}}{1 \text{ mol Ni}^{2+}}$$

$$\times \frac{0.90 \text{ g Ni actual}}{1.00 \text{ g Ni theoretical}} = 7.3 \times 10^2 \text{ g Ni}^{2+}(\text{aq})$$

23.76 $2[Cu^+(aq) + 1e^- \rightarrow Cu(s)]$ $E^{\circ}_{red} = 0.521 \text{ V}$

$\underline{\quad Cu(s) \rightarrow Cu^{2+}(aq) + 2e^- \qquad E^{\circ}_{red} = 0.337 \text{ V} \quad}$

$2Cu^+(aq) \rightarrow Cu^{2+}(aq) + Cu(s) \quad E^{\circ} = (0.521 \text{ V} - 0.337 \text{ V}) = 0.184 \text{ V}$

Rearrange Equation [20.16] for equilibrium conditions. At equilibrium, Q = K and E = 0;

$$\log K = \frac{nE^{\circ}}{0.0592}; n = 2 \qquad \log K = \frac{2(0.184)}{0.0592} = 6.2162 = 6.22; K = 1.6 \times 10^6$$

23.78   (a)   The standard reduction potential for $H_2O(l)$ is much greater than that of $Mg^{2+}(aq)(-0.83 \text{ V vs.} -2.37 \text{ V})$. In aqueous solution, $H_2O(l)$ would be preferentially reduced and no Mg(s) would be obtained.

(b) $97,000 \text{ A} \times 24 \text{ hr} \times \frac{3600 \text{ s}}{1 \text{ hr}} \times \frac{1 \text{ C}}{1 \text{ A} \bullet \text{s}} \times \frac{1 \text{ F}}{96,500 \text{ C}} \times \frac{1 \text{ mol Mg}}{2 \text{ F}} \times \frac{24.31 \text{ g Mg}}{1 \text{ mol Mg}} \times 0.96$

$$= 1.0 \times 10^6 \text{ g Mg} = 1.0 \times 10^3 \text{ kg Mg}$$

24 Chemistry of Coordination Compounds

Visualizing Concepts

24.1   *Analyze.* Given the formula of a coordination compound, determine the coordination geometry, coordination number, and oxidation state of the metal.

Plan. From the formula, determine the identity of the ligands and the number of coordination sites they occupy. From the total coordination number, decide on a likely geometry. Use ligand and overall complex charges to calculate the oxidation number of the metal.

Solve.

(a)   The ligands are $2Cl^-$, one coordination site each, and en, ethylenediamine, two coordination sites, for a coordination number of 4. This coordination number has two possible geometries, tetrahedral and square planar. Pt is one of the metals known to adopt square planar geometry when CN = 4.

(b)   CN = 4, coordination geometry = square planar

(c) $Pt(en)Cl_2$ is a neutral compound, the en ligand is neutral, and the $2Cl^-$ ligands are each –1, so the oxidation state of Pt must be +2, Pt(II).

24.4   *Analyze.* Given 5 structures, visualize which are identical to (1) and which are geometric isomers of (1).

Plan. There are two possible ways to arrange MA_3X_3. The first has bond angles of 90° between all similar ligands; this is structure (1). The second has one 180° angle between similar ligands. Visualize which description fits each of the five structures.

Solve. (1) has all 90° angles between similar ligands.

(2) has a 180° angle between similar ligands (see the blue ligands in the equatorial plane of the octahedron)

(3) has all 90° angles between similar ligands

(4) has all 90° angles between similar ligands

(5) has a 180° angle between similar ligands (see the blue axial ligands)

Structures (3) and (4) are identical to (1); (2) and (5) are geometric isomers.

24.6   *Analyze.* Given the visible colors of two solutions, determine the colors of light absorbed by each solution.

Plan. Apparent color is transmitted or reflected light, absorbed color is basically the complement of apparent color. Use the color wheel in Figure 24.24 to obtain the complementary absorbed color for the solutions.

Solve. The left solution appears yellow-orange, so it absorbs blue-violet. The right solution appears blue-green (cyan), so it absorbs orange-red.

Introduction to Metal Complexes

24.9   (a)   A *metal complex* consists of a central metal ion bonded to a number of surrounding molecules or ions. The number of bonds formed by the central metal ion is the *coordination number*. The surrounding molecules or ions are the *ligands*.

        (b)   A Lewis acid is an electron pair acceptor and a Lewis base is an electron pair donor. All ligands have at least one unshared pair of valence electrons. Metal ions have empty valence orbitals (d, s, or p) that can accommodate donated electron pairs. Ligands act as electron pair donors, or Lewis bases, and metal ions act as electron pair acceptors, or Lewis acids, via their empty valence orbitals.

24.11   *Analyze/Plan.* Follow the logic in Sample Exercises 24.1 and 24.2.   *Solve.*

        (a)   This compound is electrically neutral, and the NH_3 ligands carry no charge, so the charge on Ni must balance the –2 charge of the 2 Br^- ions. The charge and oxidation state of Ni is +2.

        (b)   Since there are 6 NH_3 molecules in the complex, the likely coordination number is 6. In some cases Br^- acts as a ligand, so the coordination number could be other than 6.

        (c)   Assuming that the 6 NH_3 molecules are the ligands, 2 Br^- ions are not coordinated to the Ni^{2+}, so 2 mol AgBr(s) will precipitate. (If one or both of the Br^- act as a ligand, the mol AgBr(s) would be different.)

24.13   *Analyze/Plan.* Count the number of donor atoms in each complex, taking the identity of polydentate ligands into account. Follow the logic in Sample Exercise 24.2 to obtain oxidation numbers of the metals.

        (a)   Coordination number = 4, oxidation number = +2

        (b)   5, +4   (c)   6, +3   (d)   5, +2

        (e)   6, +3   (f)   4, +2

24.15   *Analyze/Plan.* Given the formula of a coordination compound, determine the number and kinds of donor atoms. The ligands are enclosed in the square brackets. Decide which atom in the ligand has an unshared electron pair it is likely to donate.   *Solve.*

        (a) $4 Cl^-$ (b) $4 Cl^-, 1 O^{2-}$ (c) $4 N, 2 Cl^-$

        (d)   5 C. In CN^-, both C and N have an unshared electron pair. C is less electronegative and more likely to donate its unshared pair.

        (e)   6 O. $C_2O_4^{2-}$ is a bidentate ligand; each ion is bound through 2 O atoms for a total of 6 O donor atoms.

    (f)   4 N. en is a bidentate ligand bound through 2 N atoms.

Polydendate Ligands; Nomenclature

24.17   (a)   A monodendate ligand binds to a metal in through one atom; a bidentate ligand binds through two atoms.

    (b)   If a bidentate ligand occupies two coordination sites, three bidentate ligands fill the coordination sphere of a six-coordinate complex.

    (c)   A tridentate ligand has at least three atoms with unshared electron pairs in the correct orientation to simultanously bind one or more metal ions.

24.19   *Analyze/Plan.* Given the formula of a coordination compound, determine the number of coordination sites occupied by the polydentate ligand. The coordination number of the complexes is either 4 or 6. Note the number of monodentate ligands and determine the number of coordination sites occupied by the polydentate ligands.   *Solve.*

    (a)   *ortho*-phenanthroline, *o*-phen, is bidentate

    (b)   oxalate, $C_2O_4^{2-}$, is bidentate

    (c)   ethylenediaminetetraacetate, EDTA, is pentadentate

    (d)   ethylenediamine, en, is bidentate

24.21   (a)   The term *chelate effect* means there is a special stability associated with formation of a metal complex containing a polydentate (chelate) ligand relative to a complex containing only monodentate ligands.

    (b)   When a single chelating ligand replaces two or more monodendate ligands, the number of free molecules in the system increases and the entropy of the system increases. Chemical reactions with $+\Delta S$ tend to be spontaneous, have negative ΔG, and large positive values of K.

    (c)   Polydentate ligands can be used to bind metal ions and prevent them from undergoing unwanted chemical reactions without removing them from solution. The polydentate ligand thus hides or *sequesters* the metal ion.

24.23   *Analyze/Plan.* Given the name of a coordination compound, write the chemical formula. Refer to Table 24.2 to find ligand formulas. Place the metal complex (metal ion + ligands) inside square brackets and the counter ion (if there is one) outside the brackets. *Solve.*

    (a) $[Cr(NH_3)_6](NO_3)_3$ (b) $[Co(NH_3)_4CO_3]_2SO_4$ (c) $[Pt(en)_2Cl_2]Br_2$

    (d) $K[V(H_2O)_2Br_4]$ (e) $[Zn(en)_2][HgI_4]$

24.25   *Analyze/Plan.* Follow the logic in Sample Exercise 24.4, paying attention to naming rules in Section 24.3.   *Solve.*

    (a)   tetraamminedichlororhodium(III) chloride

    (b)   potassium hexachlorotitanate(IV)

(c) tetrachlorooxomolybdenum(VI)

(d) tetraaqua(oxalato)platinum(IV) bromide

Isomerism

24.27 *Analyze/Plan.* Consider the definitions of the various types of isomerism, and which of the complexes could exhibit isomerism of the specified type. *Solve.*

24.29 Yes. A tetrahedral complex of the form MA_2B_2 would have neither structural nor stereoisomers. For a tetrahedral complex, no differences in connectivity are possible for a single central atom, so the terms *cis* and *trans* do not apply. No optical isomers with tetrahedral geometry are possible because M is not bound to four different groups. The complex must be square planar with *cis* and *trans* geometric isomers.

24.31 *Analyze/Plan.* Follow the logic in Sample Exercises 24.5 and 24.6. *Solve.*

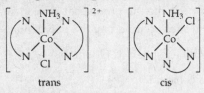

The cis isomer is chiral.

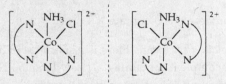

24.33 *Analyze/Plan.* Follow the logic in Sample Exercise 24.5 and 24.6. *Solve.*

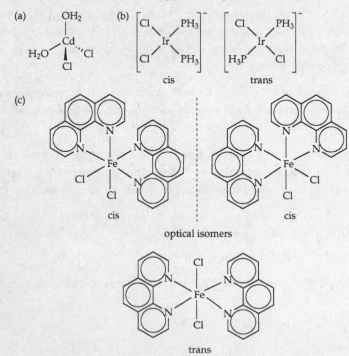

(The three isomeric complex ions in part (c) each have a 1+ charge.)

Color, Magnetism; Crystal-Field Theory

24.35   (a)   Visible light has wavelengths between 400 and 700 nm.

(b)   *Complementary* colors are opposite each other on a color wheel such as Figure 24.25.

(c)   A colored metal complex absorbs visible light of its complementary color. For example, a red complex absorbs green light.

(d) $E(J/photon) = h\nu = hc/\lambda$. Change J/photon to kJ/mol.

$$E = \frac{6.626 \times 10^{-34} \, J \bullet s}{610 \, nm} \times \frac{3.00 \times 10^8 \, m}{s} \times \frac{1 \, nm}{1 \times 10^{-9} \, m} = 3.259 \times 10^{-19} = 3.26 \times 10^{-19} \, J$$

$$\frac{3.259 \times 10^{-19} \, J}{photon} \times \frac{1 \, kJ}{1000 \, J} \times \frac{6.022 \times 10^{23} \, photons}{mol} = 196 \, kJ/mol$$

24.37   Most of the electrostatic interaction between a metal ion and a ligand is the attractive interaction between a positively charged metal cation and the full negative charge of an anionic ligand or the partial negative charge of a polar covalent ligand. Whether the interaction is ion-ion or ion-dipole, the ligand is strongly attracted to the metal center and can be modeled as a point negative charge.

24.39   (a)

(b)   The magnitude of Δ and the energy of the d-d transition for a d^1 complex are equal.

(c) $$\frac{6.626 \times 10^{-34} \, J \bullet s}{590 \, nm} \times \frac{3.00 \times 10^8 \, m}{s} \times \frac{1 \, nm}{1 \times 10^{-9} \, m} \times \frac{1 \, kJ}{1000 \, J} \times \frac{6.022 \times 10^{23} \, photons}{mol}$$

$$= 203 \, kJ/mol$$

24.41   *Analyze/Plan.* Consider the relationship between the color of a complex, the wavelength of absorbed light, and the position of a ligand in the spectrochemical series.   *Solve.*

Cyanide is a strong field ligand. The d-d electronic transitions occur at relatively high energy, because Δ is large. A yellow color corresponds to absorption of a photon in the violet region of the visible spectrum, between 430 and 400 nm. H_2O is a weaker field ligand than CN^-. The blue or green colors of aqua complexes correspond to absorptions in the region of 620 nm. Clearly, this is a region of lower energy photons than those with characteristic wavelengths in the 430 to 400 nm region. These are very general and imprecise comparisons. Other factors are involved, including whether the complex is high spin or low spin.

24.43   *Analyze/Plan.* Determine the charge on the metal ion, subtract it from the row number (3-12) of the transition metal, and the remainder is the number of d-electrons.   *Solve.*

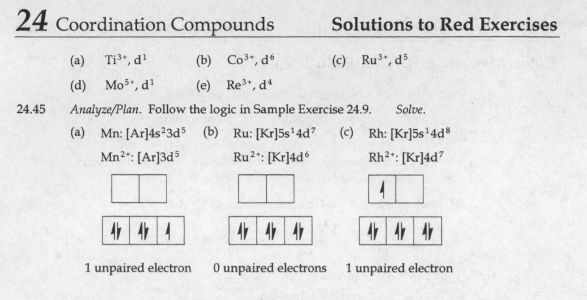

(a) Ti^{3+}, d^1 (b) Co^{3+}, d^6 (c) Ru^{3+}, d^5

(d) Mo^{5+}, d^1 (e) Re^{3+}, d^4

24.45   *Analyze/Plan.* Follow the logic in Sample Exercise 24.9.   *Solve.*

(a)   Mn: $[Ar]4s^2 3d^5$ (b)   Ru: $[Kr]5s^1 4d^7$ (c)   Rh: $[Kr]5s^1 4d^8$

       $Mn^{2+}: [Ar]3d^5$ $Ru^{2+}: [Kr]4d^6$ $Rh^{2+}: [Kr]4d^7$

1 unpaired electron   0 unpaired electrons   1 unpaired electron

24.47   *Analyze/Plan.* All complexes in this exercise are six-coordinate octahedral. Use the definitions of high-spin and low-spin along with the orbital diagram from Sample Exercise 24.9 to place electrons for the various complexes.   *Solve.*

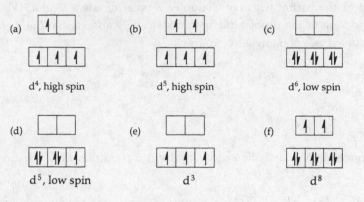

(a) d^4, high spin   (b) d^5, high spin   (c) d^6, low spin

(d) d^5, low spin   (e) d^3 (f) d^8

24.49   *Analyze/Plan.* Follow the ideas but reverse the logic in Sample Exercise 24.9.   *Solve.*

high spin

Additional Exercises

24.51 $[Pt(NH_3)_6]Cl_4$; $[Pt(NH_3)_4Cl_2]Cl_2$; $[Pt(NH_3)_3Cl_3]Cl$; $[Pt(NH_3)_2Cl_4]$; $K[Pt(NH_3)Cl_5]$

24.54   (a)   [24.53(a)] *cis*-tetraamminediaquacobalt(II) nitrate

           [24.53(b)] sodium aquapentachlororuthenate(III)

           [24.53(c)] ammonium *trans*-diaquabisoxalatocobaltate(III)

           [24.53(d)] *cis*-dichlorobisethylenediamineruthenium(II)

(b)   Only the complex in 24.53(d) is optically active. The mirror images of (a)-(c) can be superimposed on the original structure. The chelating ligands in (d) prevent its mirror images (enantiomers) from being superimposable.

24.56   (a)   In a square planar complex such as $[Pt(en)Cl_2]$, if one pair of ligands is trans, the remaining two coordination sites are also trans to each other. Ethylenediamine is a relatively short bidentate ligand that cannot occupy trans coordination sites, so the trans isomer is unknown.

(b)   A polydentate ligand such as EDTA necessarily occupies trans positions in an octahedral complex. The minimum steric requirement for a bidentate ligand is a medium-length chain between the two coordinating atoms that will occupy the trans positions. In terms of reaction rate theory, it is unlikely that a flexible bidentate ligand will be in exactly the right orientation to coordinate trans. The polydentate ligand has a much better chance of occupying trans positions, because it locks the metal ion in place with multiple coordination sites (and shields the metal ion from competing ligands present in the solution).

24.59   (a) $AgCl(s) + 2NH_3(aq) \rightarrow [Ag(NH_3)_2]^+(aq) + Cl^-(aq)$

(b) $[Cr(en)_2Cl_2]Cl(aq) + 2H_2O(l) \rightarrow [Cr(en_2)(H_2O)_2]^{3+}(aq) + 3Cl^-(aq)$

     green   brown-orange

$3Ag^+(aq) + 3Cl^-(aq) \rightarrow 3AgCl(s)$

$[Cr(en)_2(H_2O)_2]^{3+}$ and $3NO_3^-$ are spectator ions in the second reaction.

(c) $Zn(NO_3)_2(aq) + 2NaOH(aq) \rightarrow Zn(OH)_2(s) + 2NaNO_3(aq)$

$Zn(OH)_2(s) + 2NaOH(aq) \rightarrow [Zn(OH)_4]^{2-}(aq) + 2Na^+(aq)$

(d) $Co^{2+}(aq) + 4Cl^-(aq) \rightarrow [CoCl_4]^{2-}(aq)$

24.62   (a)

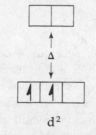

d^2

(b)   These complexes are colored because the crystal-field splitting energy, Δ, is in the visible portion of the electromagnetic spectrum. Visible light with $\lambda = hc/\Delta$ is absorbed, promoting one of the d-electrons into a higher energy d-orbital. The remaining wavelengths of visible light are reflected or transmitted; the combination of these wavelengths is the color we see.

(c) $[V(H_2O)_6]^{3+}$ will absorb light with higher energy. H_2O is in the middle of the spectrochemical series, and causes a larger Δ than F^-, a weak-field ligand. Since Δ and λ are inversely related, larger Δ corresponds to higher energy and shorter λ.

24.64   According to the spectrochemical series, the order of increasing Δ for the ligands is $Cl^- < H_2O < NH_3$. (The tetrahedral Cl^- complex will have an even smaller Δ than an octahedral one.) The smaller the value of Δ, the longer the wavelength of visible light absorbed. The color of light absorbed is the complement of the observed color. A blue complex absorbs orange light (580–650 nm), a pink complex absorbs green light (490–560 nm) and a yellow complex absorbs violet light (400–430 nm). Since $[CoCl_4]^{2-}$ absorbs the longest wavelength, it appears blue. $[Co(H_2O)_6]^{2+}$ absorbs green and appears pink, and $[Co(NH_3)_6]^{3+}$ absorbs violet and appears yellow.

24.66   (a) $[FeF_6]^{4-}$. Both complexes contain the same metal ion, Fe^{2+}; F^- is a weak-field ligand that imposes a smaller Δ and longer λ for the complex ion.

         (b) $[V(H_2O)_6]^{2+}$. Both complexes contain the same ligand, H_2O. V^{2+} has a lower charge, so the interaction with the ligand will produce a weaker field, a smaller Δ, and a longer absorbed wavelength.

         (c) $[CoCl_4]^{2-}$. Both complexes contain the same metal ion, Co^{2+}; Cl^- is a weak-field ligand that imposes a smaller Δ and a longer λ for the complex ion.

24.69   (a)

         (b)   sodium dicarbonyltetracyanoferrate(II)

         (c)   +2, 6 d-electrons

         (d)   We expect the complex to be low spin. Cyanide (and carbonyl) are high on the spectrochemical series, which means the complex will have a large Δ splitting characteristic of low spin complexes.

Integrative Exercises

24.73   In a complex ion, the transition metal is an electron pair acceptor, a Lewis acid; the ligand is an electron pair donor, a Lewis base. In carbonic anhydrase, the Zn^{2+} ion withdraws electron density from the O atom of water. The electronegative oxygen atom compensates by withdrawing electron-density from the O—H bond. The O—H bond is polarized and H becomes more ionizable, more acidic than in the bulk solvent. This is similar to the effect of an electronegative central atom in an oxyacid such as H_2SO_4.

24.75   Determine the empirical formula of the complex, assuming the remaining mass is due to oxygen, and a 100 g sample.

$$10.0 \text{ g Mn} \times \frac{1 \text{ mol Mn}}{54.94 \text{ g Mn}} = 0.1820 \text{ mol Mn}; 0.182 / 0.182 = 1$$

$$28.6 \text{ g K} \times \frac{1 \text{ mol K}}{39.10 \text{ g K}} = 0.7315 \text{ mol K}; 0.732 / 0.182 = 4$$

$$8.8 \text{ g C} \times \frac{1 \text{ mol C}}{12.0 \text{ g C}} = 0.7327 \text{ mol C}; \ 0.733 / 0.182 = 4$$

$$29.2 \text{ g Br} \times \frac{1 \text{ mol Br}}{79.904 \text{ g Br}} = 0.3654 \text{ mol Br}; \ 0.365 / 0.182 = 2$$

$$23.4 \text{ g O} \times \frac{1 \text{ mol O}}{16.00 \text{ g O}} = 1.463 \text{ mol O}; \ 1.46 / 0.182 = 8$$

There are 2 C and 4 O per oxalate ion, for a total of two oxalate ligands in the complex. To match the conductivity of $K_4[Fe(CN)_6]$, the oxalate and bromide ions must be in the coordination sphere of the complex anion. Thus, the compound is $K_4[Mn(ox)_2Br_2]$.

24.78   Calculate the concentration of Mg^{2+} alone, and then the concentration of Ca^{2+} by difference. $M \times L = \text{mol}$

$$\frac{0.0104 \text{ mol EDTA}}{1 \text{ L}} \times 0.0187 \text{L} \times \frac{1 \text{ mol Mg}^{2+}}{1 \text{ mol EDTA}} \times \frac{24.31 \text{ g Mg}^{2+}}{1 \text{ mol Mg}^{2+}} \times \frac{1000 \text{ mg}}{\text{g}}$$

$$\times \frac{1}{0.100 \text{ L H}_2\text{O}} = 47.28 = 47.3 \text{ mg Mg}^{2+}/\text{L}$$

$$0.0104 \ M \text{ EDTA} \times 0.0315 \text{ L} = \text{mol} (Ca^{2+} + Mg^{2+})$$

$$\underline{0.0104 \ M \text{ EDTA} \times 0.0187 \text{ L} = \text{mol Mg}^{2+}}$$

$$0.0104 \ M \text{ EDTA} \times 0.0128 \text{ L} = \text{mol Ca}^{2+}$$

$$0.0104 \text{ M EDTA} \times 0.0128 \text{ L} \times \frac{1 \text{ mol Ca}^{2+}}{1 \text{ mol EDTA}} \times \frac{40.08 \text{ g Ca}^{2+}}{1 \text{ mol Ca}^{2+}} \times \frac{1000 \text{ mg}}{\text{g}} \times \frac{1}{0.100 \text{ L H}_2\text{O}}$$

$$= 53.35 = 53.4 \text{ mg Ca}^{2+}/\text{L}$$

24.81 $$\frac{182 \times 10^3 \text{ J}}{1 \text{ mol}} \times \frac{1 \text{ mol}}{6.022 \times 10^{23} \text{ molecules}} = 3.022 \times 10^{-19} = 3.02 \times 10^{-19} \text{ J/photon}$$

$$\Delta E = h\nu = 3.02 \times 10^{-19} \text{ J}; \ \nu = \Delta E / h$$

$$\nu = 3.022 \times 10^{-19} \text{ J} / 6.626 \times 10^{-34} \text{ J} \cdot \text{s} = 4.561 \times 10^{14} = 4.56 \times 10^{14} \text{s}^{-1}$$

$$\lambda = \frac{2.998 \times 10^8 \text{ m/s}}{4.561 \times 10^{14} \text{ s}^{-1}} = 6.57 \times 10^{-7} \text{ m} = 657 \text{ nm}$$

We expect that this complex will absorb in the visible, at around 660 nm. It will thus exhibit a blue-green color (Figure 24.24).

25 The Chemistry of Life: Organic and Biological Chemistry

Visualizing Concepts

25.2   *Analyze/Plan.* Given structural formulas, decide which molecule will undergo addition. Consider which functional groups are present in the molecules, and which are most susceptible to addition.   *Solve.*

Addition reactions are characteristic of alkenes. Molecule (c), an alkene, will readily undergo addition.

Molecule (a) is an aromatic hydrocarbon, which does not typically undergo addition because the delocalized electron cloud is too difficult to disrupt. Molecules (b) and (d) contain carbonyl groups (actually carboxylic acid groups) that do not typically undergo addition, except under special conditions.

25.5   *Analyze/Plan.* Given ball-and-stick models, select the molecule that fits the description given. From the models, decide the type of molecule or functional group represented.

Solve. Molecule (i) is a sugar, (ii) is an organic base and a component of nucleic acids, (iii) is an amino acid, and (iv) is an alcohol.

(a)   Molecule (i) is a disaccharide composed of galactose (left) and glucose (right); it can be hydrolyzed to form a solution containing glucose. Since it is the only sugar molecule depicted, it was not necessary to know the exact structure of glucose to answer the question.

(b)   Amino acids form zwitterions, so the choice is molecule (iii).

(c)   Molecule (ii) is an organic base present in DNA (again, the only possible choice).

(d)   Molecule (iv) because alcohols react with carboxylic acids to form esthers.

Introduction to Organic Compounds; Hydrocarbons

25.7   *Analyze/Plan.* Given a condensed structural formula, determine the bond angles and hybridization about each carbon atom in the molecule. Visualize the number of electron domains about each carbon. State the bond angle and hybridization based on electron domain geometry.   *Solve.*

310

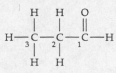

C2 and C3 both have tetrahedral electron domain geometry, 109° bond angles and sp^3 hybridization. C1 has trigonal planar electron domain geometry, 120° bond angles and sp^2 hybridization.

25.9 Carbon (of course), hydrogen, oxygen, nitrogen, sulfur, phosphorus, chlorine, (and other halogens). According to periodic trends and Figure 8.6, oxygen, nitrogen, fluorine, and chlorine are more electronegative than carbon. Sulfur has the same electronegativity as carbon.

25.11 (a) A *straight-chain hydrocarbon* has all carbon atoms connected in a continuous chain; no carbon atom is bound to more than two other carbon atoms. A *branched-chain hydrocarbon* has a branch; at least one carbon atom is bound to three or more carbon atoms.

(b) An *alkane* is a complete molecule composed of carbon and hydrogen in which all bonds are single (sigma) bonds. An *alkyl group* is a substituent formed by removing a hydrogen atom from an alkane.

(c) Alkanes are said to be *saturated* because they contain only single bonds. Multiple bonds that enable addition of H_2 or other substances are absent. The bonding capacity of each carbon atom is fulfilled with single bonds to C or H.

25.13 *Analyze/Plan.* Consider the definition of the stated classification and apply it to a compound containing five C atoms. *Solve.*

(a) $CH_3CH_2CH_2CH_2CH_3$, C_5H_{12}

(b)
$$\begin{array}{c} CH_2 \\ H_2C \quad \quad CH_2 \\ H_2C - CH_2 \end{array}$$

(c) $CH_2{=}CHCH_2CH_2CH_3$, C_5H_{10}

(d) $HC{\equiv}CCH_2CH_2CH_3$, C_5H_8 saturated: (a), (b); unsaturated: (c), (d)

25.15 *Analyze/Plan.* The general formula of an alkane is C_nH_{2n+2}. For an alkene, with 2 fewer H atoms, the general formula is C_nH_{2n}. *Solve.*

A dialkene has one more $C{=}C$ and thus two fewer H atoms than an alkene. The general formula is C_nH_{2n-2}.

25.17 *Analyze/Plan.* Follow the logic in Sample Exercise 25.3. *Solve.*

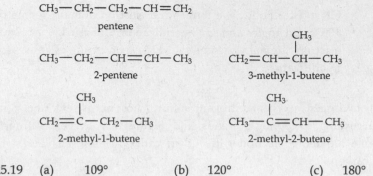

$$CH_3—CH_2—CH_2—CH=CH_2$$
pentene

$$CH_3—CH_2—CH=CH—CH_3$$
2-pentene

$$CH_2=CH—\overset{\overset{\displaystyle CH_3}{|}}{CH}—CH_3$$
3-methyl-1-butene

$$CH_2=\overset{\overset{\displaystyle CH_3}{|}}{C}—CH_2—CH_3$$
2-methyl-1-butene

$$CH_3—\overset{\overset{\displaystyle CH_3}{|}}{C}=CH—CH_3$$
2-methyl-2-butene

25.19   (a)   109°   (b)   120°   (c)   180°

25.21   *Analyze/Plan.* Follow the rules for naming alkanes given in Section 25.3 and illustrated in Sample Exercise 25.1.   *Solve.*

     (a)   2-methylhexane

     (b)   4-ethyl-2,4-dimethyldecane

     (c)

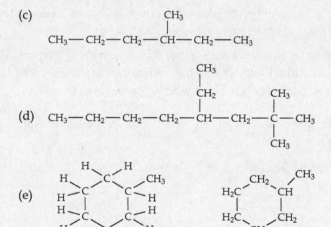

     (d)

     (e)

25.23   *Analyze/Plan.* Follow the logic in Sample Exercises 25.1 and 25.4.   *Solve.*

     (a)   2,3-dimethylheptane

     (b)   *cis*-6-methyl-3-octene

     (c)   *para*-dibromobenzene

     (d)   4,4-dimethyl-1-hexyne

     (e)   methylcyclobutane

25.25   Each doubly bound carbon atom in an alkene has two unique sites for substitution. These sites cannot be interconverted because rotation about the double bond is restricted; *geometric isomerism* results. In an alkane, carbon forms only single bonds, so the three remaining sites are interchangeable by rotation about the single bond. Although there is also restricted rotation around the triple bond of an alkyne, there is only one additional bonding site on a triply bound carbon, so no isomerism results.

25.27 *Analyze/Plan.* In order for geometrical isomerism to be possible, the molecule must be an alkene with two different groups bound to each of the alkene C atoms. *Solve.*

(a)
$$Cl-\underset{\underset{Cl}{|}}{C}=\underset{\underset{H}{|}}{C}-CH_2-CH_3, \text{ no}$$

(b)

$$\underset{H_3C}{\overset{Cl}{\diagdown}}C=C\underset{\diagdown CH_2Cl}{\overset{H}{\diagup}} \qquad \underset{Cl}{\overset{H_3C}{\diagdown}}C=C\underset{\diagdown CH_2Cl}{\overset{H}{\diagup}}$$

(c) no, not an alkene

(d) no, not an alkene

25.29 Assuming that each component retains its effective octane number in the mixture (and this isn't always the case), we obtain: octane number = 0.35(0) + 0.65(100) = 65.

Reactions of Hydrocarbons

25.31 (a) An *addition reaction* is the addition of some reagent to the two atoms that form a multiple bond. In a *substitution reaction*, one atom or group of atoms replaces (substitutes for) another atom or group of atoms. In an addition reaction, two atoms and a multiple bond on the target molecule are altered; in a substitution reaction, the environment of one atom in the target molecule changes. Alkenes typically undergo addition, while aromatic hydrocarbons usually undergo substitution.

(b) *Plan.* Consider the general form of addition across a double bond. The π bond is broken and one new substituent (in this case two Br atoms) adds to each of the C atoms involved in the π bond. *Solve.*

$$CH_3-\underset{\underset{CH_3}{|}}{CH}-\underset{\underset{H}{|}}{C}=\underset{\underset{CH_3}{|}}{C}-CH_3 + Br_2 \longrightarrow CH_3-\underset{\underset{CH_3}{|}}{CH}-\underset{\underset{Br}{|}}{C}-\underset{\underset{Br}{|}}{C}-CH_3$$

(c) *Plan.* Consider the general form of a *substitution* reaction. A Cl atom will replace one of the H atoms on the benzene ring. In the target molecule, all H atoms are equivalent, so no choice of position is required. *Solve.*

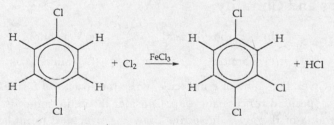

25.33 (a) *Plan.* Consider the structures of cyclopropane, cyclopentane, and cyclohexane. *Solve.*

The small 60° C—C—C angles in the cyclopropane ring cause strain that provides a driving force for reactions that result in ring opening. There is no comparable strain in the five- or six-membered rings.

(b)   *Plan.* First form an alkyl halide: $C_2H_4(g) + HBr(g) \rightarrow CH_3CH_2Br(l)$; then carry out a Friedel-Crafts reaction.   *Solve.*

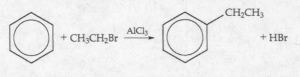

25.35   The partially positive end of the hydrogen halide, $\overset{\delta^+ \quad \delta^-}{H - X}$, is attached to the π electron cloud of the alkene cyclohexene. The electrons that formed the π bond in cyclohexene form a sigma bond to the H atom of HX, leaving a halide ion, X^-. The intermediate is a carbocation; one of the C atoms formerly involved in the π bond is now bound to a second H atom. The other C atom formerly involved in the π bond carries a full positive charge and forms only three sigma bonds, two to adjacent C atoms and one to H.

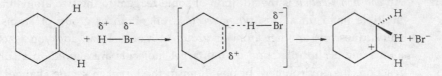

25.37   *Analyze/Plan.* Both combustion reactions produce CO_2 and H_2O:

$$C_3H_6(g) + 9/2\, O_2(g) \rightarrow 3CO_2(g) + 3H_2O(l)$$

$$C_5H_{10}(g) + 15/2\, O_2(g) \rightarrow 5CO_2(g) + 5H_2O(l)$$

Thus, we can calculate the ΔH_{comb} / CH_2 group for each compound.   *Solve.*

$$\frac{\Delta H_{comb}}{CH_2\text{ group}} = \frac{2089 \text{ kJ/mol } C_3H_6}{3 \text{ CH}_2 \text{ groups}} = \frac{696.3 \text{ kJ}}{\text{mol CH}_2} ; \frac{3317 \text{ kJ/mol } C_5H_{10}}{5 \text{ CH}_2 \text{ groups}} = 663.4 \text{ kJ/mol CH}_2$$

$\Delta H_{comb}/CH_2$ group for cyclopropane is greater because C_3H_6 contains a strained ring. When combustion occurs, the strain is relieved and the stored energy is released during the reaction.

Functional Groups and Chirality

25.39   (a)   ketone   (b)   carboxylic acid   (c)   alcohol

(d)   ester   (e)   amide   (f)   amine

25.41   *Analyze/Plan.* Given the name of a molecule, write the structural formula of an isomer that contains a specified functional group. Consider the definition of isomer, write the molecular formula of the given molecule, draw the structural formula of a molecule with the same formula that contains the specified functional group.   *Solve.*

(a)   The formula of acetone is C_3H_6O. An aldehyde contains the group $-C\overset{\displaystyle O}{\underset{\displaystyle H}{\Vert}}$

An aldehyde that is an isomer of acetone is propionaldehyde (or propanal):

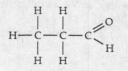

(b)   The formula of 1-propanol is C_3H_8O. An ether contains the group —O—. An ether that is an isomer of 1-propanol is ethylmethyl ether:

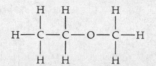

25.43   *Analyze/Plan.* Count the number of C atoms in each chain, including the carboxyl C atom. Name the chain and the acid.   *Solve.*

(a)   methanoic acid

(b)   butanoic acid

(c)   3-methylpentanoic acid

25.45   *Analyze/Plan.* In a condensation reaction between an alcohol and a carboxylic acid, the alcohol loses its —OH hydrogen atom and the acid loses its —OH group. The alkyl group from the acid is attached to the carbonyl group and the alkyl group from alcohol is attached to the ether oxygen of the ester. The name of the ester is the alkyl group from the alcohol plus the alkyl group from the acid plus the suffix -oate.   *Solve.*

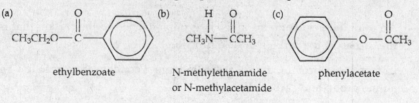

25.47   *Analyze/Plan.* Follow the logic in Sample Exercise 25.6.   *Solve.*

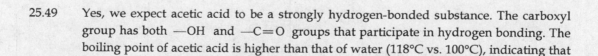

25.49   Yes, we expect acetic acid to be a strongly hydrogen-bonded substance. The carboxyl group has both —OH and —C=O groups that participate in hydrogen bonding. The boiling point of acetic acid is higher than that of water (118°C vs. 100°C), indicating that hydrogen-bonding in acetic acid is even stronger than that of water (Figure 11.10). The melting points, 16.7°C for acetic acid and 0°C for water, show a similar trend.

25.51   *Analyze/Plan.* Follow the logic in Sample Exercise 25.2, incorporating functional group information from Table 25.4.   *Solve.*

(a) CH₃CH₂CHCH₃   (b) HOCH₂CH₂OH   (c) H—C—OCH₃
with OH above the third carbon; (c) has O double-bonded above C

(d) CH₃CH₂CCH₂CH₃ (e) CH₃CH₂OCH₂CH₃
with O double-bonded above the third C in (d)

25.53   *Analyze/Plan.* Review the rules for naming alkanes and haloalkanes; draw the structures. That is, draw the carbon chain indicated by the root name, place substituents, fill remaining positions with H atoms. Each C atom attached to four different groups is chiral.   *Solve.*

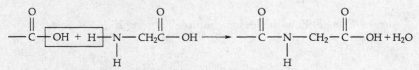

* chiral C atoms

C2 is obviously attached to four different groups. C3 is chiral because the substituents on C2 render the C1-C2 group different than the C4-C5 group.

Proteins

25.55   (a)   An α-amino acid contains an NH₂ group attached to the carbon that is bound to the carbon of the carboxylic acid function.

   (b)   In forming a protein, amino acids undergo a condensation reaction between the amino group and carboxylic acid:

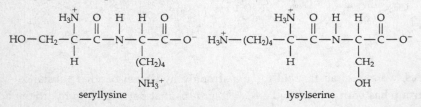

25.57   *Analyze/Plan.* Either peptide can have the terminal carboxyl group or the terminal amino group.   *Solve.*

Two dipeptides are possible:

seryllysine   lysylserine

25.59   *Analyze/Plan.* Follow the logic in Sample Exercise 25.7.   *Solve.*

(a)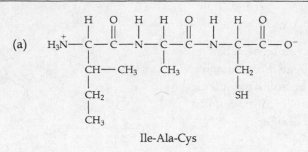

Ile-Ala-Cys

(b)   Eight: Ser-Ser-Ser; Ser-Ser-Phe; Ser-Phe-Ser; Phe-Ser-Ser; Ser-Phe-Phe;

Phe-Ser-Phe; Phe-Phe-Ser; Phe-Phe-Phe

25.61   The *primary structure* of a protein refers to the sequence of amino acids in the chain. Along any particular section of the protein chain the configuration may be helical, may be an open chain, or arranged in some other way. This is called the *secondary structure*. The overall shape of the protein molecule is determined by the way the segments of the protein chain fold together, or pack. The interactions which determine the overall shape are referred to as the *tertiary structure*.

Carbohydrates

25.63   (a)   *Carbohydrates*, or sugars, are composed of carbon, hydrogen, and oxygen. From a chemical viewpoint, they are polyhydroxyaldehydes or ketones. Carbohydrates are primarily derived from plants and are a major food source for animals.

(b)   A *monosaccharide* is a simple sugar molecule that cannot be decomposed into smaller sugar molecules by (acid) hydrolysis.

(c)   A *disaccharide* is a carbohydrate composed of two simple sugar units. Hydrolysis breaks the disaccharides into two monosaccharides.

25.65   (a)   In the linear form of galactose, the aldehydic carbon is C1. Carbon atoms 2, 3, 4, and 5 are chiral because they each carry four different groups. Carbon 6 is not chiral because it contains two H atoms.

(b)   The structure is best deduced by comparing galactose with glucose, and inverting the configurations at the appropriate carbon atoms. Recall from Solution 25.64 that both the β-form (shown here) and the α-form (OH on carbon 1 on the opposite side of ring as the CH_2OH on carbon 5) are possible.

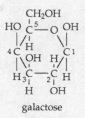

galactose

25.67   The empirical formula of glycogen is $C_6H_{10}O_5$. The six-membered ring form of glucose is the unit that forms the basis of glycogen. The monomeric glucose units are joined by α linkages.

Nucleic Acids

25.69 A *nucleotide* consists of a nitrogen-containing aromatic compound, a sugar in the furanose (five-membered) ring form, and a phosphoric acid group. The structure of deoxycytidine monophosphate is shown at right.

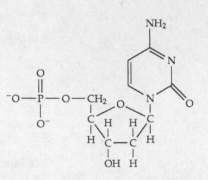

25.71 $C_4H_7O_3CH_2OH + HPO_4^{2-} \rightarrow C_4H_7O_3CH_2-O-PO_3^{2-} + H_2O$

25.73 In the helical structure for DNA, the strands of the polynucleotides are held together by hydrogen-bonding interactions between particular pairs of bases. It happens that adenine and thymine form an especially effective base pair, and that guanine and cytosine are similarly related. Thus, each adenine has a thymine as its opposite number in the other strand, and each guanine has a cytosine as its opposite number. In the overall analysis of the double strand, total adenine must then equal total thymine, and total guanine equals total cytosine.

Additional Exercises

25.75

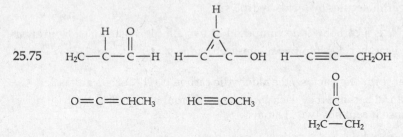

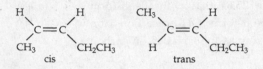

Structures with the —OH group attached to an alkene carbon atom are not included. These molecules are called "vinyl alcohols" and are not the major form at equilibrium.

25.78

H₂C—C—C—H   H—C=C—OH   H—C≡C—CH₂OH

Cyclopentene does not show cis-trans isomerism because the existence of the ring demands that the C—C bonds be cis to one another.

25.81 $H_2C=CH-CH_2OH$

$$
\begin{array}{c}
CH_2 \\
/\backslash \\
H_2C-C-H \\
| \\
OH
\end{array}
$$

(Structures with the —OH group attached to an alkene carbon atom are not included. These molecules are called "vinyl alcohols" and are not the major form at equilibrium.)

25.84   (a) Ether, $C-O-C$; alkene, $-CH=CH_2$

(b) carboxylic acid, $-\overset{O}{\overset{||}{C}}-OH$; ester, $CH_3\overset{O}{\overset{||}{C}}-O-$; aromatic,

(c) ketone, $-\overset{O}{\overset{||}{C}}-$; alkene, $-CH=CH-$; alcohol $-C-OH$

25.86   The difference between an alcoholic hydrogen and a carboxylic acid hydrogen is two-fold. First, the electronegative carbonyl oxygen in a carboxylic acid withdraws electron density from the O—H bond, rendering the bond more polar and the H more ionizable. Second, the conjugate base of a carboxylic acid, carboxylate anion, exhibits resonance. This stabilizes the conjugate base and encourages ionization of the carboxylic acid. In an alcohol no electronegative atoms are bound to the carbon that holds the —OH group, and the H is tightly bound to the O.

25.89   (a)   None

(b)   The carbon bearing the secondary —OH has four different groups attached, and is thus chiral.

(c)   The carbon bearing the —NH$_2$ group and the carbon bearing the CH$_3$ group are both chiral.

25.91   Glu-Cys-Gly is the only possible order. Glutamic acid has two carboxyl groups that can form a peptide bond with cysteine, so there are two possible structures.

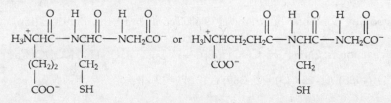

25.93   Both glucose and fructose contain six C atoms, so both are hexoses. Glucose contains an aldehyde group at C1, so it is an aldohexose. Fructose has a ketone at C2, so it is a ketohexose.

Integrative Exercises

25.95 CH_3CH_2OH CH_3-O-CH_3
     ethanol   dimethyl ether

Ethanol contains —O—H bonds which form strong intermolecular hydrogen bonds, while dimethyl ether experiences only weak dipole-dipole and dispersion forces.

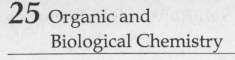

difluoromethane

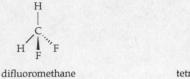

tetrafluoromethane

CH_2F_2 is a polar molecule, while CF_4 is nonpolar. CH_2F_2 experiences dipole-dipole and dispersion forces, while CF_4 experiences only dispersion forces.

In both cases, stronger intermolecular forces lead to the higher boiling point.

25.97   Determine the empirical formula, molar mass, and thus molecular formula of the compound. Confirm with physical data.

$$66.7\ g\,C \times \frac{1\ mol\ C}{12.01\ g\,C} = 5.554\ mol\ C;\ 5.554/1.388 = 4$$

$$11.2\ g\,H \times \frac{1\ mol\ H}{1.008\ g\,H} = 11.11\ mol\ H;\ 11.11/1.388 = 8$$

$$22.2\ g\,O \times \frac{1\ mol\ O}{16.00\ g\,O} = 1.388\ mol\ O;\ 1.388/1.388 = 1$$

The empirical formula is C_4H_8O. Using Equation 10.11 (MM = molar mass):

$$MM = \frac{(2.28 g/L)(0.08206\ L \bullet atm/mol \bullet K)(373K)}{0.970\ atm} = 71.9\,g/mol$$

The formula weight of C_4H_8O is 72, so the molecular formula is also C_4H_8O. Since the compound has a carbonyl group and cannot be oxidized to an acid, the only possibility is 2-butanone.

$$\overset{\displaystyle O}{\overset{\displaystyle \|}{CH_3CCH_2CH_3}}$$

The boiling point of 2-butanone is 79.6°C, confirming the identification.

25.99   The reaction is: $2NH_2CH_2COOH(aq) \rightarrow NH_2CH_2CONHCH_2COOH(aq) + H_2O(l)$

$\Delta G° = (-488) + (-237.13) - 2(-369) = 12.87 = 13$ kJ

25.103 $AMPOH^-(aq) \rightleftharpoons AMPO^{2-}(aq) + H^+(aq)$

$pK_a = 7.21;\ K_a = 10^{-pK_a} = 6.17 \times 10^{-8} = 6.2 \times 10^{-8}$

$K_a = \dfrac{[AMPO^{2-}][H^+]}{[AMPOH^-]} = 6.2 \times 10^{-8}$. When pH = 7.40, $[H^+] = 3.98 \times 10^{-8} = 4 \times 10^{-8}$.

Then $\dfrac{[AMPOH^-]}{[AMPO^{2-}]} = 3.98 \times 10^{-8}/6.17 \times 10^{-8} = 0.6457 = 0.6$